Lowering Risk by Increasing Resilience: Selected Papers from 8th International Conference on Flood Management (ICFM 8)

Lowering Risk by Increasing Resilience: Selected Papers from 8th International Conference on Flood Management (ICFM 8)

Editors

Slobodan P. Simonovic
Subhankar Karmakar
Zhang Cheng

MDPI • Basel • Beijing • Wuhan • Barcelona • Belgrade • Manchester • Tokyo • Cluj • Tianjin

Editors
Slobodan P. Simonovic
University of Western
Canada

Subhankar Karmakar
Indian Institute of
Technology Bombay
India

Zhang Cheng
China Institute of Water Resources and
Hydropower Research (IWHR)
China
Permanent Secretariat of ICFM

Editorial Office
MDPI
St. Alban-Anlage 66
4052 Basel, Switzerland

This is a reprint of articles from the Special Issue published online in the open access journal *Water* (ISSN 2073-4441) (available at: https://www.mdpi.com/journal/water/special_issues/ICFM8).

For citation purposes, cite each article independently as indicated on the article page online and as indicated below:

LastName, A.A.; LastName, B.B.; LastName, C.C. Article Title. *Journal Name* **Year**, *Volume Number*, Page Range.

ISBN 978-3-0365-5735-9 (Hbk)
ISBN 978-3-0365-5736-6 (PDF)

Cover image courtesy of Slobodan P. Simonovic

Contents

Preface to "Lowering Risk by Increasing Resilience: Selected Papers from 8th International Conference on Flood Management (ICFM 8)"

The International Conference on Flood Management (ICFM), organized every 3 years, fosters a unique brainstorming platform to exchange ideas and expertise on the multidisciplinary and multisectoral area of flood management studies.

The 8th International Conference on Flood Management (ICFM8) was initially planned from 17–19 August 2020 at University of Iowa, USA but could not be organized as a full-fledged conference due to the global pandemic situation. It was eventually agreed that the conference should be reorganized in August 2021 as a pandemic special "instead of ICFM8" webinar along with publishing of the *Book of ICFM8 Abstracts*. ICFM8 had the theme **"Lowering Risk by Increasing Resilience"** to spark interdisciplinary conversations focused on the present state of knowledge, as well as necessary future advances in research and application disciplines related to flood management, particularly the need to build resilience into future planning.

This Special Issue of *Water* includes selected research contributions to the ICFM8 that showcase key concerns and significant challenges of the future, as currently perceived by researchers, industry, policymakers, and other flood management stakeholders. The eleven papers presented herein embody a small sample of the many presentations at ICFM8. They span a variety of challenging issues related to science and technology for flood risk management; handling data and information for flood risk management; flood disaster prevention, mitigation, and adaptation; flood preparedness, response, and recovery; flood decision-making, policy, and governance; and flood resilience.

We wish to thank the ICFM8 participants who presented their research findings for wholeheartedly participating in the preparation of this Special Issue. Although the scope of *Water* is limited to selected papers submitted by the contributing participants, we are confident that many other compelling works presented at the conference found their way to archival literature elsewhere.

Last but not the least, we thank the contributing authors, reviewers, and editors of *Water* for embracing this initiative and for their committed support in publishing this topical Special Issue.

Slobodan P. Simonovic, Subhankar Karmakar, and Zhang Cheng
Editors

water

MDPI

Article

A Comprehensive Approach for Floodplain Mapping through Identification of Hazard Using Publicly Available Data Sets over Canada

Mohit Prakash Mohanty [1,2,*] and Slobodan P. Simonovic [2]

1 Department of Water Resources Development and Management, Indian Institute of Technology Roorkee, Roorkee 247 667, India
2 Department of Civil and Environmental Engineering, Western University, London, ON N6A 3K7, Canada; simonovic@uwo.ca
* Correspondence: mohit.mohanty@wr.iitr.ac.in

Abstract: Quantifying flood inundation and hazards over large regions is paramount for gaining critical information on flood risk over the vulnerable population and environment. Readily available global data and enhancement in computational simulations have made it easier to simulate flooding at a large scale. This study explores the usability of publicly available datasets in flood inundation and hazard mapping, and ensures the flood-related information reaches the end-users efficiently. Runoff from the North American Regional Reanalysis and other relevant inputs are fed to the CaMa-Flood model to generate flooding patterns for 1 in 100 and 1 in 200-year return period events over Canada. The simulated floodplain maps are overlaid on the property footprints of 34 cities (falling within the top 100 populated cities of Canada) to determine the degree of exposure during 1991, 2001 and 2011. Lastly, Flood Map Viewer—a web-based public tool, is developed to disseminate extensive flood-related information. The development of the tool is motivated by the commitment of the Canadian government to contribute $63 M over the next three years for the development of flood maps, especially in high-flood risk areas. The results from the study indicate that around 80 percent of inundated spots belong to high and very-high hazard classes in a 200-year event, which is roughly 4 percent more than observed during the 100-year event. We notice an increase in the properties exposed to flooding during the last three decades, with a signature rise in Toronto, Montreal and Edmonton. The flood-related information derived from the study can be used along with vulnerability and exposure components to quantify flood risk. This will help develop appropriate pathways for resilience building for long-term sustainable benefits.

Keywords: flood hazard; CaMa-Flood; flood map viewer; floodplain mapping; flood risk

Citation: Mohanty, M.P.; Simonovic, S.P. A Comprehensive Approach for Floodplain Mapping through Identification of Hazard Using Publicly Available Data Sets over Canada. *Water* **2022**, *14*, 2280. https://doi.org/10.3390/w14142280

Academic Editor: Chang Huang

Received: 19 June 2022
Accepted: 16 July 2022
Published: 21 July 2022

1. Introduction

In the last few decades, flood events have increased manifold leading to widespread human, environmental and economic losses [1–3]. Between 1980 and 2018, about 5997 extreme flood events resulted in 223,482 deaths and a mammoth economic loss exceeding $1 trillion [4]. A recent report by UNISDR (2015) [5] stated that flood-prone areas around the world house around 800 million residents, of which roughly one-tenth are exposed to floods annually. Apart from that, numerous scientific research signifies that the looming disaster situation is expected to become more severe in the future periods due to alterations in climatic patterns and socio-economic dynamics [6–9]. In a recent study, Dottori et al. (2018) [10] reported that with a rise in temperature up to 1.5 °C, human losses due to flooding on the global scale could rise to as high as 70–83%, including direct flood damages up to 160–240%. Under such circumstances, there is a dire need to map and quantify the flood losses to ensure optimum protection of the communities and assets.

Floodplain mapping provides qualitative (degree/severity of flood hazard) and quantitative (inundation extent, duration of flooding, inundation depth, etc.) information on flooding and is considered a viable option for minimizing flood risk. Most of the research on floodplain mapping has been demonstrated at small scales, i.e., local and regional. Considering the widespread damage, there was a growing need to extend this to larger scales at the country, region and global levels. However, there were two significant challenges: (i) substantial computational efforts and (ii) public availability of global data sets. During the last decade, considerable progress has been made in tackling these two issues. Recently, the scientific research community introduced a suite of global flood inundation models [11,12]. These sophisticated tools are designed to quantify the flood inundation dynamics using state-of-the-art algorithms over large regions. The flood hazard information is derived from the inundation parameters through suitable approaches [1,13,14]. A list of the widely used global flood models is presented in Table 1. Recently, Gaur et al. (2018, 2019) [15,16] employed CaMa-Flood to determine the timing of floods and changes in their magnitudes across Canada from 2016 to 2100. In another research, Lim et al. (2018) [17] utilized runoff observations of 11 GCMs belonging to the CMIP5 consortium as inputs to the CaMaFlood model to generate global river water depths; however, at a coarser resolution of $0.25^\circ \times 0.25^\circ$. Earlier, Winsemius et al. (2013) [18] developed a detailed framework in the GLOFRIS model to determine flood hazards while utilizing climate-related datasets. In the future, the framework may be viewed as a multi-model approach, enabling a particular component in the cascade to be used with another element. The flood hazard estimates from Winsemius et al. (2013) [18] were later downscaled to resolutions of 1 km $\times$ 1 km by Ward et al. (2013) [19], in order to account for the risk assessment.

Table 1. A list of widely used global flood models.

Name of the Model	Source
CaMa-Flood; Catchment-Based Macro-scale Floodplain model	http://hydro.iis.u-tokyo.ac.jp/~yamadai/cama-flood/, accessed on 22 March 2021
CIMA-UNEP; Centro Internazionale in Monitoraggio Ambientale and United Nations Environment Program model	https://www.preventionweb.net/organizations/8635, accessed on 10 April 2021
GLOFRIS; Global Flood Risk (model	https://www.globalfloods.eu/, accessed on 15 August 2021
JRC; Joint Research Centre model	https://ec.europa.eu/knowledge4policy/organisation/jrc-joint-research-centre_en, accessed on 16 September 2021
Fathom Global model	https://www.fathom.global/, accessed on 18 April 2021
LIS-FLOOD	http://www.bristol.ac.uk/geography/research/hydrology/models/lisflood/, accessed on 6 May 2021

On the other side, the ready availability of public data sets such as Shuttle Radar Topography Mission (SRTM) DEM, MERIT DEM [20]; hydrologic and meteorological data from reanalysis products [21,22]; GCMs [23,24]; and tide data from global tide elevation [25] have made it easier to develop comprehensive flood model set-ups to derive hazard values. Weather forecasting models use two-dimensional gridded data sets to generate the reanalysis datasets [26]. They serve as an alternative for those regions where station-level observations are scant or sparsely available. Although most studies have utilized runoff outputs from GCMs, there is a need to understand whether reanalysis observations that fuse the station level and satellite observations can also be considered efficient hydraulic inputs to these sophisticated models. Gründemann et al. (2018) [27] used the global Water Resources Reanalysis (WRR) dataset for characterizing floods over the Limpopo River basin in Southern Africa. The authors reported that the models are competent in capturing flood events over stations with a large upstream catchment area. To explore the efficacy of reanalysis products in reconstructing hydro-meteorological hazardous events at the regional scale, Senatore et al. (2020) [28] performed dynamical downscaling of two global reanalyses: ERA-Interim and ERA5 and used them in WRF-Hydro modeling system. Although the study reconstructed three heavy precipitation events, the impact at the inundation scale was not looked into further.

Among all the known natural disasters in Canada, floods are the most severe, as they constitute the highest costs for recovery [29]. Between 1970 and 2015, the total number of flood events increased significantly [30]. Therefore, identifying and quantifying flood risks is indispensable to developing robust adaptation measures and resilience mechanisms. Although past studies have attempted to derive floodplain information for the entire globe, high-resolution flood inundation and hazard information for Canada has not been deeply examined. Exhaustive information on flood hazards will help identify the severely affected regions, and together with vulnerability and exposure components, can determine the degree of flood risk. Being the second-largest country in the world, research and development on various domains of flood management are carried out at the level of each province. Because of this, a unified methodology that is nationally accepted has not been identified so far. Under these situations, it is crucial that a nationwide floodplain mapping be carried out to disseminate the information to various end-users. Recently, a Federal Floodplain Mapping Guidelines Series was launched by Public Safety Canada, which covers all components of the flood mitigation process. The extensive framework consists of four main blocks, namely: (i) Flood Hazard Identification and Priority Setting; (ii) Hydrologic and Hydraulic Procedures for Flood Hazard Delineation; (iii) Geomatics Guidelines for Floodplain mapping; and (iv) Risk-based Land-use Guide. For a holistic approach to flood management, developing a suitable information system that can provide precious flood-related information to various end-users is essential.

Based on the extensive literature review, it is now well established that quantifying flood inundation dynamics over large scales has become easier, with the increasing availability of public datasets and global flood models. However, the usability of public data sets (model inputs) freely available to the research community has not been deeply examined. Information on flood hazards, an essential component of flood risk, has not been extensively derived by past research. Most of them have either derived the overland inundation extent and depths or demonstrated how changes in river channel depths have led to riverine inundation. The exposure of properties to concurrent flooding over Canada has not been studied in detail on such a large scale. In a country like Canada, where several cities have been witnessing a significant increase in settlements and considerable investment is provided towards flood management activities, it is vital to know the degree of exposure of properties. The present study evokes the usability of publicly available datasets for carrying our floodplain mapping and deriving flood hazards at the inundation scale over Canada. With the inundation information, it also attempts to determine the exposure of properties at a decadal time scale. Lastly, it reports on developing a web-based flood information system to efficiently disseminate flood-related information to various end-users. The development of this platform is welcoming to the fact that it witnessed around 100,000 investigations and more than 11 million visitors from around the globe within a month of its release. The following section provides details of multiple datasets and the flood model used to simulate flooding patterns for different scenarios. This section also elaborates on the derivation of flood hazards from inundation values. It is followed by details on quantifying property exposure due to concurrent flooding over the last three decades. The last portion describes the development of a web-based flood information tool. Section 3 provides the results and discussions on inundation mapping, hazard modeling, validation of floodplain maps, and degree of property exposure. The last section gathers the conclusions and future perspectives of the study.

2. Materials and Methods

The proposed framework for capturing hazards through floodplain mapping and disseminating flood-related information through a web-based information system is represented in Figure 1. The framework is demonstrated over the entire of Canada, where floods are known to be the costliest and most frequently occurring among all known natural hazards. On average, around 75% of the budget from the Federal Disaster Financial Assistance Arrangements (DFAA) is utilized for bearing damages from nationwide flooding in Canada. With significant climate change impacts and changing socio-economic dynamics, flood events have escalated in the last century, and mainly severe ones have become quite common [15,31]. Under such situations, it is estimated that around $673 million, i.e., about 75% of the total DFAA, will be necessary to meet the future flood losses in Canada [32].

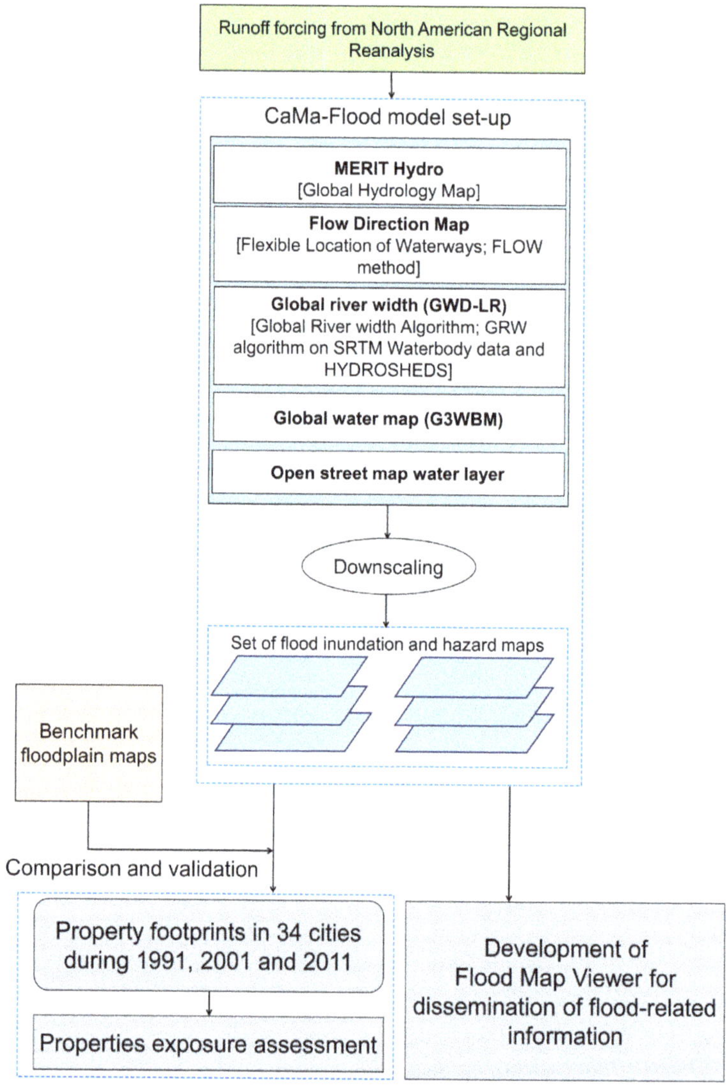

Figure 1. A proposed framework of floodplain mapping.

The extensive framework comprises four major blocks, namely: (i) Selection of readily available runoff observations; (ii) Generation of flood inundation and hazard information through the CaMa-Flood model; (iii) Quantification of decadal property exposure due to flood inundation; and (iv) Development of the Flood Map Viewer. A detailed explanation of each block is provided in the following sections.

2.1. North American Regional Reanalysis (NARR)

The NARR is a readily available product containing high-resolution estimates of various atmospheric and land surface hydrologic variables for the North American region [33]. The detailed hydro-meteorological information embedded in NARR was derived while improving NCEP-NCAR reanalysis datasets [34]. The present work uses the 3-hr runoff data from 1979 to the present with a surface resolution of $0.3^\circ \times 0.3^\circ$ as input forcing to the flood model.

2.2. Catchment-Based Macro-Scale Floodplain (CaMA-Flood) Model

The CaMa-Flood model is a well-known software used to simulate the hydrodynamics of flood waters over large regions [35]. The CaMa-Flood (version v3.6.2) considers the discretized version of river networks as hydrological units, otherwise referred to as unit catchments. The water storage within each unit catchment decides the water levels and inundation extent. This is ensured by considering the sub-grid topographic parameters of overland bathymetry and channel. For the river network map, CaMa-Flood utilizes a grid-based hybrid river network. This arrangement connects one grid to the adjoining unit catchment, resulting in a realistic sub-grid topography parameterization. The local inertial equation calculates the channel discharge and velocity [36]. On the other hand, water storage is decided by the water-balance equation.

The provision of explicit representation of flood-related outputs (water level and flooded area) is considered one of the main advantages of the CaMa-Flood model among all the other models in practice. Being a 2D model, the results derived from CaMa-Flood can be compared and validated at the 1D, i.e., river channel, and 2D, i.e., overland inundation level. Compared to the other existing models, CaMa-Flood provides high numerical efficiency for simulating inundation dynamics over large watersheds. The overland inundation is determined through a robust diagnostic scheme at the unit-catchment scale. The local inertial equation and the adaptive time step scheme optimize the computation of channel discharge and water storage [25,36].

Components of the CaMa-Flood Model

The CaMa-Flood requires a suite of geomorphological and spatial inputs for set-up, which are described in detail in the following sections.

(i) Flow Direction Map: The river network map is derived by employing the Flexible Location of Waterways (FLOW) technique [37]. This method is also utilized to generate the sub-grid-scale topographic parameters such as the length of the channel and its level, elevation of the floodplain and unit-catchment area.

(ii) Global River Width (GWD-LR): The Global Width Database for Large Rivers (GWD-LR) was developed by Yamazaki et al. (2014) [38] by utilizing the SRTM water body mask and flow direction map. This data provides the river widths from bank to bank and effective levels. It has been observed that the effective river width in this dataset is comparatively narrower when compared to a few other existing sources. However, the relative difference does not exceed 20%. Moreover, as the river widths are derived from a reliable flow direction map, the applicability of GWD-LR for large-scale inundation modeling is expected to contain few error percentages.

(iii) Global Water Map: The Global Water Body Map (G3WBM) was introduced by Yamazaki et al. (2015) [39] by implementing an automated algorithm over a set of Landsat images from the Global Land Survey (GLS) database at various temporal scales. They considered around 33,890 scenes from four GLS epochs to create a noise-free water body map. The G3WBM is also free from ice/snow gaps and cloud covers. The water body frequency was identified by overlapping multi-temporal Landsat images to distinguish the permanent water bodies from the temporary ones. It has also been observed that the G3WBM efficiently separates river channels and adjoining floodplains more precisely than other competitive datasets.

(iv) OSM Water Layer: Yamazaki et al. (2019) [40] introduced OSM Water Layer, a global surface water data by extracting water masks from the OSM. The OSM Water Layer is available freely in both PBF and GeoTiff formats. The raster version of the map has four categories, i.e., Major rivers, large rivers and lakes, minor streams, and canals.

2.3. Methodology of Floodplain Mapping

The NARR runoff dataset is downloaded and aggregated on a daily time scale. The runoff estimates within each grid for 100 and 200-year events are generated through extreme value analysis using Generalized Extreme Value (GEV) distribution. This distribution has an upper bound and a flexible tail, and is mathematically represented as

$$\mathrm{F}(x) = \exp\left\{-\left[1-\frac{k(x-\mu)}{\sigma}\right]^{\frac{1}{\xi}}\right\},\ k \neq 0 \tag{1}$$

$$\mathrm{F}(x) = \exp\left\{-\exp\left(-\frac{x-\mu}{\sigma}\right)\right\},\ k = 0 \tag{2}$$

where μ, σ, and k are the location, scale and shape parameters, respectively.

The 100 and 200-year runoff calculated at every grid is considered input to the CaMa-Flood to generate flood inundation parameters, namely inundation depth and extent. Later an efficient downscaling procedure is implemented over the coarse maps to create high-resolution maps at a resolution of 1 km. The floodplain maps for a few regions are extracted and compared with the existing regional maps of flood-prone areas.

2.4. Determination of Flood Hazard

The range of inundation depths derived through flood inundation modeling is utilized to derive flood hazard values. The discretization of hazard values is governed based on the damage to physical assets and human beings [14]. Assuming '($ɗ_n$)' is the depth of inundation associated with each pth grid in the considered domain, the value of flood hazard $\check{H}$ depends upon $ɗ_n \in \mathbf{D}\ \forall\, p \in \mathbf{P}$. Here, $\mathbf{D}$ denotes the set of all flood depths. The flood hazard $\check{H}$ may be expressed as $f(\mathrm{D})$ as described below

$$\begin{aligned}&\zeta\text{: } \mathbf{D} \rightarrow \check{H} \in \mathbf{R}^{+} \text{ such that}\\ &\eth = \zeta\,((ɗ_1), \ldots, (ɗ_n));\ \eth \in \check{H};\ p \in \mathbf{P} \text{ and } (ɗ_1), \ldots, (ɗ_n) \in \mathbf{D}\end{aligned} \tag{3}$$

where $\mathbf{R}^{+}$ represents the set of positive real numbers, while 'p' denotes the total grid cells.

Based on the hazard estimates, the values of $\zeta\,((ɗ_1), \ldots, (ɗ_n))$ are discretized into five classes, as described in Equations (4) and (5)

$$\Omega_{ɗ}\text{: } \check{H} \rightarrow \check{\mathrm{H}}_{ɗ};\ \check{\mathrm{H}}_{ɗ} = \{\hbar_{ɗ} \in \mathbf{P} : \hbar_{ɗ}\ \leq\ 5 \tag{4}$$

$$ɗ_h = \begin{cases} 1,\ 0 \leq ɗ \leq 0.2\\ 2,\ 0.2 < ɗ \leq 0.6\\ 3,\ 0.6 < ɗ \leq 1.5\\ 4,\ 1.5 < ɗ \leq 3.5\\ 5,\ ɗ > 3.5 \end{cases} \tag{5}$$

where ɗand $ɗ_h$ represent the value of flood hazard and its index for the pth grid.

2.5. Quantification of Exposure of Properties Due to Flooding

The continuous 3-hr North American Regional Reanalysis (NARR) runoff values for 1991, 2001 and 2011 were considered as hydraulic inputs to the CaMa-Flood model set-up along with the other relevant geomorphological inputs. The simulated maximum flood depth map for the three-time periods is further utilized to quantify the degree of exposure of properties. The property footprints for 1991, 2001 and 2011 are obtained from the housing statistics, Statistics Canada (www.statcan.gc.ca, accessed on 16 June 2021). Based on data availability, 34 cities are considered that fall on Canada's list of 100 most populous cities [16].

The footprints of the properties during 1991, 2001 and 2011 are overlaid on the maximum flood depth map simulated during the corresponding years to estimate the total number of properties exposed due to concurrent flooding.

2.6. Development of Flood Map Viewer

To disseminate flood-related information to various stakeholders and local communities in Canada, Flood Map Viewer (http://www.floodmapviewer.com/, accessed on 30 June 2021)—a web-based, freely accessible public tool, has been developed. The web-based tool considers entire Canada as the base map. It presents flood-related information for 100 and 200 years for various historical and climate change scenarios at a high resolution of 1 km × 1 km. Special attention is given to user-friendliness to ensure that the information can be understood by a non-technical end-user [35].

3. Results

3.1. Floodplain Maps Derived by Utilizing NARR

The 100-year and 200-year runoff are inputs to the CaMa-Flood model to generate Canada-wide floodplain maps. These high-resolution floodplain maps (resolution of 1 km × 1 km) are illustrated in Figure 2, overlaid on the MERIT DEM. The percentage of maximum water depth, an indicator of flood hazard, is presented in pie charts inside these maps. We notice that about 10.10 percent of inundated regions fall within the low-hazard category (depth below 0.2 m) during 100-year (Figure 2a) as compared to 9.68 percent for a 200-year flood event (Figure 2b). On the contrary, the cumulative percentage classes close to 80 percent for the latter, roughly 4 percent more than observed during the 100-year event. The remaining classes of flood hazards for 100-year and 200-year events constitute 8.72 and 7.89 percent (low hazard), 5.03, and 2.72 percent (medium hazard). Very-low and low flood hazard classes are mostly found over Canada's north and central regions. The efficacy of representing extreme events by NARR over Canada is well documented in the recent literature [35,41]. The observations confirm the excellent performance of NARR as a suitable input parameter, as it can capture the high-and very-high hazard spots well. In a recent article, Essou et al. (2016) [42] highlight that NARR uses a 3D-VAR assimilation approach that provides high efficiency in representing extreme events such as floods.

3.2. Validation of Floodplain Maps with Benchmark Maps

The 100-year and 200-year floodplain maps generated through computational modeling are compared with a few existing floodplain maps over six regions in Canada. The spatial flood-related information in these maps is obtained from the respective basins by considering precise and extensive input data sets such as high-resolution topography, river cross-section details, etc., while using a regional flood inundation model. As such, comparing simulated floodplain maps with the existing ones over the regions will ensure the performance of the flood model. Figure 3 compares simulated and existing floodplain maps over six areas: Lower Fraser River Basin, Grand River Basin, St. John Basin, Calgary basin, Assiniboine Basin and Red River basin. The enlarged version of the maps is provided in the Supplementary Document (Figures S1–S6). Overall, all the simulated maps perform well in representing the inundation dynamics over these regions.

While comparing the flood inundation over Fraser Basin (geographical extent ~ 11,134 km^2) for 100 years and 200 years, we notice more than 3/4th of inundated spots in the existing floodplain maps comply with the simulated ones (Figure 3a). With a 200-year map, we see an underestimation of a few areas in the western coastal region, as highlighted by red-colored grids. This is because the flood model set-up considered in this study accounts for the riverine inundation. While demonstrating the comparison over Calgary Basin (geographical extent ~1345 km^2), we notice a high similarity with the existing floodplain maps. Earlier, Sampson et al. (2015) [43] also noticed a similar behavior while comparing their simulated floodplain maps over Calgary. The authors used SRTM DEM to represent the topographical features in the flood model. The present study exhibits a higher degree of similarity with the benchmark floodplain map due to the consideration of MERIT DEM, in which significant noise corrections over SRTM DEM have been conducted.

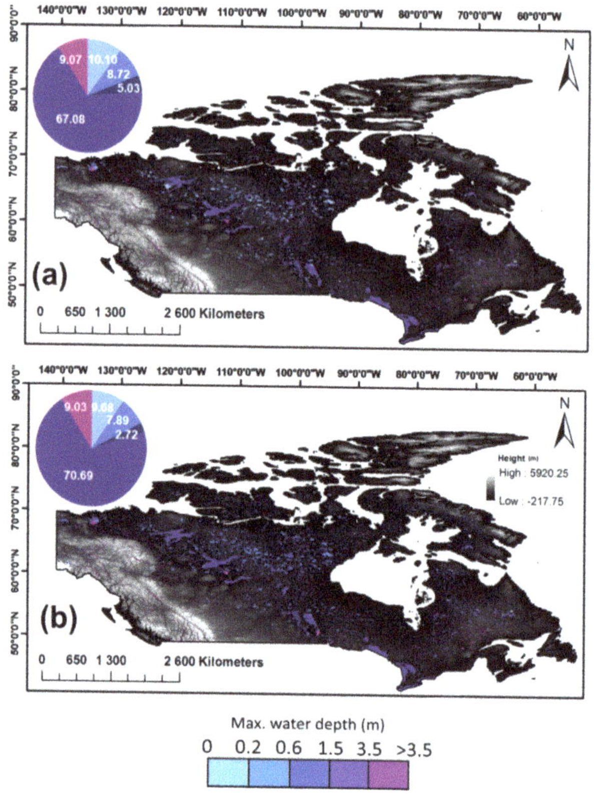

Figure 2. (**a**) 100-year and (**b**) 200-year floodplain maps derived by utilizing NARR. The inset pie-chart provides details on the percentage of area exposed to a particular degree of flood hazard.

Past research has confirmed a high degree of performance of MERIT DEM in floodplain mapping, synonymous with LiDAR DEMs over various case studies [44–47]. The Assiniboine Basin (geographical extent ~162,000 km^2) is a large river basin consisting of Qu'Appelle, Souris, and Assiniboine sub-basins. This region's past flood events in 2011 and 2014 incurred huge economic and physical losses. Blais et al. (2016) [48] regard the 2011 event as the most extreme event that Canada has witnessed. The NARR-derived floodplain maps for both scenarios match the benchmark floodplain inundated spots. A slight discrepancy is noticed over the northern and eastern parts. The possible reason for this observation is the rapid changes in the channel slope of the Assiniboine River at various chainages over the study region, which may not be precisely captured by the MERIT DEM [49]. The red river basin (geographical extent ~119,000 km^2) resides beside the Assiniboine basin. As the red river flows northwards, it follows a meandering river course, which aggravates the chances of inundation near the floodplains during extreme weather events.

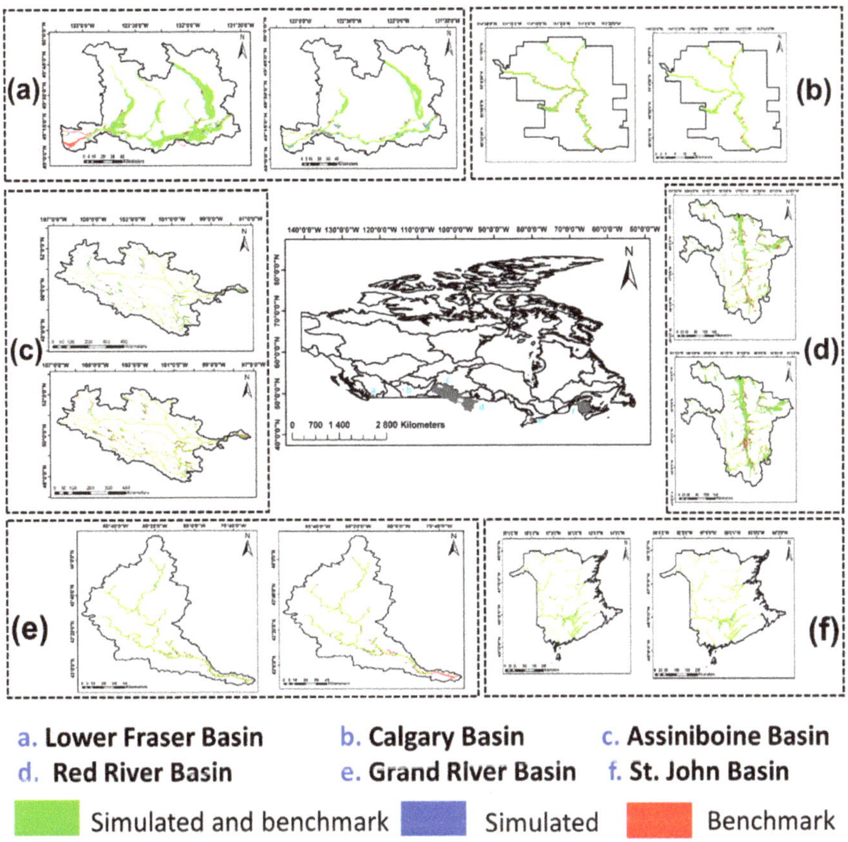

Figure 3. Comparison and validation of generated floodplain maps (both for 100 years and 200 years) against existing ones over six flood-prone regions of Canada, namely (**a**) Lower Fraser Basin, (**b**) Calgary Basin, (**c**) Assiniboine Basin, (**d**) Red River Basin, (**e**) Grand River Basin, and (**f**) St. John Basin. The enlarged version of the sub-maps is provided in the Supplementary Document (Figures S1–S6).

The flood event that occurred in 1997 over the Red river basin is often referred to as the "Flood of the Century" [50,51]. Through a visual comparison, it is clear that the simulated floodplain matches very closely with the existing flood inundated spots. The complete riverine inundation is adequately simulated over the study extent. A limited number of small stream networks are left behind. This is due to the inclusion of Global River Width data as an input, which cannot identify river widths of size less than 183 m. The Grand Basin (geographical extent ~6800 km^2) in Southwest Ontario consists of three distinct landforms: plains in the northern and western parts, moraines in the eastern and central portions, and clay in the southern. This region's major flood drivers are rapid snowmelt with concurrent rainfall and high surge. The simulated floodplain maps capture the inundated spots competitively. However, a slight underestimation is observed with the 200-year floodplain map over the basin's lower stretch. The St. John Basin (geographical extent ~12,222 km^2) is between Quebec and New Brunswick. The primary drivers of concurrent flooding occur from April to May from the runoff resulting from the melting snowpack. A major spring flooding was seen in 2008, when a relatively warmer climate resulted in a higher melting of snowpack. Although the St. John basin is a coastal region, the inundated spots over the southern stretches are simulated reasonably well by NARR

(Figure 3f). A few areas of over-predicted and under-predicted grids are noticed, possibly due to noise in MERIT DEM elevation values.

3.3. Exposure of Property to Flooding at a Decadal Time Scale

The property footprints were overlaid on the floodplain maps simulated for 1991, 2001, and 2011 with the NARR dataset. As mentioned in Section 2.5 34 cities falling in Canada's 100 most populous cities were collected based on data availability. The distribution of these cities over the provinces is as follows: two cities each in Alberta, Saskatchewan and New Brunswick; four cities in British Columbia; fifteen in Ontario; six in Quebec; and one in each in Manitoba, Nova Scotia, and Newfoundland and Labrador.

The number of properties exposed to concurrent floods in 1991, 2001 and 2011 is illustrated in Figure 4. Overall, there has been an increase in the number of properties exposed to flooding in the last three decades. Cities such as Vancouver, Victoria and Abbotsford-Mission in British Columbia showed more or less an increase in the properties exposed to flooding in 2001 and 2011. Edmonton recorded a steep rise of nearly four times in the number of properties exposed to floods in 2011 compared to 1991. Our results are supported by the recent findings by Chakraborty et al. (2021) [52]. They reported that British Columbia contains the second-highest percentage of the population exposed to flooding. Most cities in Ontario depict a rise in numbers, with the maximum being in Toronto (1997 in 1991, 2020 in 2001 and 2455 in 2011), followed by Hamilton 1224 in 1991, 1522 in 2001, and 1601 in 2011). A similar pattern is noticed in Quebec, where the number of properties exposed to flooding remained high (more than 2000) throughout the decades. The remaining cities in Quebec did not show a rapid change in the numbers over the years. Cities in New Brunswick indicate a sparse number of properties exposed to flooding, similar to Nova Scotia, Newfoundland and Labrador.

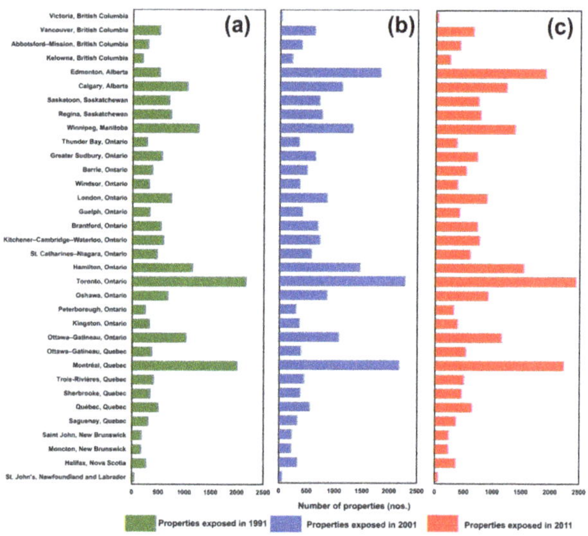

Figure 4. Number of properties exposed to flooding in (**a**) 1991, (**b**) 2001 and (**c**) 2011.

3.4. Flood Map Viewer

As highlighted in Section 2.6, Flood Map Viewer is a web-based tool developed to disseminate Canada-wide flood-related information. The tool aims to help raise awareness of the general public, professionals in the field and accountable agencies about the existing flood risk in Canada and its change under future climate conditions. The overlay of postal codes (available in the Flood Map Viewer) allows search for potential impacts up to the street and property level. Flood Map Viewer is efficient in terms of its spatial coverage, accessibility, ease of operation and visualization compared to similar web-based platforms, such as Albano et al. (2015) [53], Henriksen et al. (2018) [54], and Xu et al. 2020 [55]. Upon opening the website, the first page details various options, namely, Maps, Download, Learn More, and User Guide (Figure 5). The user can click on the 'Maps' option to explore the flood maps. At the top of this page, various possibilities for overlaying on the base map, such as Canada DEM, Drainage Basins and Postal Codes, are present (Figure 6a,b).

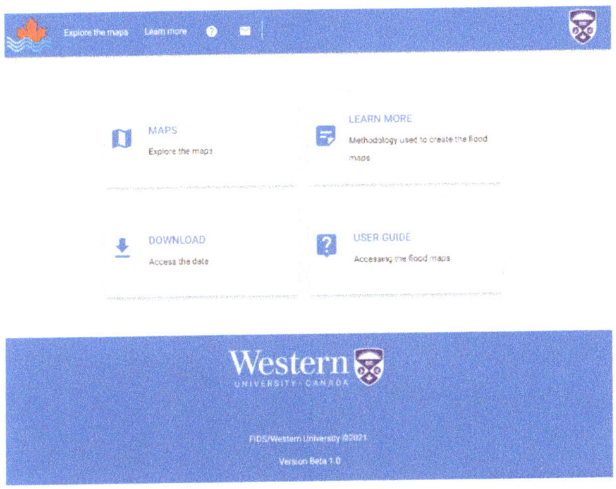

Figure 5. A screenshot of the first page of Flood Map Viewer.

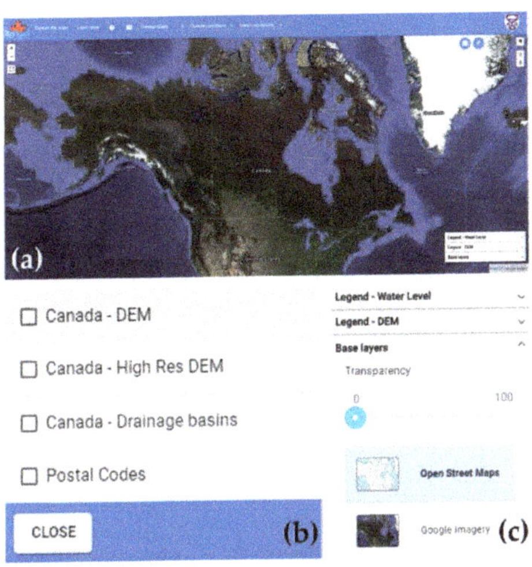

Figure 6. A screenshot of the "Maps" option in Flood Map Viewer. (**a**) depicts the main page showing the entire map of Canada, (**b**) represents the various options available for overlaying the DEM, drainage basins and postal codes over the floodplain maps, and (**c**) depicts the various options for enhancing visualization of the floodplain maps.

The second and third options provide details on the flood-related maps for the current and future conditions). Users have the flexibility to change the transparency and change the base map to an open-street map or Google Earth imagery as per their requirement (Figure 6c). The flood-related maps are available on the first page in the 'Download' option.

After clicking on this option, a set of files ranging from flood maps to spatial boundaries are available for free download. The 'Learn More' option describes the entire methodology used for floodplain mapping. The last option, 'User Guide' provides a step-by-step guide for accessing the flood-related information on the website for any user. A representative flood plain map is illustrated in Figure 7.

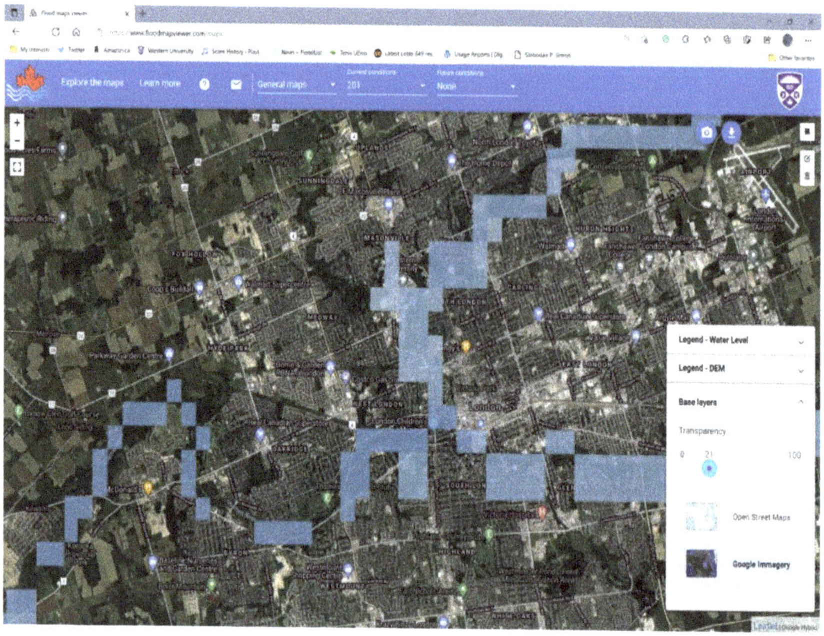

Figure 7. A representative illustration of the Flood Map Viewer (https://www.floodmapviewer.com/, accessed on 30 June 2021).

4. Discussions

This study reports on using freely available public datasets for flood inundation and hazard modeling over large regions and further shows how it can be extended to exposure assessment. The study depicts an improvement in inundation modeling as far as the resolution of the flood maps is concerned. Previous studies such as Bernhofen et al. (2018) [11], Boulange et al. (2021) [12] and Winsemius et al. (2013) [18] simulated inundation statistics at a coarser resolution, which might not be adequate further for risk characterization, and precise flood management. It is customary to obtain inundation statistics comparatively at a lower resolution with global flood models than with regional or mesoscale flood models. The reason is that the hydrological, meteorological and spatial information required to run the global flood models are available at a coarser scale. Nevertheless, the primary goal is to identify flood hot spots over Canada that have been unaddressed. It would draw attention to detailed inundation modeling at the local/regional scale. At the same time, it also provides vital information on the severity of flood hazards over the inundated sites, which is paramount for multiple stakeholders, including the Federal Government, for quantifying flood risk and implementing structural and non-structural flood control measures for flood protection.

The maximum flood depth maps for 1991, 2001 and 2011 are later utilized to characterize the exposure of properties on a nationwide scale. Although many studies have described exposure to flooding at regional scales, the present study attempts to derive it at a country-wide scale over Canada that has witnessed a rise in population and flooding in tandem in recent times. A web-based information system–the Flood Map Viewer–is developed as the final outcome to ensure holistic flood management by disseminating flood-related data. The Flood Map Viewer indicates the inundated areas during 100-year and 200-year return period events. Users can utilize a suite of options to ease the visualization of flood maps. End-users can also download the flood-related information, thus removing the barrier to sharing information between the technical flood model experts and stakeholders.

5. Conclusions and Future Perspectives

With rising flood risk over several regions in the world, water experts are identifying the utility of floodplain mapping as a sustainable tool for long-term flood management and protection. Despite most studies focusing on small regions, the possibility of expanding floodplain mapping research to broader scales was not attempted until the last few years. This has been made possible with the ready availability of global datasets and state-of-the-art numerical models for replicating dynamics of flood wave propagation. The present study explores the usability of publicly available datasets in characterizing flood hazards over Canada, and its implications for property exposure. It also emphasizes the need to develop a web-based information system as an efficient medium of communicating flood-related information to various end-users. NARR and a few other relevant datasets are considered primary inputs to CaMa-Flood, an acclaimed global flood modeling software to generate 100-year and 200-year Canada-wide floodplain maps. The simulated maps are compared and validated with the existing ones over six different flood-prone regions, based on which they are utilized to quantify the degree of exposure of properties over the last three decades. The development of the Flood Map Viewer ensures various stakeholders can easily observe and use the high-resolution flood-related information for disaster mitigation, land-use planning and resilience development. Future studies may consider recent higher resolution DEM products such as ALOS-PALSAR to test the performance of CaMa-Flood model outputs. For a holistic identification of flood hazard spots over the coastal belt, future studies may couple the existing CaMa-Flood model set-up with an ocean circulation model such as the ADvanced CIRCulation model (ADCIRC). The flood hazard information derived from the coupled set-up can be used along with vulnerability and exposure components to estimate flood risk to communities and assets. Such exhaustive information on flood risk may be considered a precious cartographic product by the disaster management experts and city planners, which would further assist in selecting suitable flood control measures for upgraded environmental development and management.

Supplementary Materials: The following supporting information can be downloaded at: https://www.mdpi.com/article/10.3390/w14142280/s1, Figure S1: Comparison of simulated and benchmark floodplain maps derived for (a) 100 and (b) 200-yr return periods for Lower Fraser Basin. Figure S2: Comparison of simulated and benchmark floodplain maps derived for (a) 100 and (b) 200-yr return periods for Calgary Basin. Figure S3: Comparison of simulated and benchmark floodplain maps derived for (a) 100 and (b) 200-yr return periods for Assiniboine Basin. Figure S4: Comparison of simulated and benchmark floodplain maps derived for (a) 100 and (b) 200-yr return periods for Red River Basin. Figure S5: Comparison of simulated and benchmark floodplain maps derived for (a) 100 and (b) 200-yr return periods for Grand River Basin. Figure S6: Comparison of simulated and benchmark floodplain maps derived for (a) 100 and (b) 200-yr return periods for St. John Basin.

Author Contributions: M.P.M., conceptualization, methodology, software, data processing and original draft preparation; and S.P.S., supervision, methodology, results review, reviewing and editing the manuscript. All authors have read and agreed to the published version of the manuscript.

Funding: The authors are grateful to the Natural Sciences and Engineering Research Council of Canada (NSERC) (No. CRDPJ 472152-14) and the Institute for Catastrophic Loss Reduction (ICLR) for funding the research.

Institutional Review Board Statement: Not applicable.

Informed Consent Statement: Not applicable.

Data Availability Statement: The NARR runoff data is available for free download at https://psl.noaa.gov/, accessed on 23 January 2021. The CaMa-Flood model and its corresponding modules can be downloaded from http://hydro.iis.u-tokyo.ac.jp/~yamadai/cama-flood/, accessed on 20 January 2021. Other relevant river and topographic-related details are available on the same web address, and can be acquired from the developer upon reasonable request. The Canada-wide floodplain maps derived by utilizing NARR are available for visualization and free download at https://www.floodmapviewer.com/ accessed on 15 July 2022.

Acknowledgments: The authors thank Dai Yamazaki for providing CaMa-Flood version v4.01. The authors acknowledge SHARCNET (www.sharcnet.ca accessed on 15 July 2022) for delivering a high super-computing facility for extensive model simulations. The existing floodplain maps over the six basins were collected from the respective river basin organizations. The authors thank the Fraser Basin Council for providing 100-year and 200-year floodplain information over the Lower Fraser Basin. The authors acknowledge Stats Canada for providing details on the property-related information.

Conflicts of Interest: The authors declare no conflict of interest.

Glossary

CaMA-Flood	Catchment-based Macro-scale Floodplain
D	Set of all flood depths
d_n	Inundation depth for a particular grid
FLOW	The Flexible Location of Waterways technique
G3WBM	Global Water Body Map
GEV	Generalized Extreme Value
GWD-LR	Global River Width
$\check{H}$	Flood Hazard
MERIT DEM	Multi-Error-Removed Improved-Terrain DEM
NARR	North American Regional Reanalysis
OSM	Open Street Map
p	Total number of grid cells in the flood model domain

References

1. Winsemius, H.C.; Aerts, J.C.J.H.; Van Beek, L.P.H.; Bierkens, M.F.P.; Bouwman, A.; Jong-man, B.; Kwadijk, J.C.J.; Ligtvoet, W.; Lucas, P.L.; van Vuuren, D.P.; et al. Global drivers of future river flood risk. *Nat. Clim. Chang.* **2016**, *6*, 381–385. [CrossRef]
2. Kinoshita, Y.; Tanoue, M.; Watanabe, S.; Hirabayashi, Y. Quantifying the effect of autonomous adaptation to global river flood projections: Application to future flood risk assessments. *Environ. Res. Lett.* **2018**, *13*, 014006. [CrossRef]
3. Menéndez, P.; Losada, I.J.; Torres-Ortega, S.; Narayan, S.; Beck, M.W. The global flood protection benefits of mangroves. *Sci. Rep.* **2020**, *10*, 4404. [CrossRef]
4. Munich, Re. NatCatSERVICE Database. Munich RE, Munich. 2018. Available online: https://www.munichre.com/en/solutions/for-industry-clients/natcatservice.html (accessed on 25 March 2022).
5. UNISDR—The United Nations Office for Disaster Risk Reduction: The Human Cost of Weather-Related Disasters 1995–2015. 2015. Available online: https://www.unisdr.org/files/46796_cop21weatherdisastersreport2015.pdf (accessed on 22 March 2022).
6. Willner, S.N.; Otto, C.; Levermann, A. Global economic response to river floods. *Nat. Clim. Chang.* **2018**, *8*, 594–598. [CrossRef]
7. Wing, O.E.; Bates, P.D.; Smith, A.M.; Sampson, C.C.; Johnson, K.A.; Fargione, J.; Morefield, P. Estimates of present and future flood risk in the conterminous United States. *Environ. Res. Lett.* **2018**, *13*, 034023. [CrossRef]
8. Vousdoukas, M.I.; Mentaschi, L.; Voukouvalas, E.; Bianchi, A.; Dottori, F.; Feyen, L. Climatic and socio-economic controls of future coastal flood risk in Europe. *Nat. Clim. Chang.* **2018**, *8*, 776–780. [CrossRef]
9. Smith, A.; Bates, P.D.; Wing, O.; Sampson, C.; Quinn, N.; Neal, J. New estimates of flood exposure in developing countries using high-resolution population data. *Nat. Commun.* **2019**, *10*, 1814. [CrossRef]
10. Dottori, F.; Szewczyk, W.; Ciscar, J.-C.; Zhao, F.; Alfieri, L.; Hirabayashi, Y.; Bianchi, A.; Mongelli, I.; Frieler, K.; Betts, R.A.; et al. Increased human and economic losses from river flooding with anthropogenic warming. *Nat. Clim. Chang.* **2018**, *8*, 781–786. [CrossRef]
11. Bernhofen, M.V.; Whyman, C.; Trigg, M.A.; Sleigh, P.A.; Smith, A.M.; Sampson, C.C.; Yamazaki, D.; Ward, P.J.; Rudari, R.; Pappenberger, F.; et al. A first collective validation of global fluvial flood models for major floods in Nigeria and Mozambique. *Environ. Res. Lett.* **2018**, *13*, 104007. [CrossRef]
12. Boulange, J.; Hanasaki, N.; Yamazaki, D.; Pokhrel, Y. Role of dams in reducing global flood exposure under climate change. *Nat. Commun.* **2021**, *12*, 417. [CrossRef]
13. Alfieri, L.; Bisselink, B.; Dottori, F.; Naumann, G.; de Roo, A.; Salamon, P.; Wyser, K.; Feyen, L. Global projections of river flood risk in a warmer world. *Earth's Future* **2017**, *5*, 171–182. [CrossRef]

14. Mohanty, M.P.; Vittal, H.; Yadav, V.; Ghosh, S.; Rao, G.S.; Karmakar, S. A new bivariate risk classifier for flood management considering hazard and socio-economic dimensions. *J. Environ. Manag.* **2020**, *255*, 109733. [CrossRef] [PubMed]
15. Gaur, A.; Gaur, A.; Simonovic, S.P. Future Changes in Flood Hazards across Canada under a Changing Climate. *Water* **2018**, *10*, 1441. [CrossRef]
16. Gaur, A.; Gaur, A.; Yamazaki, D.; Simonovic, S.P. Flooding related consequences of climate change on Canadian cities and flow regulation infrastructure. *Water* **2019**, *11*, 63. [CrossRef]
17. Lim, W.H.; Yamazaki, D.; Koirala, S.; Hirabayashi, Y.; Kanae, S.; Dadson, S.J.; Hall, W.H.; Sun, F. Long-term changes in global socio-economic benefits of flood defenses and residual risk based on CMIP5 climate models. *Earth's Future* **2018**, *6*, 938–954. [CrossRef]
18. Winsemius, H.C.; Van Beek, L.P.H.; Jongman, B.; Ward, P.J.; Bouwman, A. A framework for global river flood risk assessments. *Hydrol. Earth Syst. Sci.* **2013**, *17*, 1871–1892. [CrossRef]
19. Ward, P.J.; Jongman, B.; Weiland, F.S.; Bouwman, A.; Van Beek, R.; Bierkens, M.F.P.; Ligtvoet, W.; Winsemius, H.C. Assessing flood risk at the global scale: Model set-up, results, and sensitivity. *Environ. Res. Lett.* **2013**, *8*, 044019. [CrossRef]
20. Yamazaki, D.; Ikeshima, D.; Tawatari, R.; Yamaguchi, T.; O'Loughlin, F.; Neal, J.C.; Sampson, C.C.; Kanae, S.; Bates, P.D. A high-accuracy map of global terrain elevations. *Geophys. Res. Lett.* **2017**, *44*, 5844–5853. [CrossRef]
21. Tarek, M.; Brissette, F.P.; Arsenault, R. Evaluation of the ERA5 reanalysis as a potential reference dataset for hydrological modelling over North America. *Hydrol. Earth Syst. Sci.* **2020**, *24*, 2527–2544. [CrossRef]
22. Wang, N.; Liu, W.; Sun, F.; Yao, Z.; Wang, H.; Liu, W. Evaluating satellite-based and reanalysis precipitation datasets with gauge-observed data and hydrological modeling in the Xihe River Basin, China. *Atmos. Res.* **2020**, *234*, 104746. [CrossRef]
23. Bermúdez, M.; Cea, L.; Van Uytven, E.; Willems, P.; Farfán, J.F.; Puertas, J. A robust method to update local river inundation maps using global climate model output and weather typing based statistical downscaling. *Water Resour. Manag.* **2020**, *34*, 4345–4362. [CrossRef]
24. Toosi, A.S.; Doulabian, S.; Tousi, E.G.; Calbimonte, G.H.; Alaghmand, S. Large-scale flood hazard assessment under climate change: A case study. *Ecol. Eng.* **2020**, *147*, 105765. [CrossRef]
25. Hunter, J.R.; Woodworth, P.L.; Wahl, T.; Nicholls, R.J. Using global tide gauge data to validate and improve the representation of extreme sea levels in flood impact studies. *Glob. Planet. Chang.* **2017**, *156*, 34–45. [CrossRef]
26. Compo, G.P.; Whitaker, J.S.; Sardeshmukh, P.D.; Matsui, N.; Allan, R.J.; Yin, X.; Gleason, B.E.; Vose, R.S.; Rutledge, G.; Bessemoulin, P.; et al. The twentieth century reanalysis project. *Q. J. R. Meteorol. Soc.* **2011**, *137*, 1–28. [CrossRef]
27. Gründemann, G.J.; Werner, M.; Veldkamp, T.I. The potential of global reanalysis datasets in identifying flood events in Southern Africa. *Hydrol. Earth Syst. Sci.* **2018**, *22*, 4667–4683. [CrossRef]
28. Senatore, A.; Davolio, S.; Furnari, L.; Mendicino, G. Reconstructing flood events in Mediterranean coastal areas using different reanalyses and high-resolution meteorological models. *J. Hydrometeorol.* **2020**, *21*, 1865–1887. [CrossRef]
29. NRCan (2018) Natural Resources Canada. Federal Floodplain Mapping Framework. Available online: https://geoscan.nrcan.gc.ca/starweb/geoscan/servlet.starweb?path=geoscan/fulle.web&search1=R=308128 (accessed on 7 June 2022).
30. Public Safety Canada (2017) Floods. Available online: https://www.publicsafety.gc.ca/cnt/mrgnc-mngmnt/ntrlhzrds/fld-en.aspx\T1\textgreater{} (accessed on 10 March 2022).
31. Zadeh, S.M.; Burn, D.H.; O'Brien, N. Detection of trends in flood magnitude and frequency in Canada. *J. Hydrol. Reg. Stud.* **2020**, *28*, 100673. [CrossRef]
32. Parliamentary Budget Officer of Canada. Estimate of the Average Annual Cost for Disaster Financial Assistance Arrangements Due to Weather Events. 2016. Available online: https://www.pbo-dpb.gc.ca/web/default/files/Documents/Reports/2016/DFAA/DFAA_EN.pdf (accessed on 3 June 2022).
33. Mesinger, F.; DiMego, G.; Kalnay, E.; Mitchell, K.; Shafran, P.C.; Ebisuzaki, W.; Jovic, D.; Woollen, J.; Rogers, E.; Berbery, E.H.; et al. North American regional reanalysis. *Bull. Am. Meteorol. Soc.* **2006**, *87*, 343–360. [CrossRef]
34. Ashtine, M.; Bello, R.; Higuchi, K. Assessment of wind energy potential over Ontario and Great Lakes using the NARR data: 1980–2012. *Renew. Sustain. Energy Rev.* **2016**, *56*, 272–282. [CrossRef]
35. Mohanty, M.P.; Simonovic, S.P. Changes in floodplain regimes over Canada due to climate change impacts: Observations from CMIP6 models. *Sci. Total Environ.* **2021**, *792*, 148323. [CrossRef]
36. Bates, P.D.; Horritt, M.S.; Fewtrell, T.J. A simple inertial formulation of the shallow water equations for efficient two-dimensional flood inundation modelling. *J. Hydrol.* **2010**, *387*, 33–45. [CrossRef]
37. Yamazaki, D.; Oki, T.; Kanae, S. Deriving a global river network map and its subgrid topographic characteristics from a fine-resolution flow direction map. *Hydrol. Earth Syst. Sci.* **2009**, *13*, 2241. [CrossRef]
38. Yamazaki, D.; O'Loughlin, F.; Trigg, M.A.; Miller, Z.F.; Pavelsky, T.M.; Bates, P.D. Development of the global width database for large rivers. *Water Resour. Res.* **2014**, *50*, 3467–3480. [CrossRef]
39. Yamazaki, D.; Trigg, M.A.; Ikeshima, D. Development of a global ~90 m water body map using multi-temporal Landsat images. *Remote Sens. Environ.* **2015**, *171*, 337–351. [CrossRef]
40. Yamazaki, D.; Ikeshima, D.; Sosa, J.; Bates, P.D.; Allen, G.H.; Pavelsky, T.M. MERIT Hydro: A high-resolution global hydrography map based on latest topography dataset. *Water Resour. Res.* **2019**, *55*, 5053–5073. [CrossRef]
41. Kim, S.J.; Lee, M.; Choi, W.; Rasmussen, P.F. Utilizing North American Regional Reanalysis for climate change impact assessment on water resources in central Canada. In Proceedings of the 13th World Water Congress, Montpellier, France, 1–4 September 2008.

42. Essou, G.R.; Sabarly, F.; Lucas-Picher, P.; Brissette, F.; Poulin, A. Can precipitation and temperature from meteorological reanalyses be used for hydrological modeling? *J. Hydrometeorol.* **2016**, *17*, 1929–1950. [CrossRef]
43. Sampson, C.C.; Smith, A.M.; Bates, P.D.; Neal, J.C.; Alfieri, L.; Freer, J.E. A high-resolution global flood hazard model. *Water Resour. Res.* **2015**, *51*, 7358–7381. [CrossRef]
44. Hawker, L.; Rougier, J.; Neal, J.; Bates, P.; Archer, L.; Yamazaki, D. Implications of simulating global digital elevation models for flood inundation studies. *Water Resour. Res.* **2018**, *54*, 7910–7928. [CrossRef]
45. Liu, Y.; Bates, P.D.; Neal, J.C.; Yamazaki, D. Bare-Earth DEM Generation in Urban Areas for Flood Inundation Simulation Using Global Digital Elevation Models. *Water Resour. Res.* **2021**, *57*, e2020WR028516. [CrossRef]
46. Hao, C.; Yunus, A.P.; Subramanian, S.S.; Avtar, R. Basin-wide flood depth and exposure mapping from SAR images and machine learning models. *J. Environ. Manag.* **2021**, *297*, 113367. [CrossRef]
47. Kirezci, E.; Young, I.R.; Ranasinghe, R.; Muis, S.; Nicholls, R.J.; Lincke, D.; Hinkel, J. Projections of global-scale extreme sea levels and resulting episodic coastal flooding over the 21st century. *Sci. Rep.* **2020**, *10*, 11629. [CrossRef] [PubMed]
48. Blais, E.-L.; Greshuk, J.; Stadnyk, T. The 2011 flood event in the Assiniboine River Basin: Causes, assessment and damages. *Can. Water Resour. J. Rev. Can. Des Ressour. Hydr.* **2016**, *41*, 74–84. [CrossRef]
49. Tavares da Costa, R.; Mazzoli, P.; Bagli, S. Limitations posed by Free DEMs in watershed studies: The case of river Tanaro in Italy. *Front. Earth Sci.* **2019**, *7*, 141. [CrossRef]
50. Rannie, W. The 1997 flood event in the Red River basin: Causes, assessment and damages. *Can. Water Resour. J. Rev. Can. Des Ressour. Hydr.* **2016**, *41*, 45–55. [CrossRef]
51. Simonovic, S.P. Decision support system for flood management in the Red River Basin. *Can. Water Resour. J.* **1999**, *24*, 203–223. [CrossRef]
52. Chakraborty, L.; Thistlethwaite, J.; Minano, A.; Henstra, D.; Scott, D. Leveraging Hazard, Exposure, and Social Vulnerability Data to Assess Flood Risk to Indigenous Communities in Canada. *Int. J. Disaster Risk Sci.* **2021**, *12*, 821–838. [CrossRef]
53. Albano, R.; Sole, A.; Adamowski, J. READY: A web-based geographical information system for enhanced flood resilience through raising awareness in citizens. *Nat. Hazards Earth Syst. Sci.* **2015**, *15*, 1645–1658. [CrossRef]
54. Xu, H.; Windsor, M.; Muste, M.; Demir, I. A web-based decision support system for collaborative mitigation of multiple water-related hazards using serious gaming. *J. Environ. Manag.* **2020**, *255*, 109887. [CrossRef]
55. Henriksen, H.J.; Roberts, M.J.; van der Keur, P.; Harjanne, A.; Egilson, D.; Alfonso, L. Participatory early warning and monitoring systems: A Nordic framework for web-based flood risk management. *Int. J. Disaster Risk Reduct.* **2018**, *31*, 1295–1306. [CrossRef]

water

MDPI

Article

Flood Management Issues in Hilly Regions of Uttarakhand (India) under Changing Climatic Conditions

Mitthan Lal Kansal * and Sachchidanand Singh

Department of Water Resources Development and Management, Indian Institute of Technology, Roorkee 247667, India; ssingh1@wr.iitr.ac.in
* Correspondence: mlk@wr.iitr.ac.in

Abstract: Uttarakhand, an Indian Himalayan state in India, is famous for its natural environment, health rejuvenation, adventure, and a pilgrimage centre for various religions. It is categorised into two major regions, i.e., the Garhwal and the Kumaon, and geographically, the Bhabar and the Terai. Floods, cloudbursts, glacier lake outbursts, and landslides are the major natural hazards that cause the highest number of mortalities and property damage in this state. After becoming a full 27th state of India in 2000, the developmental activities have increased many folds, which has added to such calamities. This study briefly summarises the major incidents of flood damage, describes the fragile geology of this Himalayan state, and identifies the natural as well as the anthropogenic causes of the flood as a disaster. It also highlights the issue of climate change in the state and its adverse impact in the form of extreme precipitation. Besides these, it reviews the challenges involved in flood management and highlights the effective flood risk management plan that may be adopted to alleviate its adverse impacts.

Keywords: flood management issues; natural disasters; climate change; government policies

Citation: Kansal, M.L.; Singh, S. Flood Management Issues in Hilly Regions of Uttarakhand (India) under Changing Climatic Conditions. *Water* **2022**, *14*, 1879. https://doi.org/10.3390/w14121879

Academic Editors: Slobodan P. Simonovic, Subhankar Karmakar and Zhang Cheng

Received: 11 February 2022
Accepted: 7 June 2022
Published: 10 June 2022

1. Introduction

The occurrence of natural disasters and climate change are the two complex and most challenging issues which have long-term impacts on any country's environmental, economic, and social well-being. Researchers have focused on forecasting extreme hydro-meteorological events, such as cloudbursts and extreme precipitation, and their increased intensity and frequency, especially in the mountainous regions [1]. The mountainous region's characteristics such as fragile geology and restricted or marginal accessibility further escalate such hazards into a disaster [2]. Flash flood, along with the landslides in mountainous regions, is the most catastrophic natural disaster that causes the loss of lives and damage to infrastructure and other economic activities [3]. In the last 30 years, flood alone has contributed to about 50% of the total affected 80 million people due to natural disasters. On the economic front, flood damage is worth more than USD 11 million around the world [4]. It is feared that precipitation extremes and the ensuing flooding are expected to become even more frequent in most parts of Asia, Africa, and South and Southeast Asia in the next decades due to climate change. Moreover, these countries face significant difficulty acquiring land use, topography, and hydro-meteorological information, contributing to major uncertainty in flood management studies [5]. Among Asian nations, India has the most flood-related fatalities and is highly exposed to disaster risks. In India, planning and enacting a climate response strategy is even more difficult as the country has vast geographic heterogeneity [6]. India ranks 31st out of 191 countries in the INFORM Risk Index (Index for Risk Management), 2020 [7]. In 2021, India ranked 121st out of 182 countries under the Notre Dame Global Adaptation Initiative (ND-GAIN) Index [8], representing the vulnerability to climate change.

The Indian Himalayan Region (IHR), with diverse demographic, economic, social, environmental, and political systems, plays an important role in the country's nexus of water, energy, and food. It has an inimitable topographical and climatic setting, making it prone to numerous hydro-meteorological disasters such as floods, cloudbursts, glacier lake outbursts, and landslides [9,10]. In the last 30–40 years, the frequency and severity of natural hazards have risen due to various anthropogenic and changing climatic conditions [11]. The anthropogenic activities such as the construction of dams, roads, deforestation, etc., have further aggravated the disasters which have disrupted the Himalayan ecosystem in the various states such as Uttarakhand [12]. The climate change in the Himalayas has resulted in irregular precipitation, temperature rise, drying up of perennial rivers, depletion of natural resources, and an increase in the frequency and intensity of flash floods [13]. Uttarakhand, an important state in the IHR, has undergone a considerable expanse of flood disasters. The major natural disasters in Uttarakhand include the disasters of 1970, 1986, 1991, 1998, 2001, 2002, 2004, 2005, 2008, 2009, 2010, 2012, 2013, 2016, 2017, 2019, 2020, and 2021 [14,15]. It may be noticed that the severity and frequency of flood incidents have increased in recent years.

Flood risk management is crucial as it provides optimal utilisation and exploitation of land and water resources that bring prosperity and sustainable development to a nation [16]. In India, flood risk management in plains has been emphasised, and policies have been framed to mitigate its ill effects. However, flood risk management in hilly regions is still in the infancy stage, particularly due to complex and tough terrain with limited accessibility and a low level of monitoring [17,18]. Experience with the last few decadal floods has shown that the structural measures alone could not ensure adequate security against such disasters. Therefore, an effective flood management strategy to safeguard these disasters is essential. This study briefly summarises the major flood damage incidents in the state and identifies its natural and anthropogenic causes. Further, the study reviews the challenges involved in flood management and highlights the flood risk management plan that may be adopted to alleviate its adverse impacts.

2. Material and Methods

2.1. Study Area Physiography

The IHR is one of the world's youngest mountain ranges and is characterised by fragile and sensitive ecosystems concerning geological hazards, geodynamics, topography, biodiversity, and water resources status [19]. It spreads over 11 states and 2 union territories and covers 500,000 sq. km. It stretches approximately 2500 km from the Hindu Kush in the north-west to Myanmar in the south-east [20].

The state of Uttarakhand, a part of IHR, is located between latitudes 28.44° and 31.28° north and longitudes 77.35° and 81.01°east. Uttarakhand was created out of northern Uttar Pradesh in the year 2000. Uttarakhand is well-known for its thriving spiritual and religious tourism industry, ecological diversity and richness, and a cultural ethos rooted in traditions. The state hosts numerous pilgrimage sites, namely Panch Kedar for Hindus, Piran Kaliyar for Muslims, Hemkund Sahib for Sikhs, Digambar Jain temple for Jains, and Christ Church for Christians, the hotspot of tourism activities. The state is divided into two regions, Garhwal (the western region) and Kumaun (the eastern region), as shown in Figure 1, having 95 development blocks in 13 districts.

Many perennial Indian rivers originate in Uttarakhand. There are six major river basins in the states, namely, Alaknanda, Bhagirathi, Ganga, Kali, Ramganga, and Yamuna, as shown in Figure 2. The major basins of Garhwal region are Alaknanda, Bhagirathi, Yamuna, and Ganga and major tributaries are Bhilangna, Dhauliganga-Garhwal, Mandakini, and Tons. The major basins of Kumaon region are Kali and Ramganga and major tributaries are Saryu, Gori Ganga, and Pindar. Some basic characteristics of the major river basin of Uttarakhand estimated from geographic information system (GIS) analysis are given in Table 1.

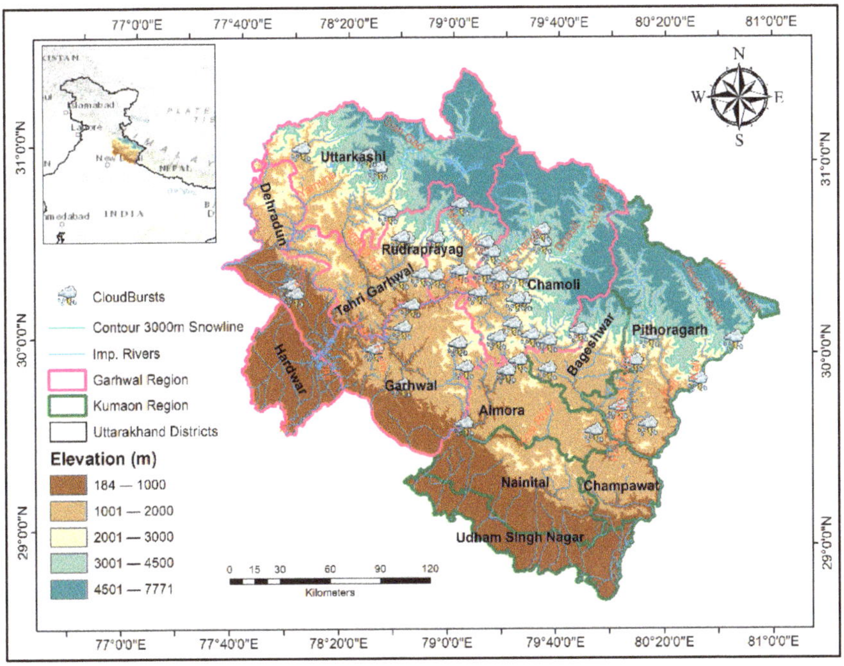

Figure 1. Map of Uttarakhand showing district, regions, and locations of cloudbursts.

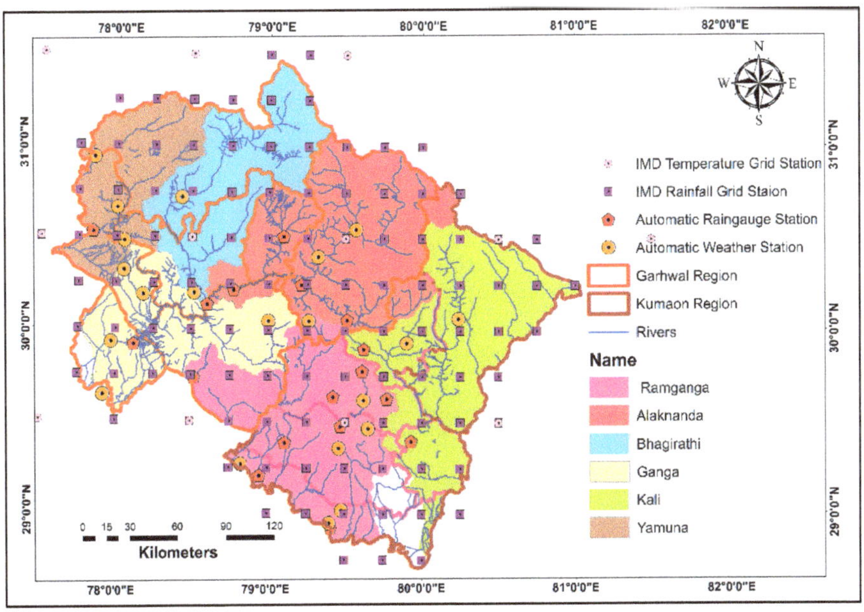

Figure 2. The spatial distribution of IMD-ARG and IMD-AWG along with the major river basins in the state of Uttarakhand.

Table 1. The basic characteristics of major rivers of Uttarakhand estimated from GIS analysis.

S. No.	Name	Area (sq. km.)	Mean Annual Rainfall (mm)	Mean Slope (Degrees)	Mean Elevation (m)	Elevation Range (m)	River Length (km)
1	Alaknanda	11,083	1035	28	3402	459–7785	195
2	Bhagirathi	7323	1011	27	3451	459–7054	190
3	Yamuna	5462	1175	24	2182	360–6278	170
4	Ganga	7282	1240	14	936	217–3076	125
5	Ramganga	11,319	1354	13	939	190–3089	185
6	Kali	11,014	1193	26	2552	185–7070	252

The state is divided into five transverse zones: (a) To the south of the Himalayan Frontal Fault (HFF)—the Terai. (b) Between the Shivalik (Outer Himalayan) range and the Main Boundary Fault (MBF)—the Doons. (c) Between the Main Central Thrust (MCT) and the MBF, along with the ridges up to around 3000 m high—the Middle Himalaya. (d) The area north of the MCT, which includes the perennially snow-covered peaks at heights of up to 8000 m—Inner Himalaya. (e) To the north of the snow-covered hills—the Trans Himalaya. Uttarakhand is a geological part of the Western Himalayas that can be classified into five morphological regions of varying physiographical and geological features. The regions from South to North comprise the Outer (sub-Himalayas), then the Lower Himalayas (Lesser Himalayas), followed by the Greater Himalayas, the Tethys (Tibet), and the Trans Himalaya [21].

2.2. Climatic Characteristics

The climate in the state ranges widely from the Terai region's sub-tropical humid atmosphere to the tundra-like atmosphere in the Greater Himalayas. The climate and landscape vary greatly with altitude (186–7619 m), from the largest glaciers in the higher elevation to the subtropical forests in the lowest elevation. Substantial annual precipitation falls as snow and feeds the Himalayans at high altitudes (>3000 m) [22].

The lowlands along the border of Uttar Pradesh are covered by the drier Terai-Savanna grasslands, the moist deciduous forest of the Upper Gangetic Plains, and form the Bhabhar belt. However, due to the extensive agricultural practices, most of these lowland forests have been cleared [21]. Snow and bare ice occupy the highest altitudes.

This mountainous state occupies 53,483 sq. km., i.e., 1.63% of India's total land area. Out of the 13 districts, Uttarkashi occupies the maximum area of 8039 sq. km. (15%), while Champawat covers the minimum area of 1634 sq. km. (3%). Uttarakhand experiences four types of seasons, namely, Monsoon from June to September (JJAS), Post-monsoon from October to November (ON), Cold weather from December to February (DJF), and Pre-monsoon (Hot Weather) from March to May (MAM). The estimated normal annual precipitation in the state is 1446 mm. Rain usually starts at the end of April and lasts until September. The monsoon rainfall amounts to 78% of annual rainfall, making this state prone to flood incidents during monsoon. Bageshwar receives the maximum normal annual rainfall (1872 mm), out of which 81% falls in the monsoon season. Haridwar receives the minimum normal annual rainfall (1107 mm) of all.

Further, due to differences in elevation, position, slope, and other topographical features throughout the state, the state's temperatures vary widely. During March and April, temperatures rise until they peak in May and June, when the mean maximum temperature in the southern and valley regions of the state is between 34 and 38 °C, and the mean minimum temperature is between 20 and 24 °C. Temperatures can reach 42 °C in the valleys and southern half of the state and 30 °C in two-kilometre-high elevations on some days. Day and night temperatures begin to dip around the end of September and reach the lowest in January and early February. January is the coldest month, with an average maximum temperature of around 20 °C and an average lowest temperature of approximately 6 °C in the southern section and river valleys. The mean maximum and mean minimum temperatures range from 10 to 12 °C at altitudes of 2 km [23]. Table 2 summarises the district-wise details of the area, rainfall, temperature, and elevation characteristics in the various seasons. Figure 3 shows the state's district-wise climatic characteristics regarding normal rainfall and temperature.

Table 2. The district-wise climatic characteristics and topography of the state.

District	Area (sq. km.)	Temperature (°C)						Normal Rainfall (mm)					Elevation (m)	
		Min. Temp.			Max. Temp.			ON	DJF	MAM	JJAS	Annual	Min.	Max.
		Min.	Max.	Mean	Min.	Max.	Mean							
Almora	3088	1.9	19.2	11.0	13.7	26.7	21.6	32	118	188	1238	1576	519	2619
Bageshwar	2267	−4.8	14.5	4.9	6.4	20.3	14.8	37	128	192	1515	1872	714	6513
Chamoli	7821	−13.3	8.5	−2.8	−3.0	15.1	7.1	40	150	229	1286	1705	714	7619
Champawat	1634	4.2	20.5	12.9	15.5	28.1	23.1	28	92	116	1114	1351	268	2199
Dehradun	3055	2.1	20.9	12.0	14.6	29.6	23.4	25	127	170	931	1253	283	2962
Garhwal	5444	4.4	22.0	14.0	16.3	31.3	24.9	29	107	156	1070	1361	254	3049
Hardwar	2372	6.6	25.6	17.0	19.0	36.8	28.8	19	93	91	904	1107	214	874
Nainital	4124	5.0	22.0	14.3	16.5	31.2	24.9	31	101	124	1099	1355	212	2522
Pithoragarh	7228	−4.2	8.6	3.0	−2.9	15.1	7.1	27	122	134	1184	1467	428	6985
Rudraprayag	1821	−7.8	12.8	2.2	3.9	18.8	12.6	44	144	262	1400	1851	584	6869
Tehri Garhwal	3854	−0.6	18.4	9.2	11.7	25.7	20.2	30	131	204	1031	1396	339	6392
Udham Singh Nagar	2737	7.3	24.8	17.2	19.1	35.6	28.3	32	78	81	1209	1400	186	421
Uttarkashi	8039	−7.0	8.6	−4.0	−3.5	15.9	7.3	30	144	203	875	1252	711	6990
Overall	53,483	–	–	–	–	–	–	31	123	172	1120	1446	186	7619

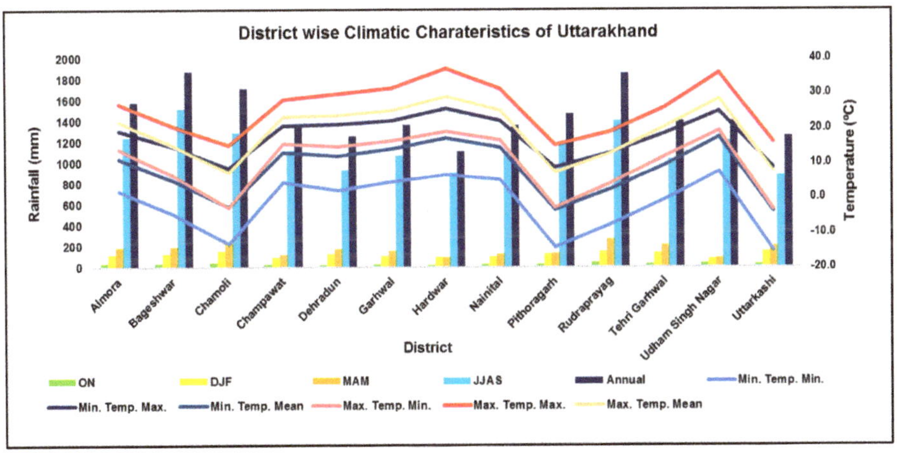

Figure 3. The climatic characteristics of various districts in Uttarakhand.

2.3. Flash-Flood Incidents

Hydro-meteorological disasters have increased rapidly over the last 20 years [24], as displayed in Table 3. Chamoli, Pithoragarh, Rudraprayag, and Uttarkashi are the most affected districts. Therefore, to alleviate the future occurrences of such disasters and suggest probable mitigation strategies, it is necessary to identify various issues and challenges faced during flood management.

Table 3. The list of major flash-flood locations in the state of Uttarakhand.

S. No.	Date	Region	District	Damage	Reference
1	7 February 2021	Tapovan	Chamoli	More than 200 people were killed or missing, and several hydel plants suffered	[25,26]
2	18 August 2019	Makudi	Uttarkashi	17 people died	https://sandrp.in/ (accessed on 25 January 2022)
3	16 July 2018	Tharali	Chamoli	55 houses, 10 vehicles, 2 ropeways, 1 road bridge washed, 2 cattle died, mini-hydro projects affected	https://sandrp.in/ (accessed on 25 January 2022)
4	14 August 2017	Dharchula	Pithoragarh	16 people	https://www.skymetweather.com (accessed on 25 January 2022)
5	28 May 2016	Kemra	Tehri	120 houses, 100 animals	Millennium post, 28 May 2016
6	15 August 2014	Purala	Pauri Gharwal	16 people reported dead	[13]
7	16 June 2013	Kedarnath	Rudraprayag	10,000 people, 365 houses	[27,28]
8	13 September 2012	Ukhimath	Chamoli	66 people	[15]
9	3 August 2012	Pandrasu ridge	Uttarkashi	35 people, 436 livestock lost, 591 houses	[29]
10	11 August 2001	Phata	Rudraprayag	27 people, 64 animals, 22 houses	[30]

2.4. Hydroclimatic and Topographic Dataset

The daily gridded IMD precipitation datasets at 0.25° resolution and maximum temperature datasets at 1° resolution from 1990 to 2019 were used in this study. Both datasets were obtained from https://www.imdpune.gov.in/ (accessed on 25 January 2022). Around 103 rainfall and 13 temperature grid points were utilised to cover the entire state, as shown in Figure 2. The daily rainfall and temperature datasets of the fifth generation "European Centre for Medium-Range Weather Forecasts (ECMWF) Atmospheric Reanalysis (ERA5) [31]" for the period 1990–2019 from Google Earth Engine [32] were also utilised.

The elevation map was prepared from Shuttle Radar Topography Mission (SRTM) Digital Elevation Model (DEM) acquired from https://portal.opentopography.org (accessed on 25 January 2022) at 30 m resolution. The locations of the India Meteorological Department's (IMD) Automatic Weather Station (AWS) and Automatic Rain Gauges (ARG) were obtained from http://aws.imd.gov.in:8091 (accessed on 25 January 2022). The satellite imageries of the flood events were acquired from Google Earth Pro. The climatic characteristics of the state were also obtained from the IMD annual report, and State Disaster Management Action plan on climate change [6,33,34].

2.5. The Proposed Methodology for Flood Risk Management

The methodology adopted in this study is divided into three parts, A, B and C, as shown in Figure 4. Part A summarises the primary causes of the disasters in the state and analyses the climatic variation by incorporating rainfall and temperature datasets. This part also discusses the topographic influences causing the two major disasters in the state.

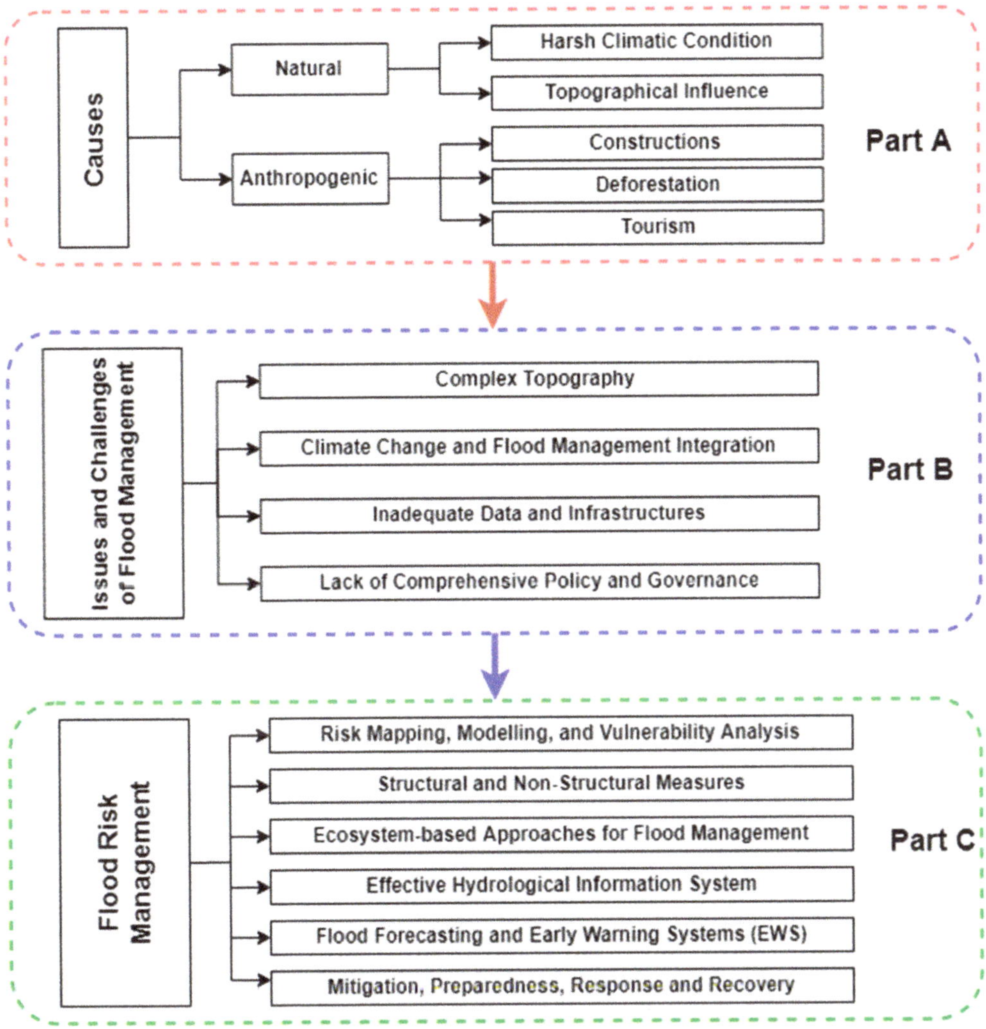

Figure 4. The proposed methodology for flood risk management.

Part B reviews the issues and challenges involved in the flood management of the region. Part C provides the basic approach for flood risk management in the Himalayan state to reduce the likelihood of flood incidents for people living in flood-prone areas.

2.6. Methodology for Climatic Variation Assessment

The methodology for climatic (rainfall and temperature) assessment is shown in Figure 5. The daily rainfall datasets are aggregated into seasonal datasets to assess the rainfall trends and variations. Then, the histogram of the rainfall value is plotted to assess the distribution. Furthermore, the Mann–Kendall and Theil–Sen slope method (Equations (1)–(5)) is utilised to estimate each district's trend. Non-parametric tests, such as the Mann–Kendall (MK) [34,35] test, are frequently used to identify patterns in time series of hydrological changes. For instance, Y_i ranked from i = 1, 2...N − 1, and Y_j ranked from j = i + 1, 2 . . . N. Each data point Y_i is used as a reference point and is compared with all other data points Y_j such that,

$$\mathrm{Sgn}\,(Y_j - Y_i) = \begin{cases} +1, > (Yj - Yi) \\ 0, = (Yj - Yi) \\ -1, < (Yj - Yi) \end{cases} \tag{1}$$

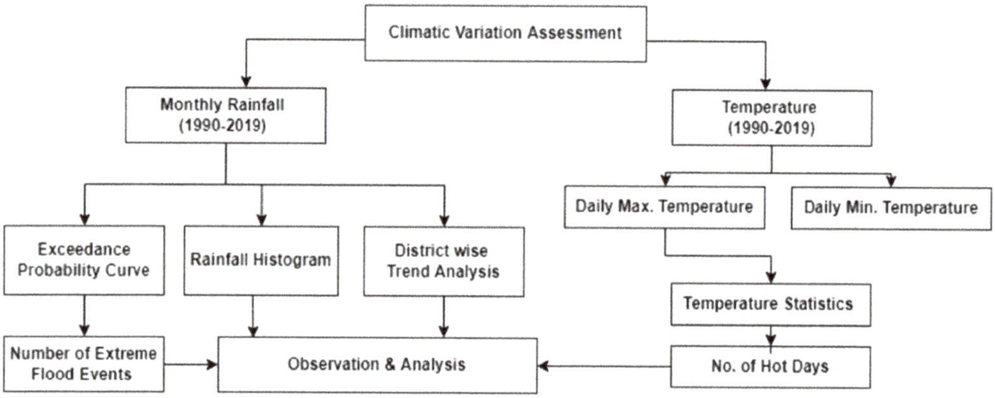

Figure 5. Methodology for climatic variation assessment.

The Kendall statistic S is estimated as

$$S = \sum\nolimits_{i=1}^{N-1} \sum\nolimits_{j=i+1}^{N} \mathrm{Sgn}(Yj - Yi) \tag{2}$$

The variance of the statistic S is defined by

$$\mathrm{Var}(S) = \frac{N(N-1)(2N+5) - \sum_{i=1}^{N} t_i(i)(i-1)(2i+5)}{18} \tag{3}$$

In which t_i denotes the number of ties up to sample i. The test statistic $\mathbf{Z_c}$ is estimated as

$$Z_c = \begin{cases} \frac{S-1}{\sqrt{\mathrm{Var}(S)}}, & S > 0 \\ 0, & S = 0 \\ \frac{S+1}{\sqrt{\mathrm{Var}(S)}}, & S < 0 \end{cases} \tag{4}$$

and follows a standard normal distribution. If Z_c is positive, the trend is upward; the negative Z_c shows a downward trend. A statistical significance of 95%, i.e., *p*-value of less than 0.05, is considered for the trend to exist. Additionally, for the determination of slope when there is a linear trend, Sen's slope method is used. Moreover, the magnitude of a time series trend was evaluated by a simple non-parametric procedure developed by Sen [35]. The trend is calculated by

$$\beta = \mathrm{Median}\left(\frac{Yj - Yi}{j - i}\right), \; j > i \tag{5}$$

where β is Sen's slope estimate. β > 0 indicates an upward trend in a time series and vice versa.

Now, the rainfall events of the monsoon season (JJAS) are considered to estimate the number of extreme monthly flood events. The probability of exceedance, i.e., the Weibull method [36], is applied to estimate the suitable threshold. For this estimation, it is necessary to sort all the rainfall data in decreasing order and assign a serial order number ranging from 1 to n. The formula is given by,

$$Probability\ of\ exceedance\ (PE) = \frac{m}{(n+1)} \quad (6)$$

where m = rank order and n = number of events.

The probability of exceedance value of 5% is selected as a threshold to estimate the number of extreme flood events. It is assumed that a high amount of rainfall would indicate a flood event. Moreover, the daily maximum temperature statistics of 30 years are utilised to analyse the number of hot days every year. Mean plus one Standard Deviation is selected as a threshold for hot days count for the analysis.

2.7. *Methodology for Topographic Influence Estimation*

For the analysis of topographic influence, the visual image interpretation technique was employed [37]. The information of area, shape, texture, tone, size, pattern, and association of the incidents was taken as the image interpretation elements. Moreover, the analysis used Google Earth imagery of pre and post incidents information gleaned from literature and the DEM of the region to estimate the basic physiography of the affected watersheds. Two case studies were investigated. The first case study examined the Kedarnath flood incident of 2013 [38–41]. The second case study assessed the Chamoli flash flood of 2021 [26,42].

3. Results and Discussion

3.1. *Part A—Causes of the Flood Disasters in the State*

The Himalayan's young, fragile, steep gradients are vulnerable to climate change [19]. In addition, these mountain areas' ecological and social-economic systems are becoming increasingly susceptible due to swelling human density. Moreover, the disasters disrupt society's functioning and adversely damage infrastructure, properties, life, and the environment. Furthermore, they aggravate other distress such as social conflict, financial crisis, environmental degradation, and diseases. Both natural and anthropogenic causes trigger these disasters in the mountain ecosystem.

3.1.1. Natural Causes

The two major natural causes of disaster in the state are the harsh climatic conditions causing extreme rainfall and the complex undulating topography causing rapid flood characteristics. During the southwest monsoon season, the Uttarakhand area experiences significant to extremely heavy rainfall.

The following are the primary meteorological conditions that contribute to high rainfall in the area:

(1) From the Bay of Bengal, the low-pressure areas rise and migrate through the central part of India and then recurve northwards and north-eastwards and cause high rainfall in the Uttarakhand's foothill. Furthermore, depressions from the Arabian sea cross north Maharashtra and south Gujarat coasts, reach the Kumaun-Garhwal region and cause severe cloudbursts, flash floods, and landslides.

(2) Flooding due to heavy rainfall occurs in the central and eastern Himalayas when the monsoon's axis shifts to the Himalayan foothills from the Northern Indian plains in July and August (the setting in the 'break situation').

(3) Western/Extra-tropical disturbances, originating from Caspian and the Mediterranean Sea in the far west and moving towards north India through Afghanistan, Iran, and Pakistan, cause snow and rain during the winter season over the western Himalayas.

This rainfall from Western disturbances occurs four to five times per month during monsoon and six to seven times during the winters.

Results for Climatic Variation Assessment

A detailed seasonal rainfall trend analysis (Equations (1)–(5)) for each district is carried out to study the temporal variation in rainfall, and the results are displayed in Table 4. The increasing trend is represented by '+ve', decreasing trend by '-ve', and the no-trend by '0'. The observations made are as follows,

(1) In the monsoon season (JJAS), 9 out of 13 districts, namely, Almora, Bageshwar, Chamoli, Champawat, Dehradun, Garhwal, Hardwar, Rudraprayag, and Tehri Garhwal, show an increasing rainfall trend.
(2) In post-monsoon (ON), no significant trend is observed.
(3) In the cold season (DJF), an increasing trend is observed for Chamoli, Pithoragarh, and Rudraprayag districts, while in the other districts, no trend is observed.
(4) In the hot weather season (MAM), all the districts except Uttarkashi show an increasing rainfall trend.
(5) Out of 13 districts, 8 districts, namely Almora, Bageshwar, Chamoli, Champawat, Hardwar, Pithoragarh, Rudraprayag, and Tehri Garhwal, exhibit an increasing annual rainfall trend.

Table 4. The trend analysis of seasonal and annual rainfall for each district.

District	Season				Annual
	JJAS	ON	DJF	MAM	
	Trend, p-Value, z Value				
Almora	+ve, 0.003, 2.98	0, 0.652, 0.45	0, 0.733, 0.34	+ve, 0.004, 2.85	+ve, 0.09, 1.7
Bageshwar	+ve, 0.01, 2.57	0, 0.084, 1.73	0, 0.266, 1.11	+ve, 0.001, 3.23	+ve, 0.048, 1.97
Chamoli	+ve, 0.021, 2.31	0, 0.077, 1.77	+ve, 0.01, 2.59	+ve, 0, 4.13	+ve, 0.049, 1.97
Champawat	+ve, 0.011, 2.53	0, 1, 0	0, 0.967, 0.04	+ve, 0.008, 2.67	+ve, 0.001, 3.18
Dehradun	+ve, 0.021, 2.31	0, 0.363, 0.91	0, 0.119, 1.56	+ve, 0, 3.6	0, 0.034, 2.12
Garhwal	+ve, 0.01, 2.57	0, 0.375, 0.89	0, 0.586, 0.55	+ve, 0.007, 2.72	0, 0.055, 1.92
Hardwar	+ve, 0.049, 1.97	0, 0.719, −0.36	0, 0.182, 1.34	+ve, 0.004, 2.85	+ve, 0.119, 1.56
Nainital	0, 0.058, 1.89	0, 1, 0	0, 0.965, −0.04	+ve, 0.021, 2.3	0, 0.043, 2.02
Pithoragarh	0, 0.075, 1.78	0, 0.132, 1.51	+ve, 0.046, 2	+ve, 0.027, 2.21	+ve, 0.048, 1.98
Rudraprayag	+ve, 0.011, 2.53	0, 0.076, 1.77	+ve, 0.038, 2.08	+ve, 0, 3.83	+ve, 0.033, 2.13
Tehri Garhwal	+ve, 0.011, 2.53	0, 0.163, 1.39	0, 0.125, 1.53	+ve, 0.001, 3.3	+ve, 0.013, 2.49
Udham Singh Nagar	0, 0.095, 1.67	0, 0.824, 0.22	0, 1, 0	+ve, 0.003, 2.99	0, 0.002, 3.09
Uttarkashi	0, 0.058, 1.89	0, 0.103, 1.63	0, 0.08, 1.75	0, 0.091, 1.69	0, 0.067, 1.83

Overall, it is observed that the rainfall trend is mostly increasing in the pre-monsoon and monsoon seasons. Chamoli and Rudraprayag districts show the maximum increasing rainfall trend. This increasing rainfall trend during the monsoon season results in frequent flash floods and cloudbursts in the state. The results obtained are consistent with the state IMD's annual report [34].

The rainfall histogram (Figure 6a) prepared from 103 grid points for 30 years, considering months from June to September, at 100 mm bin size, shows that the majority (24.6%) of the monthly rainfall falls in the range 100–200 mm. Around 20% of monthly rainfall lies in the range of 0–100 mm, 19.6 % lies in the range 200–300 mm, 14.1% lies in the range 300–400 mm, 9.8% lies in the range 400–500 mm, and only 11.8% has rainfall >500 mm.

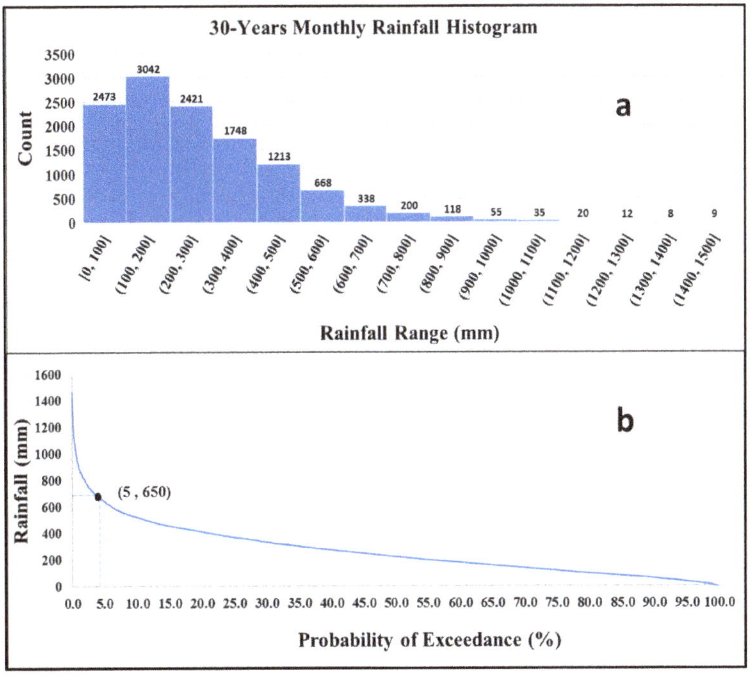

Figure 6. (**a**) The histogram and (**b**) the probability of exceedance curve of the monthly rainfall (June to September) obtained for the years 1990–2019.

The probability of exceedance curve (Figure 6b) was prepared using Equation (6), and considering the value of 5%, the value of rainfall depth threshold was estimated to be 650 mm. Therefore, all the monthly rainfall of more than 650 mm is considered an extreme flood. The result shows that the maximum number (288) of total extreme flood events is in July, 242 in August, 51 in September, and 34 in June in the past 30 years. This also demonstrates that the region is highly vulnerable in July and August. To understand the trend of the flood incidents, the Mann–Kendall test was carried out. It is observed that there is no statistically significant trend in the monsoon season (June to September) (Figure 7). However, a very small increasing trend in July and August and a small decreasing trend in June and September may be due to random fluctuating components in the time series.

The daily maximum temperature statistics revealed that the region has a normal maximum temperature of 30.1 °C and a standard deviation of 6.2 °C. Therefore, considering the threshold temperature as 36.3 °C (mean + one std. dev.), the values of no. of hot days obtained are shown in Figure 8. It is observed that the years 1999, 2002, 2009, 2010, 2012, 2014, and 2019 experienced more than 70 hot days in a year. To understand the trend, the Mann–Kendall test was performed. It is observed that statistically, there is no significant trend in the number of hot days during the last three decades. The very small increasing trend may be due to natural random fluctuating components in the time series. These increasing hot day events contribute to the glaciers' mass loss, permafrost degradation, and increased snowmelt in the Himalayas.

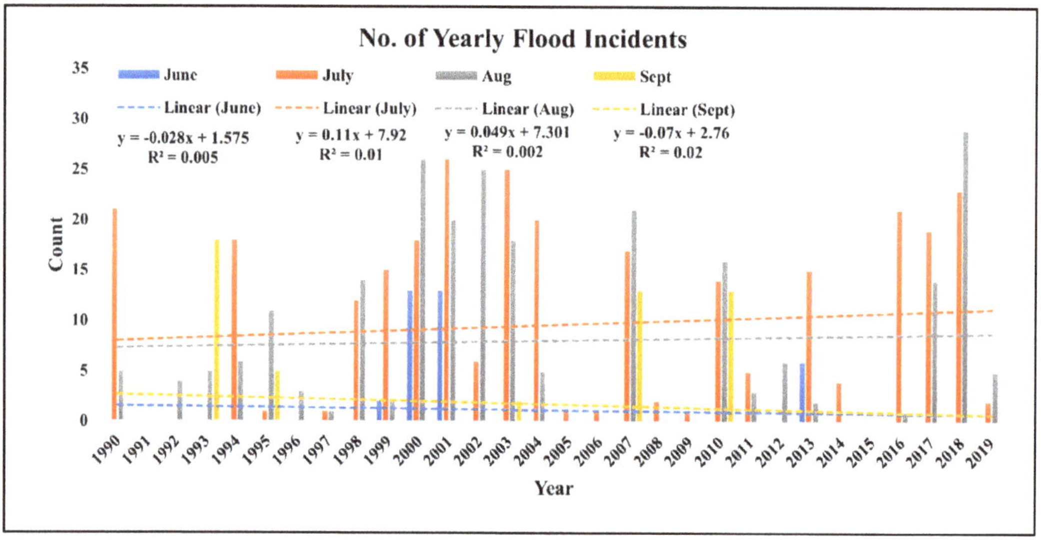

Figure 7. The time series of the number of flood incidents.

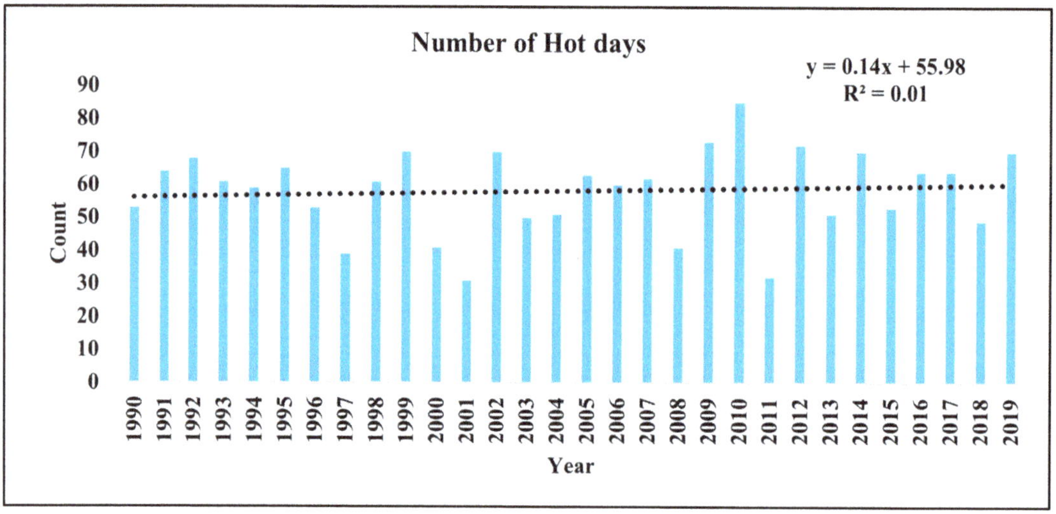

Figure 8. The time series of the number of hot days.

Results for Topographic Influence Assessment

The floods developed in a mountainous region differ from those generated in flood plains lying in low elevation zones. In hilly regions of Uttarakhand, floods are sudden and often termed flash floods. These flash floods are usually generated by extreme events such as heavy rainfall, cloudbursts, glacial lake outbursts, and dams created due to the blockage of rivers [43]. Such incidents are localised, occur with little to no warning, and pose significant damage in downstream areas. A wide range of variables, including hydrological characteristics, such as high-intensity rainfall and huge flow velocity, high frequency of occurrence, high peak flood, and basic characteristics such as the steepness of slope, elevation, and shape of the area, the degree to which the water carries sediments, and the vulnerability of the populations to the flowing floods, can further stimulate their impact.

In 2013, an extreme event in the form of a cloudburst combined with flash floods and landslides occurred in the Mandakini River basin. More than 10,000 people were presumed dead as a result of this catastrophe. Over 135 highways, 150 bridges, and 2000 houses/buildings were affected [39,40]. The basin length of Mandakini is about 66 km and covers 1982 sq. km. The relative relief ranges from 177 to 1416 m, in which 53% of the relied ranges from 500 to 750 m (moderate-high). Additionally, around 68% of the region has a slope of 25–35° [44]. Such topographical conditions of the basin make it prone to flash floods. The pre- and post-flash-flood scenario near the Kedarnath temple depicting the debris deposition and damage to the roads is shown in Figure 9.

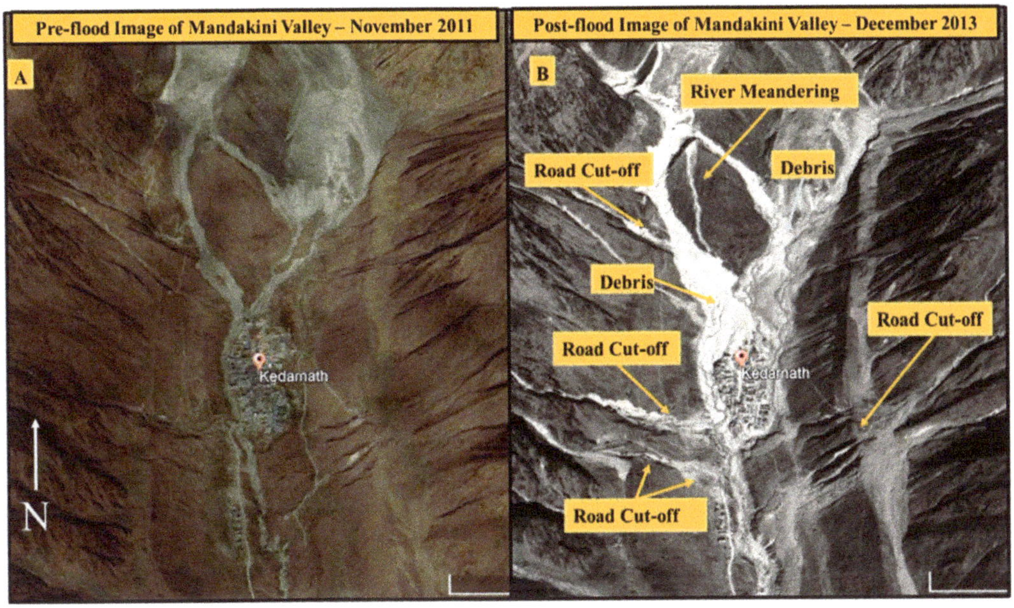

Figure 9. The Google Earth images showing, (**A**) Pre-flood (November 2011), (**B**) Post flood (December 2013) impact of the extreme flash-flood event in the Kedarnath region of the Mandakini river basin.

Apart from extreme weather-induced disasters, the region also suffers from landslides and avalanches. In 2021, the Chamoli region experienced a flash-flood event resulting from an avalanche near the Ronti Glacier (Figure 10) at an elevation of 4064 m. At the confluence of the Rishi and Dhauliganga rivers, the height is 1908 m, while Nanda Devi, at 7817 m, is the highest point in the region. The floods, while travelling downstream, demolished the bridge near Raini village, damaged the Rishiganga and the National Thermal Power Corporation (NTPC) Tapovan Hydel plant, and killed over 200 people [45]. The velocity and flood discharge at 89 m upstream of the Rishiganga Dam is estimated to be 24 m/s and 12,448 cumecs. The length of the rockslide surface is about 1182 m, having a maximum width of 755 m and a mean slope of 62° [46].

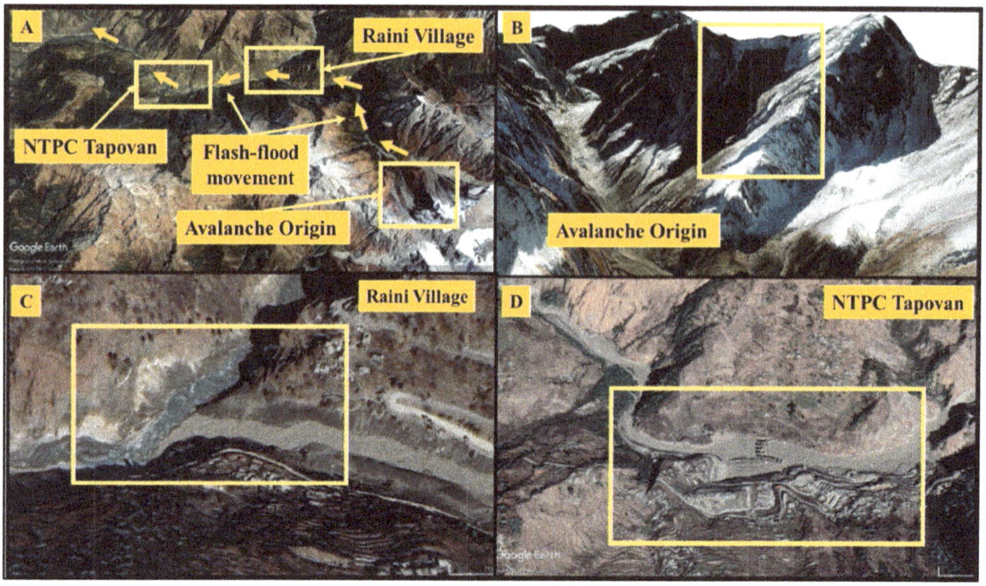

Figure 10. The Google Earth imagery shows, (**A**) An overall view of the incident, (**B**) The actual location of the avalanche near Ronti Glacier, (**C**) The confluence of Dhauliganga and Rishiganga, near Raini village, and (**D**) The inundated NTPC Tapovan sites.

3.1.2. Anthropogenic Causes

The increasing population growth in these mountain regions hampers the ecological and socio-economic systems. Numerous anthropogenic activities, such as the construction of roads, buildings, hydropower projects, abundant tourism, forest encroachment, etc., pose a serious threat to the climate and mountain ecosystem [40]. Blasting and massive cutting of the rocks, which generates high-intensity vibrations, is the primary method of clearing mountains for development. New roads are continually being developed, and old highways are being expanded and maintained in the Indian Himalayas. Such activities cause steepening of the hill slopes and further lead to slope instability. The unplanned development over the riverbed has essentially replaced the agricultural land, allowing the river habitat to decline.

The elimination of natural vegetation, and the expansion of impervious areas, lead to a reduced water storage capacity and, as a result, a higher flood peak and a shorter time to peak [47]. Certain anthropogenic ritual activities such as cremation ceremonies cause heavy pressure on the forest ecosystem as they result in deforestation and lead to air and water pollution. Morbidity among agricultural workers is exacerbated by these developments, which have a greater impact on the economy [48].

Greenhouse gas (GHG) emissions in the valley have increased significantly due to uncontrolled human activities such as tourism and vehicle traffic. The AR6 IPCC reports that solar radiation is absorbed by anthropogenic aerosols such as black carbon, which increases moist static energy, warms the lower troposphere, and further increases convection inhibition, suppressing light rainfall [49]. Aerosols intensify convection in the Himalayan foothills during the pre-monsoon season, resulting in regional convergence; this phenomenon is known as the 'Elevated-Heat-Pump' mechanism [50]. Intense monsoon rainfall in northern India and western Nepal in 2013, which caused landslides and one of the worst floods in history, has been further linked to increased GHG and aerosols [38]. A direct association between an increase in tourism and the landslide's frequency has also been observed [51]. The major anthropogenic activities in the state of Uttarakhand has been discussed [52] and are shown in Figure 11.

Figure 11. The impact of anthropogenic activities on the degradation of river ecosystem in Uttarakhand. (**A**) (Source: https://sandrp.in/2020/02/24/uttarakhand-road-widening-work-in-almora-damages-traditional-water-sources/ (accessed on 6 June 2022)), (**B**) (Source: https://sandrp.in/2021/09/17/uttarakhand-cloud-bursts-around-vyasi-hep/ (accessed on 6 June 2022)), (**C**) (Source: https://news.euttaranchal.com/uttarakhand-bans-quarrying-ganga tributaries-following-ngt-order (accessed on 6 June 2022)), (**D**) (Source: https://news.euttaranchal.com/no-mining-within-10-km-national-parks-uttarakhand-high-court (accessed on 6 June 2022)), (**E**) (Source: https://www.downtoearth.org.in/news/urbanisation/undone-by-rampant-mining-illegal-buildings-41450 (accessed on 6 June 2022)), (**F**) (Source: https://www.hindustantimes.com/cities/dehradun-news/uttarakhand-char-dham-yatra-to-kedarnath-halted-amid-heavy-rain-imd-warns-of-thunderstorms-101653301156476.html (accessed on 6 June 2022)).

3.2. Part B—Issues and Challenges of Flood Management in the State of Uttarakhand

Flood hazards in Himalayan Rivers are diverse, necessitating a multi-faceted and multi-scaled approach to preparedness. It is necessary to monitor, forecast, and disseminate information about these processes and phenomena. These initiatives must cross traditional disciplinary boundaries and involve scientists, policymakers, and members of civil society working together at all levels of government and administration to be effective. The four major challenges of flood management in Uttarakhand are: (1) complex and steep topography that controls the flow path and velocity, (2) lack of comprehensive policy and governance, which is essential for effective risk governance, (3) inadequate data and infrastructure, that is necessary for estimation and evaluation of flood-generating mechanisms, and (4) climate change and flood management integration, which is required for a long-term solution to flooding risks and hazards.

3.2.1. Complex Topography of the Mountainous Terrain

The complex and steep topography of the hilly regions of Uttarakhand links to their unusual sharp atmospheric changes (i.e., moisture, precipitation, radiation, temperature, pressure), soil, vegetation, and hydrological conditions over short distances. These sharp gradients throughout the terrain control the form of precipitation, intensity and frequency, groundwater interactions, biodiversity, and soil moisture, which sequentially lead to high rates of flood variability over short distances [17,53,54]. The main impact in hilly terrain is undercutting check dams, river damming by debris, riverbanks collapse and erosion, debris flows and deposits, channel displacement, clogging bridges, scour, and inundations of low-lying areas. Floods associated with geophysical flows, such as glacial dam burst floods, river avulsions, and glacial lake outburst floods (GLOF), are often extremely disruptive and lack enough evidence and models for understanding and controlling them in mountain regions [55,56]. When combined with limited hydro-meteorological statistics, such complexities lead to partial knowledge of the flood generation process, particularly with localised extreme events such as cloudbursts and glacier lake outbursts, landslides, and other convective storms. In the Himalayas, the cloudbursts are the least understood weather patterns [15,28]. The terms thunderstorms and cloudbursts have been used for decades interchangeably. However, the cloudbursts are hard to track as they are scattered and mostly occur on remote mountain slopes with limited rain gauges [24]; they are often only reported if lives are lost [57]. Given the degree to which the infrastructure and human resources are harmed, it is crucial to underline the processes that produce cloudbursts and establish a model for proper predictions.

3.2.2. Lack of Comprehensive Policy and Governance on Flood Mitigation

The key gaps in current policy are the lack of rigidity of the legislative framework, the inability to follow a legal strategy, the question of authorisation of power, and the local government's role in policy processes. In the Disaster Mitigation, Management and Prevention Act (2005), the separation of responsibilities and the functions of local authorities was remarkably integrated but not implemented as a legislative framework [10]. Land management policies and preparation may better mitigate floods by reducing risk and hazard exposure. Such laws include construction codes, land use laws, environmental protection acts, emergency management legislation, and water policy, which govern human actions and practices focused on environmental and ecological conservation, flood prevention, and control measures. Building law, for example, specifies the safe distance of houses from the rivers, construction size, and floor height to reduce waterlogging and flood damage. In flood control, institutional aspects and effective governance play a critical role. Policy planning alone is not adequate to handle the flood without enough preparation.

Furthermore, for local-level planning and proper land use management, the participation of the community is also required. There is also a requirement for strong coordination among the institutes and flood management boards for effective risk governance. A strong rehabilitation and resettlement policy for re-establishing the socio-economic stability of the vulnerable communities is much needed in the state.

3.2.3. Inadequate Data and Infrastructure

The Himalayan regions' basic climate and hydrological data are scarce, which greatly disrupts the prediction, estimation, and evaluation of devastating flood-generating climate events, flood warnings, and other life-saving management systems [17]. Due to the remoteness, lack of connectivity, inadequate communication networks, and other infrastructure, it is challenging to develop response systems and instrumentation in hilly mountain areas. Primary data collection issues include transportation logistics to the gauging sites, loss of facilities during extreme flood events, and the expense of installation and maintenance of gauges. In addition, in river basins with steep valley walls, wireless coverage, and radio telemetry are frequently inaccessible with highly restricted choices for transmitting long-term hydrographic data and early warning systems. Remote sensing systems can address the difficulties of in situ measurements. However, mountainous places might have significant constraints (e.g., resource-intensive processing of data through remote sensing, limited data documentation, the cloud cover impacts, or other data biases) and complex topography, including insufficient spatial and temporal resolutions [58–60]. Due to the difficulties in data gathering and the insufficiency of information, the hydrology of hilly mountain areas is still not understood fully.

In Uttarakhand, the flood forecasting (FF) and early-warning system (EWS) are expanding. The district offices do have sirens but have a small range of 2 km, which is a significant shortcoming considering the geographic area that would be addressed. Data, however, is transmitted with the help of different media, such as televisions, mobiles, radio, and newspapers. With the speed of urbanisation and development witnessed in the villages, various patterns were observed in the individuals. Many people used radio, and those who had television did not wish to watch weather reports. Rather, many people, despite having TV access, did not have the set-up boxes that made them work. As per the India Water Resources Information System portal, the state of Uttarakhand has 85 flood forecasting sites of the Central Water Commission, as shown in Figure 12, along various rivers of Uttarakhand. However, most rivers in this region come under the category of "Classified river"; thus, the information about their flood discharge, velocities, and related warning is unknown to the public. Furthermore, due to a limited number of communication towers in remote locations and the unavailability of good quality real-time data, the whole set-up of the warning system fails during heavy rainfall and cloudbursts, which leads to a huge loss of lives and properties.

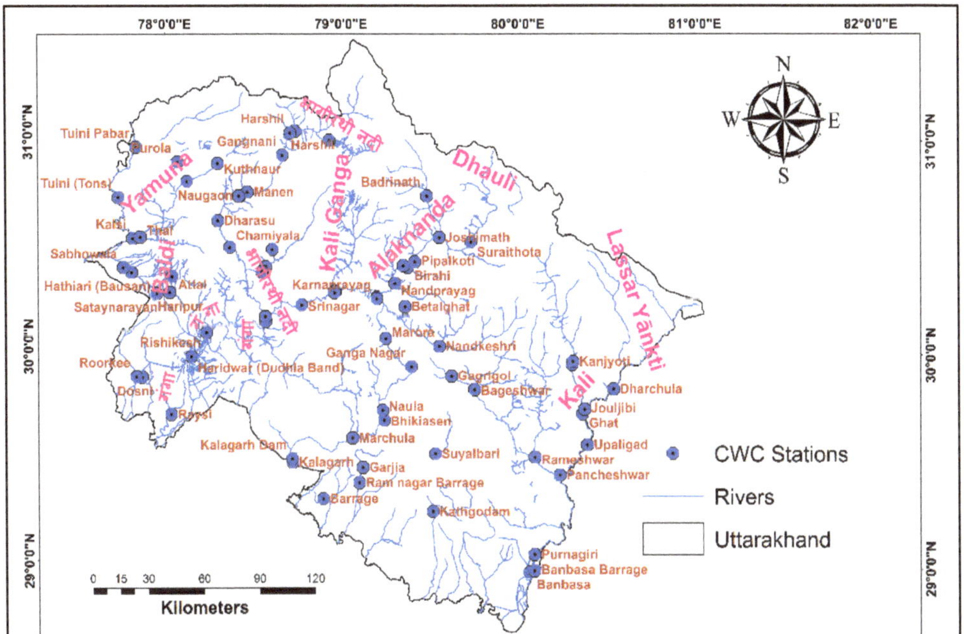

Figure 12. The locations of various river monitoring stations of the Central Water Commission.

3.2.4. Climate Change and Flood Management Integration

The Himalayan glaciers are the source of several river systems in India, but they are affected by human-caused global warming and climate change. The AR6, IPCC 2021, states that significant changes in temperature and precipitation have occurred in the Himalayas since the 1960s, including glacier mass loss, permafrost degradation, snowmelt, and increased glacier runoff [61]. Moreover, studies show that the glaciers in the Himalayan regions have unpredictable retreat and volume change rates due to microclimate and physiography variations [62]. Out of over 5000 glaciers, only 11 glaciers in the Indian Himalayas are being examined for their mass balance. Nearly 100 glaciers are regularly tracked for glacier snout shifts [48]. Decreasing glacier area and magnitude can briefly raise runoff before the glacier has retreated to such a degree that it no longer leads to runoff, temporarily raising flood levels and eventually causing landslides, GLOFs, and other geophysical flows. Due to the rapid response of landscapes, ecosystems, vegetation, water, snow and glaciers, and soil to environmental change, mountain areas are exclusively vulnerable to climate change [63–65]. Thus, incorporating climate change into flood management could be an effective long-term solution to flooding risks and hazards.

Furthermore, the land erosion in the state is likely due to a combination of excessive rainfall and prolonged drought periods. The vulnerability of mountain areas against climate change indicates that mountain populations become highly susceptible because snowpack, glacial cover, and weather cycles shift and increase the severity and intensity of several mountain flood processes. Future flood management strategies must integrate glacier shifts, monsoonal rains, and snowfall patterns.

3.3. *Part C—Flood Risk Management Plan*

Flood risk management aims to reduce the likelihood of flood incidents for people living in flood-prone areas. It involves "the development of policies and strategies and plans for implementation and associated means of review". The hilly regions of Uttarakhand are mostly affected by the flash-floods situation because of steep slopes and high drainage density. Thus, implementing flood reduction measures requires a holistic approach throughout the pre-flood, post-flood, and post-flood stages [66,67]. During the pre-flood stage, the recommendations are as follows: implementation of disaster contingency planning and flood risk management for all causes of flood, preventing inapt development within the flood plains, constructing physical flood defence infrastructure, implementation of proper warning and forecast systems, proper land use planning and public communications. Then, at the 'during-flood' stage, the following measures could be followed: issuing a warning to the appropriate authorities, forecasting the future river flow conditions, and identifying the probability of flood-forming zones. At the 'post-flood' stage, measures adopted are as follows: reconstruction and restoration of damaged infrastructure, providing relief to the affected, regeneration and retrieval of the economic activities, and improving and reviewing the flood management activities for future occurrences. Thus, for effective flood management, coordinated development and management of flood plains for maximising the benefits in a balanced manner could be ensured via the following approach:

3.3.1. Risk Mapping, Modelling, and Vulnerability Analysis

Mapping the specific hazards, locating important facilities, mobilising resources during and after an event, and mapping the physical constraints of an incident site to mitigate the risks of possible hazardous occurrences would help in hazard zoning, vulnerability assessment, and identifying the risk areas. Furthermore, involving local knowledge, perceptions, and socio-cultural values would help develop know-how, resources, and self-help capacities.

3.3.2. Structural and Non-Structural Measures

A combined approach of structural and non-structural measures should be adopted to control flash-flood hazards effectively, as shown in Figure 13. The structural measures generally include the catchment interventions and water control works, such as embankment protection, channel modifications and improvements and diversions, and river training works such as Check dams and the Gabion wall. The non-structural measure is divided into two classes: risk acceptance and risk reduction. The risk acceptance class adopts certain strategies for flash-flood insurance and emergency response systems. In the risk reduction approach, the prevention and mitigation strategies are adopted.

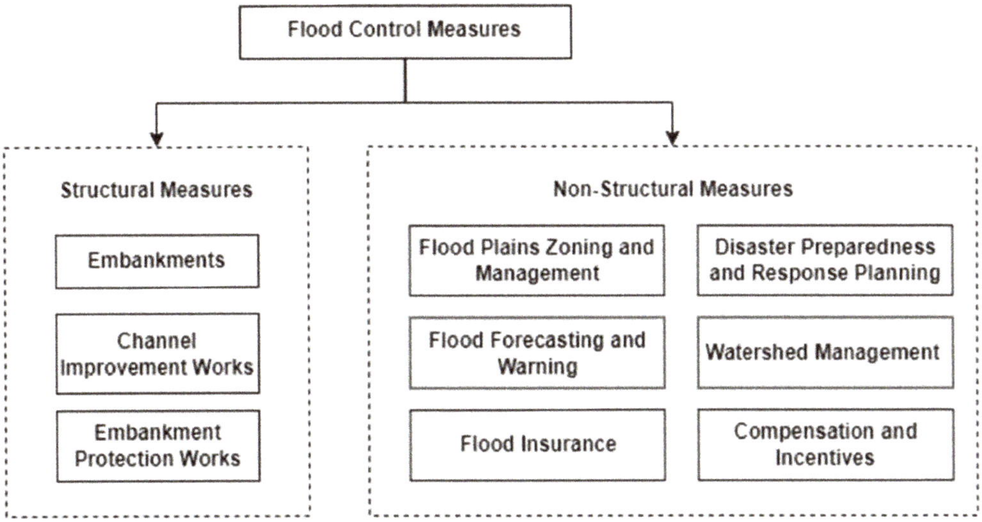

Figure 13. The flood control measures.

3.3.3. Ecosystem-Based Approaches for Flood Management

Researchers and practitioners have recommended that Nature-Based Solution (NBS), a novel approach toward catastrophe risk reduction, water security, and climate change resilience, is more effective and sustainable than traditional strategies [68]. Both urban and rural areas can benefit from ecosystem flood management services. The ecosystem services can lower the hazard caused by heavy precipitation. The wetlands can prevent floods by storing and temporarily slowing them down. Additionally, the roots of the wetland vegetations stabilise the riverbanks and hill slopes by holding the soils. The runoff fraction reduces the vegetated buffer zones, and other natural processes such as infiltration, evapotranspiration, water storage in aquifers, etc., reduce the runoff fraction [69,70]. Thus, a large-scale NBS is ideal in extreme events due to its ability to make more space for water, hence reducing the risk through retention, slowdown, and infiltration [71].

3.3.4. Effective Hydrological Information System

The management of flood data and related information necessitates the implementation of an effective Hydrological Information System. It could be done by maintaining the hydrometric networks and utilising its datasets for hydrological and hydrodynamic modelling, which would help generate flood scenarios and forecast floods during extreme events. Additionally, the interaction of land use land cover with climate change and hydrology should be adopted in the research [72–74]. Furthermore, increasing inter-sectoral communication through enhancing human and institutional capacities must be promoted. Moreover, the development of shared information portals and platforms for facilitating knowledge exchange between stakeholders should be espoused.

3.3.5. Flood Forecasting and Early Warning Systems (EWS)

For effective flood control, accurate and timely flood forecasts and early warnings are continually improved and upgraded [75]. An early warning system employs a flood sensor coupled to a transmitter to detect increasing water levels in local water basins (rivers, lakes) or flood defence structures such as dams and embankments. When the water reaches a crucial level, a signal is wirelessly relayed to the receiver. The flood warning is subsequently communicated to concerned agencies and vulnerable people through mobile phones [76,77]. By incorporating local participation, the critical flood levels could be determined. IMD and CWC are developing observation sites, weather stations, computational simulation tools, computer infrastructure, and communication networks to improve the forecasting framework to deliver more reliable data that can produce greater socio-economic benefits. The spatial distribution of IMD–AWS and ARG in Uttarakhand is already shown in Figure 2. The locations of the flood forecast station of the CWC are shown in Figure 12. For example, approximately 11 CWC gauging sites in the Alaknanda river basin can be viewed at https://ffs.tamcnhp.com/main/site (last accessed on 10 April 2022). In an extreme rainfall event, real-time flood level information upstream and downstream of the Alaknanda river could aid in formulating appropriate disaster response and management strategies.

3.3.6. Mitigation, Preparedness, Response, and Recovery

The whole system is broken into four steps for effective flood management: Mitigation, Preparedness, Response, and Recovery [33].

Mitigation aims to reduce or remove the risk of flooding substantially before its occurrence by developing a strategy to minimise the effects of disaster on a community, facility, or organisation through raising public knowledge of disaster risk management. The main components of the mitigation strategy, therefore, are,

- Land-use planning and management.
- Retrofitting of the existing structures.
- Risk assessment and vulnerability analysis.
- Performance and conceptual design of the structures.
- Building bylaws and codes.

Preparedness activities are structured to create a sense of readiness for an emergency disaster. The key initiative for disaster preparedness is community participation in contributing to the response process by being involved in the training and capacity building process. The effective flood management response involves providing immediate emergency relief and search and rescue assistance. The actual work of this process depends on the characteristics of the community, but the primary goal is to meet citizens' basic needs before the rehabilitation begins. In the hilly regions, the response phase forms an important criterion for successfully preventing the adverse impact of natural hazards, as the warning durations are less. In this regard, incident management systems and tools for analysing the consequences of disasters built on geographic information systems (GIS) may substantially aid rapid response and emergency management. A useful application of geospatial technology during a disaster may be a map showing public access to emergency services (such as electricity and transportation). The final stage, known as the recovery phase, includes assessing the damage, rebuilding, and helping the afflicted. Restoring normalcy to disaster-affected areas is the overall objective of this stage. Conducting a quick damage assessment forms a top objective to expedite the rescue effort. Rapid damage assessment and relief operations can be supplemented by measures such as coordinating the spatial inventory of damage, assuaging and educating the public, and averting reoccurrence.

4. Conclusions

The Indian Himalayan Regions are the home of the major Indian population and create an important nexus among water, energy, and food for society, biodiversity, and life, but they are often subject to catastrophic natural disasters, especially floods. The study highlights the natural causes, such as harsh climatic conditions and topographical influence and anthropogenic causes, such as construction activities, deforestation, and tourism that have aggravated the extreme flood conditions. Further, the study identifies the challenges of flood management, viz., complex topography, weak policy and governance, lack of adequate data and infrastructure, and the need for climate change and flood management integration. The study suggests adopting an effective flood risk management plan to assess the multi-institutional, land use planning, mitigation policy execution, and mitigation strategies in pre-, during-, and post-flood scenarios with a special focus on the districts, viz., Chamoli, Pithoragarh, Rudraprayag, and Uttarkashi.

Furthermore, adopting nature-based solutions, such as wetlands, is proposed to counter extreme flood events. Additionally, a combined structural and non-structural measures approach is recommended to manage flash floods effectively. Furthermore, the study commends a need for collaborative governance between local, state, and central policymakers and an integrated communication network for improved flood control in hilly regions. Additionally, the work recommends installing and upgrading flood forecasts and early warning systems for the region's accurate and timely flood forecasts. Adopting an effective flood management system in hilly regions requires an infrastructural and technical contribution to long-term monitoring and assessment. Moreover, comprehensive model-based studies of climate changes and environmental effects and process-based hydrology studies in altitudinal and mountainous gradients should be espoused.

Author Contributions: Conceptualisation—M.L.K. and S.S.; formal analysis and writing—S.S.; supervision and guidance—M.L.K. All authors have read and agreed to the published version of the manuscript.

Funding: The second author is thankful to the Ministry of Education, Government of India, for the fellowship grant for carrying out the research at the Department of Water Resources Development & Management at IIT Roorkee. The APC is waived-off.

Institutional Review Board Statement: Not applicable.

Informed Consent Statement: Not applicable.

Data Availability Statement: Not applicable.

Acknowledgments: The authors are thankful to the Ministry of Education, Government of India for the financial support provided through the Indian Institute of Technology, Roorkee, for carrying out this research.

Conflicts of Interest: The authors declare no conflict of interest.

References

1. Munpa, P.; Kittipongvises, S.; Phetrak, A.; Sirichokchatchawan, W.; Taneepanichskul, N.; Lohwacharin, J.; Polprasert, C. Climatic and Hydrological Factors Affecting the Assessment of Flood Hazards and Resilience Using Modified UNDRR Indicators: Ayutthaya, Thailand. *Water* **2022**, *14*, 1603. [CrossRef]
2. Wester, P.; Mishra, A.; Mukherji, A.; Shrestha, A.B.; Change, C. *The Hindu Kush Himalaya Assessment*; Wester, P., Mishra, A., Mukherji, A., Shrestha, A.B., Eds.; Springer International Publishing: Cham, Switzerland, 2019; ISBN 978-3-319-92287-4.
3. Murray, V.; Ebi, K.L. IPCC Special Report on Managing the Risks of Extreme Events and Disasters to Advance Climate Change Adaptation (SREX). *J. Epidemiol. Community Health* **2012**, *66*, 759–760. [CrossRef] [PubMed]
4. Bhatt, G.D.; Sinha, K.; Deka, P.K.; Kumar, A. Flood Hazard and Risk Assessment in Chamoli District, Uttarakhand Using Satellite Remote Sensing and GIS Techniques. *Int. J. Innov. Res. Sci. Eng. Technol.* **2014**, *3*, 15348–15356. [CrossRef]
5. Ashraf, S.; Luqman, M.; Iftikhar, M.; Ashraf, I.; Hassan, Z.Y. Understanding Flood Risk Management in Asia: Concepts and Challenges. In *Flood Risk Management*; Hromadka, T., Rao, P., Eds.; IntechOpen: Rijeka, Croatia, 2017.
6. Raj, J. *Uttarakhand Action Plan on Climate Change: Transforming Crisis into Opportunity*; Government of Uttarakhand: Dehradun, India, 2014.
7. European Commission. INFORM Index for Risk Management. India Country Profile. Available online: https://drmkc.jrc.ec.europa.eu/inform-index (accessed on 6 June 2022).
8. Chen, C.; Noble, I.; Hellmann, J.; Coffee, J.; Murillo, M.; Chawla, N. Notre Dame Global Adaptation Initiative. Available online: https://gain.nd.edu/our-work/country-index/rankings/ (accessed on 6 June 2022).
9. Lindell, M.K.; Arlikatti, S.; Huang, S.K. Immediate behavioral response to the June 17, 2013 flash floods in Uttarakhand, North India. *Int. J. Disaster Risk Reduct.* **2019**, *34*, 129–146. [CrossRef]
10. Dash, P.; Punia, M. Governance and disaster: Analysis of land use policy with reference to Uttarakhand flood 2013, India. *Int. J. Disaster Risk Reduct.* **2019**, *36*, 101090. [CrossRef]
11. Dimri, A.P.; Kumar, D.; Choudhary, A.; Maharana, P. Future changes over the Himalayas: Maximum and minimum temperature. *Glob. Planet. Chang.* **2018**, *162*, 212–234. [CrossRef]
12. Geneletti, D.; Dawa, D. Environmental impact assessment of mountain tourism in developing regions: A study in Ladakh, Indian Himalaya. *Environ. Impact Assess. Rev.* **2009**, *29*, 229–242. [CrossRef]
13. Mishra, P.K.; Thayyen, R.J.; Singh, H.; Das, S.; Nema, M.K.; Kumar, P. Assessment of cloudbursts, extreme rainfall and vulnerable regions in the Upper Ganga basin, Uttarakhand, India. *Int. J. Disaster Risk Reduct.* **2021**, *69*, 102744. [CrossRef]
14. Das, S.; Ashrit, R.; Moncrieff, M.W. Simulation of a Himalayan cloudburst event. *J. Earth Syst. Sci.* **2006**, *115*, 299–313. [CrossRef]

15. Dimri, A.P.; Chevuturi, A.; Niyogi, D.; Thayyen, R.J.; Ray, K.; Tripathi, S.N.; Pandey, A.K.; Mohanty, U.C. Cloudbursts in Indian Himalayas: A review. *Earth Sci. Rev.* **2017**, *168*, 1–23. [CrossRef]
16. Wheater, H.; Evans, E. Land use, water management and future flood risk. *Land Use Policy* **2009**, *26*, 251–264. [CrossRef]
17. Tullos, D.; Byron, E.; Galloway, G.; Obeysekera, J.; Prakash, O.; Sun, Y.H. Review of challenges of and practices for sustainable management of mountain flood hazards. *Nat. Hazards* **2016**, *83*, 1763–1797. [CrossRef]
18. Li, W.; Lin, K.; Zhao, T.; Lan, T.; Chen, X.; Du, H.; Chen, H. Risk assessment and sensitivity analysis of flash floods in ungauged basins using coupled hydrologic and hydrodynamic models. *J. Hydrol.* **2019**, *572*, 108–120. [CrossRef]
19. Saha, S.K.; Kumar, A.S. Northwest Himalayan Ecosystems: Issues, Challenges and Role of Geospatial Techniques. In *Remote Sensing of Northwest Himalayan Ecosystems*; Navalgund, R.R., Kumar, A.S., Nandy, S., Eds.; Springer: Singapore, 2019; pp. 3–14. ISBN 978-981-13-2127-6.
20. Satendra; Gupta, A.K. *Uttarakhand Disaster 2013*; Naik, V.K., Roy, T.K.S., Sharma, A.K., Dwivedi, M., Eds.; National Institute of Disaster Management: New Delhi, India, 2014; ISBN 9789382571148.
21. Pandey, V.K.; Mishra, A. Trends of Hydro-Meteorological Disaster in Uttarakhand, India. *Int. J. Curr. Res.* **2018**, *4*, 1–7.
22. Singh, P.; Kumar, N. Effect of orography on precipitation in the western Himalayan region. *J. Hydrol.* **1997**, *199*, 183–206. [CrossRef]
23. Climatological Publication Section. *Climate of Uttarakhand*; India Meteorological Department: Pune, India, 2014.
24. Singh, S.; Kansal, M.L. Cloudburst—A Major Disaster in The Indian Himalayan States. In *Civil Engineering for Disaster Risk Reduction*; Kolathayar, S., Pal, I., Chian, S.C., Mondal, A., Eds.; Springer: Singapore, 2022; pp. 115–126. ISBN 978-981-16-5312-4.
25. Pandey, P.; Chauhan, P.; Praveen, C.M.B.; Suresh, K.T. Cause and Process Mechanism of Rockslide Triggered Flood Event in Rishiganga and Dhauliganga River Valleys, Chamoli, Uttarakhand, India Using Satellite Remote Sensing and in situ Observations. *J. Indian Soc. Remote Sens.* **2021**, *3*, 1011–1024. [CrossRef]
26. Singh, S.; Kansal, M.L. Chamoli flash-flood mapping and evaluation with a supervised classifier and NDWI thresholding using Sentinel-2 optical data in Google earth engine. *Earth Sci. Inform.* **2022**, *15*, 1073–1086. [CrossRef]
27. Mishra, A.; Srinivasan, J. Did a cloud burst occur in Kedarnath during 16 and 17 June 2013? *Curr. Sci.* **2013**, *105*, 1351–1352.
28. Kumar, A.; Gupta, A.K.; Bhambri, R.; Verma, A.; Tiwari, S.K.; Asthana, A.K.L. Assessment and review of hydro-meteorological aspects for cloudburst and flash flood events in the third pole region (Indian Himalaya). *Polar Sci.* **2018**, *18*, 5–20. [CrossRef]
29. Gupta, V.; Dobhal, D.P.; Vaideswaran, S.C. August 2012 cloudburst and subsequent flash flood in the Asi Ganga, a tributary of the Bhagirathi river, Garhwal Himalaya, India. *Curr. Sci.* **2013**, *105*, 249–253.
30. Joshi, V.; Kumar, K. Extreme rainfall events and associated natural hazards in Alaknanda valley, Indian Himalayan region. *J. Mt. Sci.* **2006**, *3*, 228–236. [CrossRef]
31. Copernicus Climate Change Service. (C3S) ERA5: Fifth Generation of ECMWF Atmospheric Reanalyses of the Global Climate; Copernicus Climate Change Service Climate Data Store (CDS). Available online: https://cds.climate.copernicus.eu/cdsapp#!/home (accessed on 6 June 2022).
32. Gorelick, N.; Hancher, M.; Dixon, M.; Ilyushchenko, S.; Thau, D.; Moore, R. Google Earth Engine: Planetary-scale geospatial analysis for everyone. *Remote Sens. Environ.* **2017**, *202*, 18–27. [CrossRef]
33. DMMC. *State Disaster Management Action Plan for the State of Uttarakhand*; Disaster Mitigation & Management Centre Uttarakhand Secretariat: Uttarakhand, India, 2012.
34. Guhathakurta, P.; Bandgar, A.; Menon, P.; Prasad, A.K.; Sable, S.T.; Sangwan, N. *Climate Research and Services Observed Rainfall Variability and Changes over Uttarakhand State*; India Meteorological Department: Pune, India, 2020; Volume 52.
35. Sen, P.K. Estimates of the Regression Coefficient Based on Kendall's Tau. *J. Am. Stat. Assoc.* **1968**, *63*, 1379–1389. [CrossRef]
36. Ciupak, M.; Ozga-zieli, B.; Tokarczyk, T.; Adamowski, J. A Probabilistic Model for Maximum Rainfall Frequency. *Water* **2021**, *13*, 2688. [CrossRef]
37. Asokan, A.; Anitha, J.; Ciobanu, M.; Gabor, A.; Naaji, A.; Hemanth, D.J. Image processing techniques for analysis of satellite images for historical maps classification—An overview. *Appl. Sci.* **2020**, *10*, 4207. [CrossRef]
38. Cho, C.; Li, R.; Wang, S.Y.; Yoon, J.H.; Gillies, R.R. Anthropogenic footprint of climate change in the June 2013 northern India flood. *Clim. Dyn.* **2016**, *46*, 797–805. [CrossRef]
39. Chevuturi, A.; Dimri, A.P. Investigation of Uttarakhand (India) disaster-2013 using weather research and forecasting model. *Nat. Hazards* **2016**, *82*, 1703–1726. [CrossRef]
40. Kansal, M.L.; Shukla, S.; Tyagi, A. Probable Role of Anthropogenic Activities in 2013 Flood Disaster in Uttarakhand, India. In Proceedings of the World Environmental Water Resources Congress Water without Borders, Portland, Oregon, 1–5 June 2014; pp. 924–937. [CrossRef]
41. Das, P.K. 'The Himalayan Tsunami'—Cloudburst, Flash Flood & Death Toll: A Geographical Postmortem. *IOSR J. Environ. Sci. Toxicol. Food Technol.* **2013**, *7*, 33–45. [CrossRef]
42. Shrestha, A.B.; Steiner, J.; Nepal, S.; Maharjan, S.B.; Jackson, M.; Rasul, G.; Bajracharya, B. Understanding the Chamoli Cause, process, impacts, context of rapid infrastructure development. *Cryosphere* **2021**, 1–15. Available online: https://www.icimod.org/article/understanding-the-chamoli-flood-cause-process-impacts-and-context-of-rapid-infrastructure-development/ (accessed on 6 June 2022).
43. Sene, K. *Hydrometeorology Forecasting and Applications*; Springer: Dordrecht, The Netherlands, 2010; ISBN 978-90-481-3402-1.

44. Kumar, A.; Negi, M.S. Physiographic study of Mandakini valley (Rudraprayag District) Garhwal Himalaya by morphometric analysis and Geospatial Techniques. *Int. J. Geomat. Geosci.* **2016**, *7*, 285–298.
45. Thayyen, R.J.; Mishra, P.K.; Jain, S.K.; Wani, J.M.; Singh, H. Hanging glacier avalanche (Raunthigad-Rishiganga) and debris flow disaster of 7th February 2021, Uttarakhand, India: A Preliminary assessment. *Nat. Hazards* **2021**, 1–37. [CrossRef]
46. Pandey, V.K.; Kumar, R.; Singh, R.; Kumar, R.; Rai, S.C.; Singh, R.P.; Tripathi, A.K.; Soni, V.K.; Ali, S.N.; Tamang, D.; et al. Catastrophic ice-debris flow in the Rishiganga River, Chamoli, Uttarakhand (India). *Geomat. Nat. Hazards Risk* **2022**, *13*, 289–309. [CrossRef]
47. Singh, S.; Kansal, M.L. Flash Flood Hazard mapping using Satellite Images and GIS: A Case Study of Alaknanda River Basin. In *Flash Floods: Challenges and its Management*; The Institution of Engineers Centenary Publication: Kolkata, India, 2020; pp. 77–83.
48. Thadani, R.; Singh, V.; Chauhan, D.; Dwivedi, V.; Pandey, A. *Climate Change in Uttarakhand: Current State of Knowledge and Way Forward*; Bishen Singh Mahendra Pal Singh: Dehradun, India, 2015; pp. 1–66.
49. Wang, H.; Easter, R.C.; Zhang, R.; Ma, P.L.; Singh, B.; Zhang, K.; Ganguly, D.; Rasch, P.J.; Burrows, S.M.; Ghan, S.J.; et al. Aerosols in the E3SM Version 1: New Developments and Their Impacts on Radiative Forcing. *J. Adv. Model. Earth Syst.* **2020**, *12*, 1–36. [CrossRef]
50. D'Errico, M.; Cagnazzo, C.; Fogli, P.G.; Lau, W.K.M.; Hardenberg, J.; Fierli, F.; Cherchi, A. Indian monsoon and the elevated-heat-pump mechanism in a coupled aerosol-climate model. *J. Geophys. Res. Atmos.* **2015**, *120*, 8712–8723. [CrossRef]
51. Prasad, A.S.; Pandey, B.W.; Leimgruber, W.; Kunwar, R.M. Mountain hazard susceptibility and livelihood security in the upper catchment area of the river Beas, Kullu Valley, Himachal Pradesh, India. *Geoenviron. Disasters* **2016**, *3*, 1. [CrossRef]
52. Joshi, L.M.; Singh, A.K.; Kotlia, B.S. Rivers of Uttarakhand Himalaya: Impact of Floods in the Pindar and Saryu Valleys. In *The Indian Rivers*; Scientific and Socio-Economic Aspects; Singh, D.S., Ed.; Springer: Singapore, 2018; pp. 413–427. ISBN 978-981-10-2984-4.
53. Haeberli, W.; Beniston, M. Climate change and its impacts on glaciers and permafrost in the Alps. *Ambio* **1998**, *27*, 258–265. [CrossRef]
54. Beniston, M. Climatic Change in Mountain Regions: A Review of Possible Impacts. In *Climate Variability and Change in High Elevation Regions: Past, Present & Future*; Advances in Global Change Research; Diaz, H.F., Ed.; Springer: Dordrecht, The Netherlands, 2003; Volume 15, pp. 5–31. [CrossRef]
55. Ziersen, J.; Clauson-Kaas, J.; Rasmussen, J. The role of Greater Copenhagen utility in implementing the city's Cloudburst Management Plan. *Water Pract. Technol.* **2017**, *12*, 338–343. [CrossRef]
56. Mergili, M.; Emmer, A.; Juřicová, A.; Cochachin, A.; Fischer, J.T.; Huggel, C.; Pudasaini, S.P. How well can we simulate complex hydro-geomorphic process chains? The 2012 multi-lake outburst flood in the Santa Cruz Valley (Cordillera Blanca, Perú). *Earth Surf. Process. Landf.* **2018**, *43*, 1373–1389. [CrossRef]
57. Thayyen, R.J.; Dimri, A.P.; Kumar, P.; Agnihotri, G. Study of cloudburst and flash floods around Leh, India, during August 4–6, 2010. *Nat. Hazards* **2013**, *65*, 2175–2204. [CrossRef]
58. Anders, A.M.; Roe, G.H.; Hallet, B.; Montgomery, D.R.; Finnegan, N.J.; Putkonen, J. Spatial patterns of precipitation and topography in the Himalaya. *Spec. Pap. Geol. Soc. Am.* **2006**, *398*, 39–53. [CrossRef]
59. Bookhagen, B. Appearance of extreme monsoonal rainfall events and their impact on erosion in the Himalaya. *Geomat. Nat. Hazards Risk* **2010**, *1*, 37–50. [CrossRef]
60. Shrestha, D.; Singh, P.; Nakamura, K. Spatiotemporal variation of rainfall over the central Himalayan region revealed by TRMM Precipitation Radar. *J. Geophys. Res. Atmos.* **2012**, *117*, 1–14. [CrossRef]
61. Masson-Delmotte, V.; Zhai, P.; Pirani, S.L.; Connors, C.; Péan, S.; Berger, N.; Caud, Y.; Chen, L.; Goldfarb, M.I.; Gomis, M.; et al. *Climate Change 2021: The Physical Science Basis. Contribution of Working Group I to the Sixth Assessment Report of the Intergovernmental Panel on Climate Change*; Cambridge University Press: Cambridge, UK, 2021.
62. Krishnan, R.; Shrestha, A.B.; Ren, G.; Rajbhandari, R.; Saeed, S.; Sanjay, J.; Syed, A.; Vellore, R.; Xu, Y.; You, Q.; et al. Unravelling Climate Change in the Hindu Kush Himalaya: Rapid Warming in the Mountains and Increasing Extremes. In *The Hindu Kush Himalaya Assessment*; Wester, P., Mishra, A., Mukherji, A., Shrestha, A.B., Eds.; Springer International Publishing: Cham, Switzerland, 2019; p. 41. ISBN 9783319922881.
63. Huggel, C. Assessment of Glacial Hazards based on Remote Sensing and GIS Modeling. *Schriftenr. Phys. Geogr. Glaziologie Geomorphodynamik* **2004**, *44*, 87.
64. Terzi, S.; Torresan, S.; Schneiderbauer, S.; Critto, A.; Zebisch, M.; Marcomini, A. Multi-risk assessment in mountain regions: A review of modelling approaches for climate change adaptation. *J. Environ. Manag.* **2019**, *232*, 759–771. [CrossRef] [PubMed]
65. de Jong, C. Challenges for mountain hydrology in the third millennium. *Front. Environ. Sci.* **2015**, *3*, 1–13. [CrossRef]
66. Wang, Z.; Lai, C.; Chen, X.; Yang, B.; Zhao, S.; Bai, X. Flood hazard risk assessment model based on random forest. *J. Hydrol.* **2015**, *527*, 1130–1141. [CrossRef]
67. Chan, F.K.S.; Chuah, C.J.; Ziegler, A.D.; Dąbrowski, M.; Varis, O. Towards resilient flood risk management for Asian coastal cities: Lessons learned from Hong Kong and Singapore. *J. Clean. Prod.* **2018**, *187*, 576–589. [CrossRef]
68. Vojinovic, Z.; Alves, A.; Gómez, J.P.; Weesakul, S.; Keerakamolchai, W.; Meesuk, V.; Sanchez, A. Effectiveness of small- and large-scale Nature-Based Solutions for flood mitigation: The case of Ayutthaya, Thailand. *Sci. Total Environ.* **2021**, *789*, 147725. [CrossRef]

69. Gallay, I.; Olah, B.; Gallayová, Z.; Lepeška, T. Monetary valuation of flood protection ecosystem service based on hydrological modelling and avoided damage costs. An example from the Čierny hron river basin, Slovakia. *Water* **2021**, *13*, 198. [CrossRef]
70. Hillard, U.; Sridhar, V.; Lettenmaier, D.P.; McDonald, K.C. Assessing snowmelt dynamics with NASA scatterometer (NSCAT) data and a hydrologic process model. *Remote Sens. Environ.* **2003**, *86*, 52–69. [CrossRef]
71. Thorslund, J.; Jarsjö, J.; Jaramillo, F.; Jawitz, J.W.; Manzoni, S.; Basu, N.B.; Chalov, S.R.; Cohen, M.J.; Creed, I.F.; Goldenberg, R.; et al. Wetlands as large-scale nature-based solutions: Status and challenges for research, engineering and management. *Ecol. Eng.* **2017**, *108*, 489–497. [CrossRef]
72. Sridhar, V.; Jin, X.; Jaksa, W.T.A. Explaining the hydroclimatic variability and change in the Salmon River basin. *Clim. Dyn.* **2013**, *40*, 1921–1937. [CrossRef]
73. Sridhar, V.; Kang, H.; Ali, S.A. Human-induced alterations to land use and climate and their responses for hydrology and water management in the Mekong River Basin. *Water* **2019**, *11*, 1307. [CrossRef]
74. Seong, C.; Sridhar, V.; Billah, M.M. Implications of potential evapotranspiration methods for streamflow estimations under changing climatic conditions. *Int. J. Climatol.* **2018**, *914*, 896–914. [CrossRef]
75. Yang, S.N.; Chang, L.C. Regional inundation forecasting using machine learning techniques with the internet of things. *Water* **2020**, *12*, 1578. [CrossRef]
76. Chang, L.C.; Chang, F.J.; Yang, S.N.; Tsai, F.H.; Chang, T.H.; Herricks, E.E. Self-organizing maps of typhoon tracks allow for flood forecasts up to two days in advance. *Nat. Commun.* **2020**, *11*, 1983. [CrossRef]
77. Chang, F.; Hsu, K.; Chang, L. *Flood Forecasting Using Machine Learning Methods*; MDPI: Basel, Switzerland, 2019; ISBN 978-3-03897-549-6.

water

MDPI

Article

Uncertain Accelerated Sea-Level Rise, Potential Consequences, and Adaptive Strategies in The Netherlands

Jos van Alphen [1,*], Marjolijn Haasnoot [2] and Ferdinand Diermanse [2]

[1] Staff Delta Programme Commissioner, 2511 WB The Hague, The Netherlands
[2] Deltares, 2600 MH Delft, The Netherlands; marjolijn.haasnoot@deltares.nl (M.H.); ferdinand.diermanse@deltares.nl (F.D.)
* Correspondence: jos.van.alphen@deltacommissaris.nl

Abstract: Recent observations and publications have presented the possibility of a high and accelerated sea-level rise (SLR) later this century due to ice sheet instability and retreat in Antarctica. Under a high warming scenario, this may result in a sea level in 2100 that is up to 2 m higher than present and 5 m in 2150. The large uncertainties in these projections significantly increase the challenge for investment planning in coastal strategies in densely populated coastal zones such as the Netherlands. In this paper, we present the results of two studies that were carried out within the framework of the Dutch Delta Programme. The first study showed that it is not only the absolute SLR that presents a challenge but also the annual rate of rise. The latter impacts the lifetime of constructions such as barriers and pumping stations. When the rate of sea-level rise increases up to several centimeters per year, the intended lifetime of a flood defense structure may be reduced from a century to several decades. This new challenge requires new technologies, experiments, strategies, and governance. The second study explored different strategies for the long term to adapt to high SLR (>1 m) and assessed the consequences thereof on adaptation and developments in the coming 2–3 decades. We believe that strategic choices have to be made regarding the permanent closure of estuaries, the pumping or periodic storage of high river discharges, agriculture in an increasingly saline coastal area, and the maintenance of the coastline by beach nourishments. These strategic choices have to be complemented by no-regret measures such as spatial reservations for future sand extraction (for beach nourishments) and future expansion of flood defenses, water discharge, and water storage. In addition, it is advised to include flexibility in the design of new infrastructure.

Keywords: sea-level rise; adaptation; flood risk; water resources; low-lying coasts

Citation: van Alphen, J.; Haasnoot, M.; Diermanse, F. Uncertain Accelerated Sea-Level Rise, Potential Consequences, and Adaptive Strategies in The Netherlands. *Water* **2022**, *14*, 1527. https://doi.org/10.3390/w14101527

Academic Editors: Slobodan P. Simonovic, Subhankar Karmakar, Zhang Cheng, Athanasios Loukas and William Llovel

Received: 22 January 2022
Accepted: 6 May 2022
Published: 10 May 2022

1. Introduction

Coastal regions, especially deltas, have many benefits, such as supply of fresh water, sediments and nutrients from debouching rivers, wealthy and productive ecosystems that guarantee food by fisheries, aquaculture and agriculture, flat land to be developed, and access to international trade. Therefore, these areas attract people to live, work, and recreate, leading to continuous high investments in assets and infrastructure. As a result, coastal cities and communities are home to about 10% of the world's population [1], especially in megacities such as Amsterdam, Buenos Aires, Hamburg, London, Miami, New York, Sydney, and Tokyo and rapidly expanding cities such as Bangkok, Dhaka, Ho Chi Minh City, Jakarta, Lagos, Manila, Mumbai, Shanghai, and Yangon. On the other hand, these areas are vulnerable to coastal erosion, floods and droughts, and pollution, and frequently suffer from related damage, casualties, and disruption.

Sea-level rise (SLR) is expected to impose large challenges to coastal zones. Without population change and additional adaptation, the population exposed to 100 yr coastal floods is projected to triple at 1.4 m SLR relative to the 2020 level. By 2100, the value of global assets within the 100 yr coastal floodplain is projected to vary between USD 7.9 and 14.2 trillion, depending on the emission scenario [2].

Between 1901 and 2018, the global mean sea level increased by 0.20 m. Between 1901 and 1971, the average rate of SLR was 1.3 mm/year, increasing to 1.9 mm/year between 1971 and 2006, and further increasing to 3.7 mm/year between 2006 and 2020 [3]. This SLR acceleration is human-made, and will probably continue this century [3]. Future climate change and SLR projections are driven by emissions and/or concentrations from illustrative Representative Concentration Pathways (RCPs) and Shared Socio-economic Pathways (SSPs) scenarios, respectively [2]. Depending on the greenhouse gas emission scenario, climate change may result in a SLR in 2100 (relative to 1995–2014) of 0.28–0.55 (assuming an extremely low emission scenario SSP1–1.9) to 0.98–1.88 m (assuming an extremely high emission scenario SSP5-8.5). Due to deep uncertainty in ice-sheet processes in Greenland and Antarctica, a SLR of 2 m in 2100 and 5 m in 2150 cannot be ruled out [3].

These large uncertainties in the projections significantly increase the challenge for investment planning in coastal management strategies in low-lying densely populated coastal zones such as the Netherlands. Planning and implementation of coastal defense projects may take many decades. E.g., after the 1953 storm, surge completion of the Delta-works dams, sluices, and storm surges barriers in the southwestern part of the Netherlands took 45 years. This long lead time for planning and implementation, as well as the large potential consequences of coastal flooding, makes it crucial to consider uncertain but possible high-end sea-level rise projections already for nearby investment decisions in water management and land use. To adapt to uncertain climate change, the Netherlands follows an adaptive approach with short-term actions and long-term options to be implemented depending on how the future unfolds [4].

This paper describes the approach of the Dutch government in dealing with projections of uncertain but potential accelerated SLR. We present the consequences regarding coastal management, flood protection, fresh water supply, and water drainage. The challenge of SLR may be dealt with by a wide variety of strategies, ranging from advancing seawards with a new higher coastal defense, maintaining and elevating the present coast, accommodation of water, and adapting land use to retreat from places at high risk [5]. This paper presents examples of these strategies that are being developed for the Netherlands. In this way, this paper illustrates the enormous potential impact of accelerating SLR on densely populated coastal areas. Hence, it also underlines the need to reduce greenhouse gas emissions in order to mitigate SLR and avoid these dramatic consequences. Moreover, it contributes to the debate on the urgency and strategies to adapt on SLR, as advocated in the recent IPCC WG2 report [2].

2. The Netherlands and the Delta Programme

The Netherlands is situated in the delta of the rivers Rhine, Meuse, Scheldt, and Ems along the North Sea coast. The coastal region consists of sandy beaches and dunes, estuaries, intertidal areas, and low-lying polders. These polders are densely populated, including the capital Amsterdam, government seat The Hague, and mainport Rotterdam. About 26% (10,500 km^2) of the Netherlands' territory is below mean sea level, and about 60% is vulnerable to floods from the North Sea, rivers, or lakes (Figure 1). This flood-prone area contains about 60% of the population of 17 million and produces about 60% of the gross domestic product (GDP) of about EUR 800 billion. Because of the large number of inhabitants and high value of assets, the Netherlands has a high level of flood protection, provided by a comprehensive system of dams, seawalls, storm surge barriers, dikes, dunes, pumps, sluices, and regular beach nourishments. Two major floods in 1916 and 1953 resulted in closing of the major tidal inlets and estuaries from the sea, which altogether shortened the original Dutch coastline of 1200 km to about 300 km. This flood protection and water management system is complemented by a well-developed governance system of district water boards, national agencies (Rijkswaterstaat), funding, legislation [6], and well-experienced knowledge institutes and private parties (e.g., dredgers). Maintenance and regular upgrading of this flood defense system to changing conditions costs about EUR 1.2 billion annually [7].

As a result of growing concern about the potential impact of climate change and SLR on the Dutch delta, in 2010, the government initiated the Delta Programme. The aim of the Delta Programme is to maintain the Dutch delta as an attractive place to live, work, and recreate for present and future generations [8]. The Delta Programme has a long-term time horizon, up to 2100. To make the inherent uncertainty in climate and socio-economic developments manageable, an adaptive approach has been adopted, including scenarios, adaptive strategies, and periodic (every 6 years) review [4]. A 'Signal' group of independent experts advises the Delta Commissioner on an annual basis on relevant new developments and knowledge that might influence the goals and implementation of the Delta Programme [9].

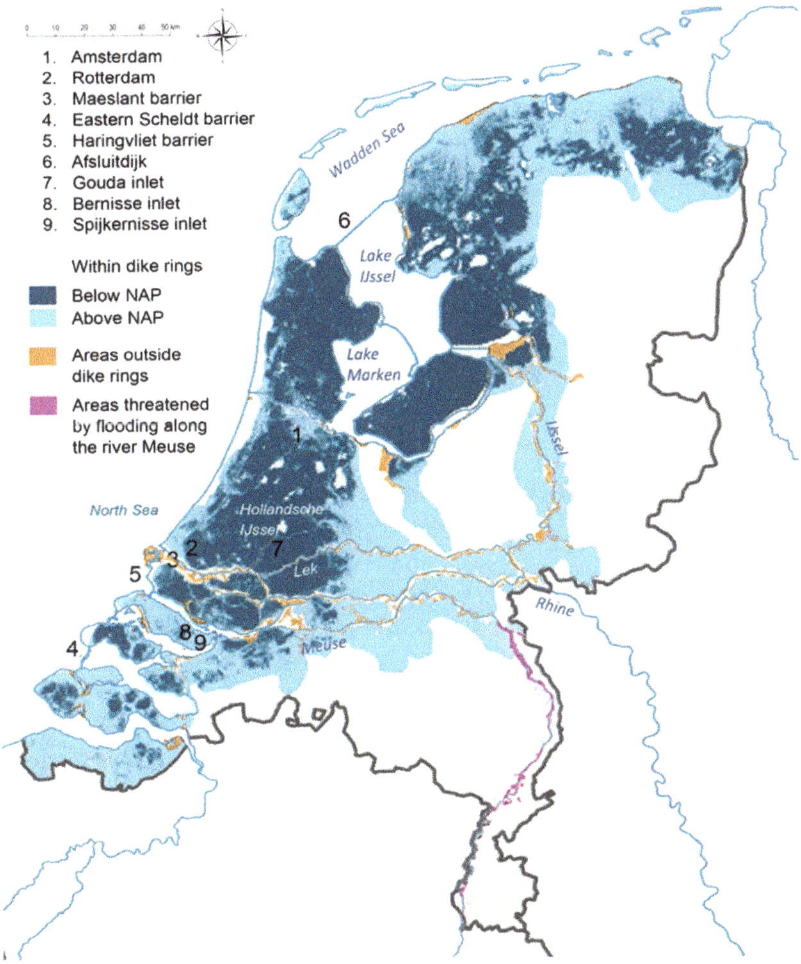

Figure 1. Flood-prone areas in the Netherlands, present situation. NAP is ordnance datum, which is about mean sea level. Dike rings are areas protected by flood defenses. (Reprinted with permission from Haasnoot et al., 2020 [10]).

In 2014, the Delta Commissioner proposed strategies and measures to prepare for climate change, related to a 1.5 and 3.5 °C global warming scenario and anticipating a SLR of 0.3–1.0 m in 2100 (relative to 1990) [11]. This was in line with the most recent national climate change scenarios of the Royal Dutch Meteorological Institute [12]), which were based on 5th IPCC Assessment Report [13]. In 2017, the Signal group advised the Delta Commissioner to pay more attention to a possible acceleration of SLR and its potential consequences for the Dutch delta [14].

The Dutch delta is extremely vulnerable to SLR and an acceleration may have a major impact on the livability of the delta for future generations. This acceleration may unfold in the second half of this century, so present measures are sufficient until at least 2050. However, in SLR-related flood-prone areas, up to 2050, an estimated EUR 600 billion of investments is planned in urban developments, infrastructure, sustainable energy, and climate adaptation [15]. These types of investments predominantly involve projects with a lifetime of 50–100 years, which also determine future land use, and may trigger or block future adaptation options. Considering a potential acceleration of SLR within this lifetime, it becomes necessary to analyze whether these planned investments increase future risk, how this can be avoided, and which strategic choices have to be made. To answer these questions, in 2017, the Delta Commissioner initiated a study to assess the potential impact of accelerated SLR on the Dutch delta, followed in 2019 by an inventory of potential strategies and measures to deal with this accelerated SLR.

3. Materials and Methods

This paper describes the approach and main results of the SLR impact assessment on the Dutch delta and the inventory of potential strategies to deal with this threat.

Whether and how a natural delta adapts to SLR depends on the balance between the amount of sediment supplied to the delta by rivers, waves, and currents and the volume that is necessary for intertidal areas, riverbeds, and land to keep pace with SLR. Apart from the Wadden Sea, the Dutch delta is not natural anymore: rivers have been embanked and most floodplains have been turned into polder systems (and hence cut off from natural sediment supply), the coastline is artificially maintained with beach nourishments, the inlets and estuaries have been closed with dams or protected by storm surge barriers, and former lakes have been drained and turned into polders, with ground surface levels up to 6 m below the mean sea level. The impact of SLR on the Dutch delta is established on three essential elements of delta management: coastline maintenance, flood protection, and fresh water supply.

Our inventory was based on a scenario-neutral approach [16], based on [17], wherein we assessed the consequences conditional on the magnitude and rate of SLR (up to 3 m and 60 mm/year) on the present strategies. Scenarios could then be used to assess when this may occur. For example, in an RCP2.6 or RCP4.5 emission scenario, 1 m SLR may occur beyond or around 2100. In an RCP8.5 scenario, also including a contribution of melting Antarctic land ice, 1 m SLR may already be reached around 2070 and meet 1.95 m in 2100 [18].

The assessment of SLR consequences was primarily based on model computations and simulations, see [10] for more details. Necessary volumes of beach nourishments were estimated by using the area of the marine coastal zone (4000 km^2) multiplied by different rates of SLR. Conceptual 0-dimensional water balance models were used to get a first impression of different combinations of storage and pumping capacities to manage river floods. Two-dimensional hydrodynamic model (WAQUA [19]) computations established hydraulic loads to assess related dike reinforcement efforts and storm surge barrier closure frequencies. Additionally, probabilistic model computations were carried out to assess impacts of sea-level rise on flood risks. The slow response of groundwater salinization to SLR was investigated by a regional 3D variable density groundwater flow model [20]. The results of this model provided input data for the National Water Model, a combination of groundwater, subsurface water, and surface water simulation models as well as water demand models to assess (fresh) water availability and demands in the Netherlands [21]. To estimate the effect of salinization on fresh water intake locations along the main rivers, a 3D hydraulic model supplied data on salt intrusion at values of 0, 2, and 4 m SLR.

These model computations provided an overview of critical consequences of SLR on our present water supply and flood management infrastructure (see Figure 2). For example, at 1 m SLR, the Eastern Scheldt storm surge barrier closure frequency may be increased from the current 1.5 to 45 times a year, if current closure criteria remain unchanged. This will have major consequences on the required maintenance effort, but also on the preservation of the intertidal habitats behind the barrier.

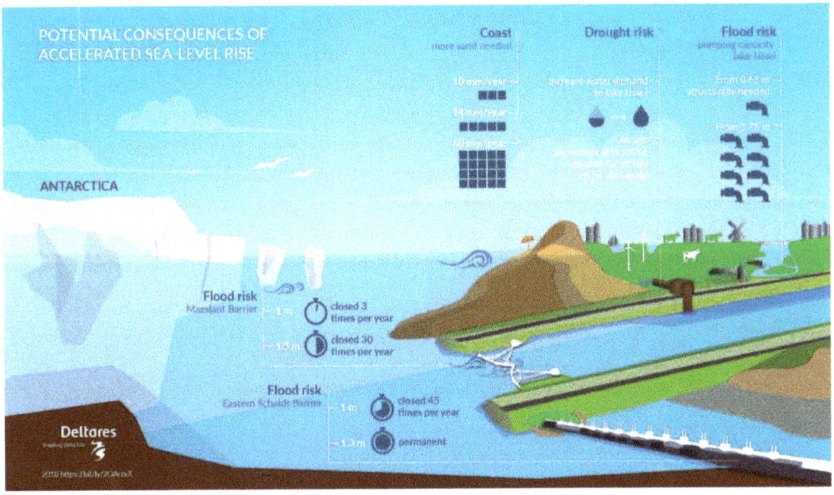

Figure 2. The consequences of accelerated SLR for the main elements of Dutch water management.

For the inventory of future strategies, the generic IPCC adaptation strategies (protect, accommodate and retreat) [4] were used as a starting point and slightly altered for the specific situation in the Netherlands. This alteration was based on an inventory of 189 plans to deal with SLR that were collected since 2008 [22]. An indicator list was developed to 'score' strategies on criteria such as technical and societal feasibility, impacts on flood protection and water resources, adaptability, and relation with other transitions. An expert elicitation workshop and a multi-day student hackathon were organized to provide a quick-scan assessment of adaptation measures and strategies and to provide further inspiration on potential alternative adaptation strategies. Finally, adaptation pathways were designed to connect the various strategies and measures in sequence and to identify potential triggers for a shift of strategies and illuminate lock-ins that may block future adaptation needs, e.g. urban developments along to be raised flood defenses. Recommendations for low-regret actions were formulated to prepare for adaptations in the years to follow. Examples are pilots to develop new techniques to increase beach nourishment volumes and spatial reservations for future expansion of water-related infrastructure.

4. Results

4.1. The Potential Impact of Sea-Level Rise on the Dutch Delta

An accelerated SLR may affect the Dutch delta in different ways. The consequences of a potential accelerated SLR on present water management structures and strategies according to the adopted 'what if' scenarios are described below. Figure 2 illustrates the potential consequences of SLR on coastal maintenance, water demand, pumping capacity, and closure frequency of storm surge barriers. Figure 3 summarizes some critical consequences ('tipping points') of SLR magnitude (left) or SLR rate (right) on flood defense and water supply infrastructure and coastal zone management. The most important locations are shown in Figure 1.

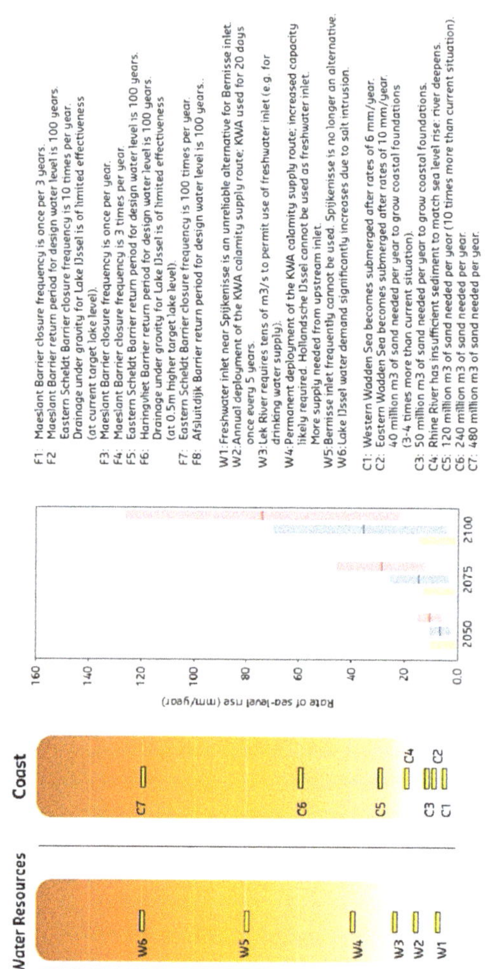

Figure 3. The consequences of accelerated SLR on measures related to flood risk, water management, and coastal maintenance. Left panel is sea-level height, right panel is SLR rate. The colored bars represent different SLR scenarios for the Netherlands: yellow = Delta scenarios 2014, blue: RCP 4.5, and pink: RCP8.5. The horizontal yellow bars in the three central columns represent the tipping points in present flood risk, water management, and coastal management policies, referred to by codes and explained below the diagram (Reprinted with permission from Haasnoot et al., 2020 [10]).

Maintenance of the sandy coastline requires increasing volumes of sand nourishments to compensate for erosion and to keep up with SLR. This volume is proportional to the annual rate of sea-level rise. The global averaged SLR accelerated to about 4 mm/year the last decade, but along the Dutch coast, this acceleration is difficult to assess due to large interannual variability in wind climate and thermal expansion. [Le Bars, personal communication]. With this annual SLR of 2 mm/year, the volume of sand nourishments on eroded beaches and erosive tidal channels is 12 million m^3/year on average. An accelerated SLR may increase this volume at the end of the century with more than a factor 10 (K7 in Figure 3). The relatively shallow Dutch continental shelf (depth of 20–40 m) consists of vast sandy deposits originating from glacial times. The increase in potential demand for beach nourishments would make spatial reservations for sand extraction necessary, taking account of the rapid development of large offshore wind energy farms.

Large parts of the coastal area consist of intertidal basins, such as the Wadden Sea, which is a UNESCO heritage site. Preservation of the 4700 km^2 intertidal habitats such as shoals and mudflats requires sediment supply that enables these areas to keep pace with SLR. Otherwise, these areas may lose surface area and eventually drown. This can have large consequences on ecology, fisheries, and recreation, and also on flood protection, since these shallow areas reduce wave action and hydraulic loads on the flood defenses. For the Wadden Sea, it is expected that natural sediment supply by tidal action may fall short when SLR exceeds an annual rate of 6–10 mm/year (which may be around 2050) [23]. (see also K1 and K2 in Figure 3).

Flood protection against storm surges is provided by a system of dams, seawalls, and storm surge barriers. With accelerating SLR and without improvements of the flood defenses, the current design conditions for these flood defenses may be met more frequently. The level of protection may drop and the flooding probability may increase. Storm surge barriers are of special interest, since SLR increases the closure frequency. E.g., the Maeslant barrier, which protects Rotterdam, is designed to close 1/10 per year (once every decade). At 1 m SLR, it may need to be closed 3 times per year (V4 in Figure 3), and at 1.5 m SLR, it may be closed 30 times a year. This may induce a larger maintenance effort. Furthermore, the barrier was not designed to close that often and may therefore meet its end-of-life time earlier than was originally intended. More importantly, it also increases the probability of a closure coinciding with a flood wave on the river, increasing riverine flood risk in the area that is protected by the barrier. To maintain the present level of flood protection, an upgrade of the storm surge barrier becomes necessary before the designed end-of-life of the barrier, and/or large dike improvement works.

Excessive river discharge and rainfall is discharged by gravity from large lakes and canals into the North Sea. With SLR, the interval during which gravity drainage into the sea is possible is gradually shortened, and consequently, the natural discharge capacity is reduced. To maintain the required discharge capacity, the sluices have to be widened and/or complemented with pumps. When SLR exceeds 0.65 m above the present level, the low tide along the Afsluitdijk may equal the level of Lake Ijssel. Then, discharge under gravity is no longer possible and drainage should be entirely performed by pumps. This requires a pumping capacity of about 1200–3200 m^3/s, depending on the accepted lake level fluctuations. In the 'what if' scenarios, this may become necessary from 2070 onward. Artificial drainage of the river discharge of Rhine and Meuse through their present river outlets near Rotterdam may require even larger pumping capacities.

In the coastal region, there is a continuous inward saline influx in river mouths and through ground water into the deep lying polders. Presently, the influx through surface waters is counterbalanced by the outflow of fresh riverine water, except during periods of low discharge. Polders that are suffering from upwelling saline groundwater are flushed with fresh water that is transferred from upstream intakes, such as Gouda, which serves 1100 km^2 of polder area. The Gouda intake supplies about 80 Mm^3 per year on average, mainly during summer. With increasing SLR, the salinization influx may increase. Fresh water intakes for drinking water, agriculture, and industrial purposes may close more frequently or have to be transferred to more upstream fresh water locations. To maintain the present agricultural land use and crop types, the fresh water demand for flushing may increase, which requires expansion of the present supply route and supply volumes. For the Gouda intake, the required fresh water supply may increase to 120 Mm^3 in 2100 and 185 Mm^3 per year in 2200.

Most existing large flood and water management structures in the Netherlands are designed for a lifetime of 100–200 years, i.e., according to the present SLR of 2 mm/year, for an SLR of 0.20–0.40 m. An accelerated SLR (e.g., up to 30 mm/year) may reduce this lifetime significantly. Figure 4 illustrates that in the RCP8.5 scenario, the lifetime of a 50 cm SLR design may be reduced from 65 years presently to only 10 years at the end of the century. To solve this challenge, we must think of more robust measures that can handle a large SLR (to achieve a long lifetime), or measures that are adaptive (to gradually expand, reinforce, or increase in height when necessary), or more fundamentally, entirely different strategies of delta management.

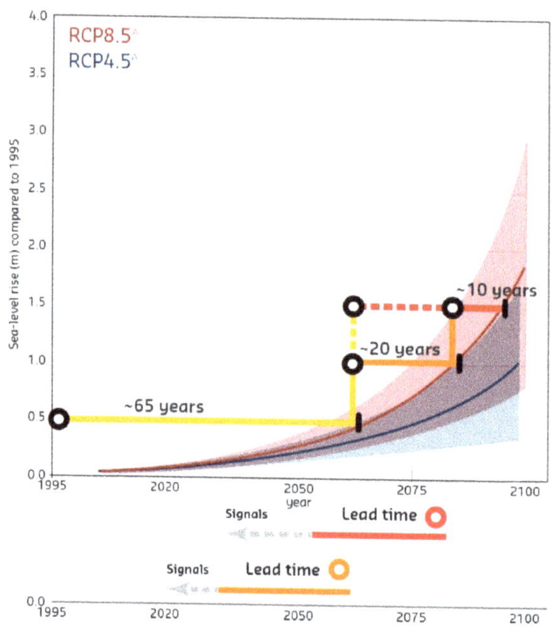

Figure 4. The effect of accelerated SLR on the lifetime of investments. In the present situation, a design for 50 cm holds 65 years. The potential acceleration later this century reduces the lifetime of a 50 cm design to about 10 years (Reprinted with permission from Haasnoot et al., 2020 [10]).

The important message of these observations is that from 2050 onward, large uncertainty appears in the expected SLR, depending on what will occur with Antarctic land ice loss, but also that the potential consequences for the Dutch delta may be huge and may accumulate on all aspects that are relevant for delta management in a rather short time, i.e., several decades.

4.2. Strategies to Cope with Sea-Level Rise

Experiences with previous Dutch major water management programmes show that policy development, plan preparation, decision making, and implementation may take several decades: the Deltaworks programme, initiated after the devastating 1953 flood, took about 45 years to complete, and the Room for the River programme, initiated after the 1993 and 1995 flood events of the Rhine and Meuse Rivers, lasted 20 years until finalization. With these experiences in mind, and potential consequences of accelerated SLR emerging from 2050 onward, studies and preparations for new strategies should start now to be ready when this might become necessary.

In [5], three different types of strategies were distinguished:

- Retreat, which involves reducing the exposure to coastal hazards by moving people, assets, and activities landward to higher ground. This choice can be motivated by excessive economic or environmental impacts of protection or long-term adaptation needs;
- Accommodate, which implies that people continue to use the land at risk but do not attempt to prevent the land from being flooded. This option includes creating emergency flood shelters, elevating buildings on piles or mounds, converting agriculture to fish farming, or growing flood or salt tolerant crops;
- Protect, which involves hard structures such as seawalls and dikes, as well as nature-based solutions such as dunes and vegetation, to protect the land from the sea so that existing land uses can continue. Protection can be provided by strengthening the existing coastline or by advancing a higher new coastline.

With growing attention for SLR in the last decade, many ideas to cope with SLR have been proposed for the Dutch delta [22]. Inspired by the abovementioned IPCC typology for the Dutch delta, these strategies are now categorized as follows (see Figure 5, [24]):

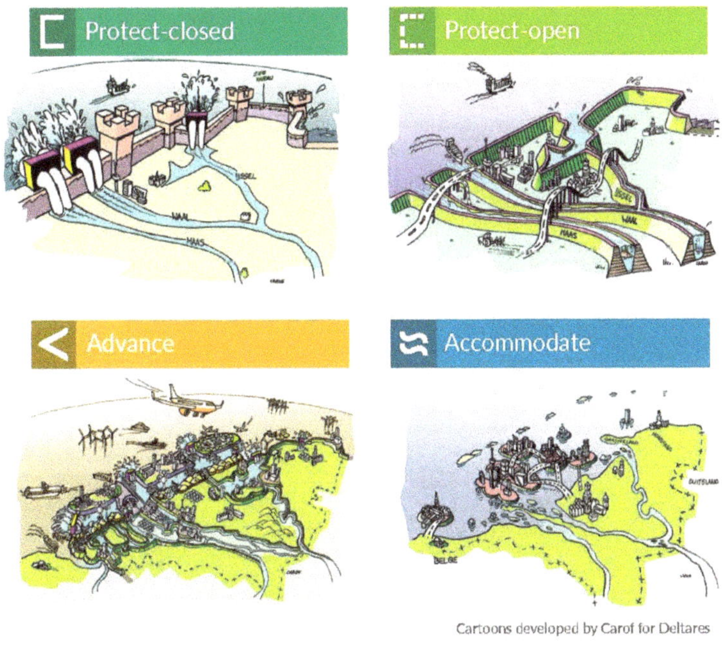

Figure 5. Potential types of strategies for the Netherlands to adapt to accelerated SLR (copyright of Carof, for Deltares).

Protect ('freeze'): continue the present policy to preserve the territory of the Netherlands by increasing volumes of beach nourishments, and raising dikes and seawalls. Present land use can be continued, but requires increasing efforts in flood protection and water management. Regarding the estuaries, two variants of protection emerge:

(a) Close them off from the sea completely by dams, navigation locks, and sluices. This 'protect-closed' strategy protects the inland area against SLR and the related salinization. However, handling river discharge requires enormous pumping stations, with a capacity of several thousand m^3/s, in combination with large areas for temporary flood storage.

(b) Maintain the open connection with the sea. In this 'protect-open' strategy, rising sea level extends its influence upstream along the rivers, flooding unembanked (harbor) areas and requiring extensive dike improvement programmes to maintain flood protection standards. In addition, a large inland area may be affected by salinization of surface waters.

The 'accommodate' strategy is based on the 'living with water' concept. For the Dutch delta, this could be a combination of 'retreat' ('flee', horizontal retreat) and 'accommodate' (vertical retreat, by creating mounds, raising buildings, or creating ring dikes to protects urban areas, or by introducing floating solutions). In this strategy, flood defenses are maintained to the present height, but not raised to rising sea level. As a result, low-lying areas may experience more frequent flooding (e.g., from presently 1:10.000 to 1:10 years in the future). This strategy may have a large societal impact since large parts of the country must be re-organized or abandoned and millions of inhabitants must adapt or migrate to higher ground.

The 'advance' ('fight') strategy is new compared to the original IPCC types. In this strategy, the present coastline is extended seaward to build a more robust coastal flood defense. It can be considered as a special kind of protect strategy, also creating 100–500 km^2 of new land. A more radical alternative is the construction of a new (closed) coastline 10–20 km offshore, protecting the present coastline and creating a brackish reservoir in which river discharge during river flood events can be buffered before it is pumped across this new coastline. The present land use can be continued; moreover, the new coastline creates 500 km^2 for urban and industrial developments, recreation, and nature, and possibly, a new airport. The sand volumes necessary to expand the present shoreface or to construct an offshore coastline vary between 10 and 20 billion m^3.

These different strategies all involve prolonged and massive investments in construction and maintenance and have profound impacts on present and future land use. Especially, the 'living with water' strategy may cause large societal and political debate, since the Dutch culture has been for a thousand years based on protection against floods and reclaiming land from the sea. On the other hand, elements of this 'amphibious' strategy already exist locally, but are less known and widespread. For example, to provide additional flood discharge and storage capacity, in the former Noordwaard, polder houses have been removed or relocated to outside the flood plain or to mounds [25] (see Figure 6). Along parts of the coasts (e.g., Wadden Sea and Westerschelde), managed realignment is implemented to restore salt marshes and to aid coastal defense [26].

Policy development, decision making, elaboration, and implementation of the measures can take several decades. However, there is still a lot of uncertainty regarding future SLR, and starting a debate on the appropriate strategy right now may seem much too early when assuming a best-case scenario in which global warming is kept to a maximum of 2 °C, or even when considering the likely range of projections with a higher global warming scenario. On the other hand, the planned investments up to EUR 600 billion in housing, infrastructure, sustainable energy, and agriculture in flood-prone areas may substantially increase the future risk potential, and create a further 'lock in' of continued developments in low-lying flood-prone polders when fundamental choices are avoided.

The adaptive approach of the Delta Programme tries to find a balance between 'too little too late' and 'too much too early', to determine the measures that are necessary (or 'low regret') for now, maintain open options for future additional measures, and monitor developments in order to timely decide on accelerating present strategies or implementing additional measures [4]. Adaptation pathways connect present investment agendas with future perspectives by different combinations of measures, identifying where a change of strategy is still possible and how to avoid 'lock in' [27]. 'Lock-in' situations can be avoided by low-regret measures, including spatial reservations for potential future measures such as sand extraction, water storage, and flood defense reinforcements.

Figure 6. Relocated houses on flood-free mounds in Noordwaard flood storage and bypass area (picture from Brabants Dagblad).

Adaptive strategies can benefit from flexible measures, which are easy to alter, speed up, or slow down depending on the measured climate change. Examples are 'nature-based' solutions such as beach nourishments. Robust measures that can cope with high-end scenarios may be preferred when the adjustment costs are relatively high compared to a more robust initial design [28]. This is especially the case with investments in 'hard' infrastructure such as locks, seawalls, and dams. An interesting hybrid example is the expansion of the pumping station in IJmuiden, where the expensive ground works and foundation was already constructed in a robust way for the worst-case scenario (SLR + 0.85 m in 2100), while the pumps are only placed for the best-case scenario (SLR + 0.35 m in 2100), creating flexibility to place additional pumps when necessary [29]

Potential adaptation pathways are presented in Figure 7 and can be explained as follows:

- Until 2050, SLR projections for different emission scenarios vary only little. Therefore, a continuation of the present protect strategy seems logical as a first step. A complicated societal debate is avoided, choices can be postponed until more knowledge about SLR becomes available, and present land use can be continued by increasing the present technical measures in flood protection and water management, presumably up to a sea-level rise of 2 m and a related rate of SLR. The increasing volumes involved in beach nourishments (up to 60 million m^3/year) can be dredged on the Dutch continental shelf. This strategy largely depends on scaling up present sand extraction and beach nourishments operations of generally 1–5 million m^3 per project. An interesting pilot is the so-called 'Sand motor', a 1.28 km^2 beach nourishment of 21.5 million m^3, costing about EUR 70 million. After 10 years of monitoring, it was concluded that the 'Sand motor' acted as a 'feeder beach' from which the sand is transported to adjacent dunes and beaches by natural processes such as wind, tidal currents, and waves [30]. It also illustrated the 'benefits of scale' effect, since such larger volumes enable the use of more efficient large vessels and nourishment techniques, and reduces the costs of material mobilization and demobilization costs for ground works. As a consequence, the usual beach nourishment costs were almost halved.
- In addition to these kind of pilots, low-regret measures can be taken, such as spatial reservations for sand extraction in the North Sea, to avoid conflicts with other marine users, including wind farm developers. Nevertheless, a large loss of intertidal areas

in the Wadden Sea cannot be prevented if SLR rates consistently exceed 6 mm/year (western Wadden Sea) or 10 mm/year (Eastern Wadden sea). Nor can the salt intrusion be fully avoided. In this way, the protection pathway is continued, with the risk of large transfer cost for future generations, if a shift to other strategies is needed.

- With rising sea levels, the first dilemma that may emerge is whether to permanently close the estuaries or to keep them open to facilitate free river discharge. To keep both options open in the coming decades, it is necessary to avoid new developments in unembanked areas (or at least make them flood-proof), and reserve space for future dike improvements. A first inventory shows that future dike raising requires 12–17 m additional horizontal space per meter raised. This requires valuable space, especially in urban areas such as Rotterdam, which can only be used for temporary activities. Furthermore, in the closed option, drainage of river discharge requires large pumping capacity, up to several thousand m^3/s or even more. This pumping requirements can be reduced when flood waters can be temporarily stored. This storage capacity can be found in present water systems, or in creating new systems in the coastal zone, attached to the river mouths. Model simulations can help find the optimum combination of dike raising, storage volumes, and pumping capacity.
- Regarding agriculture and horticulture, a gradual transition to salt (and drought)-tolerant crops is low-regret, since salinization of the coastal area is expected in both the retreat and protection strategies.
- With continued SLR and increasing technical efforts, a fundamental choice between 'advance' or 'retreat' becomes inevitable. The 'advance' strategy can be seen as a next step in the present protective strategy, heavily relying on technical measures such as flood defenses and pumps. Therefore, some experts advocate to start with this strategy already, and avoid short-term expenses on maintenance and on upgrading the present flood and water management infrastructure. The retreat strategy marks a significant 'change of mind', which may be reached after a long societal debate, or become reality when natural disaster creates a 'fait accompli'.

Worldwide, many examples exist of large-scale technical measures that may be necessary to handle a SLR of 2 m or even more. Examples are the Japanese 10 m high tsunami seawalls (costs up to USD 30 million/km [31]) and even more expensive 'super levees' such as the 36 m high and 33 km long Seamangum closure dike in South Korea (cost USD 1.7 billion, [32], the 300 m^3/s Mubarak pumping station in the Aswan dam, large-scale land reclamations by sand nourishments for urban development in Palm Jumeirah, Dubai (cost USD 12 billion for 5.6 km^2) and Eko City, Lagos (USD 6 billion for 10 km^2) [33]. Additionally, examples exist of (managed) retreat or 'living with water' measures, such as mounds and shelters in the tsunami-affected coastal zone of Japan, cyclone shelters in Bangladesh, and raised buildings in the Mississippi delta, but these areas are often less populated than the Dutch delta.

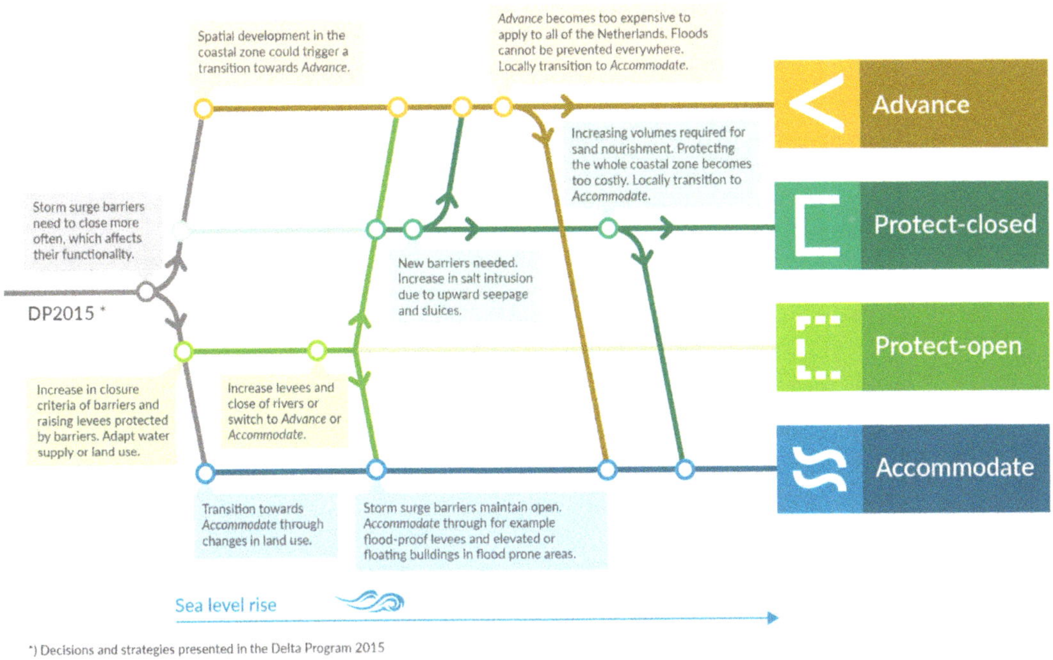

Figure 7. Adaptation pathways for the Netherlands to adapt to accelerated SLR. DP2015 represents the present strategies, adopted in 2014 [11].

4.3. The First Step

A choice for one of these strategies has large consequences for present and future land use, requires a solid preparation and societal debate, and is not necessary at the present rate of SLR, as there is still time to prepare. Furthermore, the Netherlands does not necessarily have to choose only one of the four strategies. A regional differentiation of the strategies, in which some areas receive extra protection while in other areas a retreat or accommodate strategy is applied, is a viable option as well, especially when there are large differences in population density and flood hazard.

To support a timely preparation on SLR, in 2019, the Delta Commissioner and Minister of Infrastructure and Water Management jointly initiated a 6-year comprehensive research programme that consists of the following tracks [34]:

- research aimed to reduce the uncertainty regarding SLR. This requires international cooperation on polar research, but also on monitoring SLR in coastal seas;
- develop a method to signal a potential acceleration of SLR timely and with sufficient confidence. This involves statistical research on SLR data and projections;
- establish the potential effects on the flood risk and water management of the Dutch delta and related infrastructure and establish the lifetime of the present policies. Detailed model computations for different SLR scenarios must be performed to predict future coastal erosion, necessary beach nourishment volumes, hydraulic loads and required flood defense strengthening, storage and pumping capacities, and increase of fresh water supply. Special attention is paid to the discharge of river water with the rising sea-level [35], which is a major challenge for many deltas worldwide;
- explore long-term options which might become necessary when present policies become short. In this situation, the present territory and land use of the Netherlands might be affected by regular or permanent flooding and salt intrusion. In cooperation with other authorities and NGOs, it must be explored whether present investment

agendas on, e.g., housing, infrastructure, and sustainable energy may occupy land that might be necessary for future strategies to deal with SLR, and how these agendas can be aligned to create synergy with these future strategies;

- prepare the implementation of the necessary policies and measures by:
  - collecting expertise and knowledge on large-scale long-term societal transitions;
  - starting pilots to gain experience with the possibilities to scale up present techniques and methods (e.g., building with nature);
  - propose spatial reservations for future sand extraction, flood protection, water discharge, storage, and supply; and
  - introducing the adaptive approach in the design of projects that are planned for the coming years in housing, infrastructure, and replacement of present (aging) infrastructure.

This research programme is a joint cooperation of several ministries, water management authorities, provinces, municipalities, knowledge institutes and universities, private parties, NGOs, and neighboring countries, under the leadership of the Delta Programme Commissioner and Minister of Infrastructure and Water Management. The results will become available in 2026, to give input to the next 6-year evaluation of the Delta Programme.

5. Discussion and Conclusions

SLR is becoming a worldwide threat for coastal areas, especially because of its potentially accelerating rate. Whether and when this will occur is still uncertain, and depends largely on the success of CO_2 reduction measures, global warming, and the response of Antarctica's land ice. The exploration of adaptation pathways helps to manage this uncertainty by enabling timely adaptation and limiting regret of investments. As sea levels rise faster, adaptive measures must be implemented at an increasing frequency or to a higher sea level. This is challenging for investments with a long lifetime or for measures that need decades to prepare and implement in an orderly way, such as coastal defenses or managed retreat. With accelerating SLR, the time required may become a limit. Therefore, long-term planning and accelerated implementation, particularly in the next decade, is important to close the gaps between present and necessary adaptation [2].

Using the Dutch approach as an inspiration for other urbanized coastal zones threatened by SLR, it is recommended to start a comprehensive programme consisting of:

- research into the behavior of Antarctic ice, combined with the development of a method to signal SLR acceleration in a timely way with sufficient confidence;
- research into the consequences of accelerated SLR on present water management infrastructure such as flood defenses, storm surge barriers, locks and pumping stations, and fresh water intake locations. How will SLR affect their lifetime, maintenance, replacement, or reconstruction? Consequently, what will be the effect on the land use that this affected infrastructure serves? Which strategic choices in water management and land use strategies become inevitable? E.g., is the present type of fresh-water-based agriculture in the coastal zone sustainable with increasing salinization due to SLR? Or do we continue with new urban development in low-lying coastal areas which require increasing efforts to maintain dry?;
- the exploration of long-term strategies (e.g., 'protect', 'advance', 'accommodate', and 'retreat') that might become necessary when present policies start to fail. Vice versa, identify necessary short-term measures to keep these future strategies open and avoid regret investments. An important no-regret measure is the spatial reservation for future sand extraction (for beach nourishments) and for future expansion of flood defenses, water discharge, and water storage.
- flexible measures, which are important to bridge the coming decades of uncertainty in SLR. Nature-based solutions, such as beach nourishments or stimulating natural sedimentation on flood defense forelands, are relatively cheap and more flexible to adapt to changing conditions than concrete structures. In addition, they have beneficial side effects on, e.g., nature restoration and recreation. Pilots may be helpful

to improve the effectiveness of these measures and establish their 'scaling up' potential, as illustrated by the Sand motor mega beach nourishment [30].

- adaptive design. When decisions have to be made regarding the (re)construction of long-lifetime robust infrastructure, an adaptive design should be promoted, to make future expansion possible without excessive costs.

The first results of this Dutch sea-level rise programme illustrate the large potential impact of accelerating SLR on densely populated coastal areas. Although adaptive options can be formulated, the implementation will require large and long-lasting efforts and cause strong societal and political debates. This underlines the need to reduce greenhouse gas emissions in order to mitigate SLR, avoid these dramatic consequences, and reduce adaptation efforts.

Author Contributions: Conceptualization, J.v.A., M.H. and F.D.; methodology, M.H. and F.D.; software, F.D. and M.H.; formal analysis and investigation, F.D. and M.H.; writing—original draft preparation, J.v.A., M.H. and F.D.; writing—review and editing, J.v.A.; visualization, M.H. All authors have read and agreed to the published version of the manuscript.

Funding: This research received no external funding.

Institutional Review Board Statement: Not applicable.

Informed Consent Statement: Not applicable.

Conflicts of Interest: The authors declare no conflict of interest.

References

1. Global Commission on Climate Adaptation. *Adapt Now: A Global Call for Leadership on Climate Resilience*; Global Center on Adaptation: Rotterdam, The Netherlands; World Resources Institute: Washington, DC, USA, 2019.
2. IPCC. Summary for Policymakers. In *Climate Change 2022: Impacts, Adaptation and Vulnerability. Contribution of Working Group II to the Sixth Assessment Report of the Intergovernmental Panel on Climate Change*; Cambridge University Press: Cambridge, UK, 2022. Available online: https://www.ipcc.ch/report/ar6/wg2/ (accessed on 9 May 2022).
3. IPCC. Summary for Policymakers. In *Climate Change 2021: The Physical Science Basis. Contribution of Working Group 1 to the Sixth Assessment Report of the Intergovernmental Panel on Climate Change*; Cambridge University Press: Cambridge, UK, 2021. Available online: https://www.ipcc.ch/report/ar6/wg1/#SPM (accessed on 20 March 2022).
4. Bloemen, P.J.T.M.; Hammer, F.; Van der Vlist, M.J.; Grinwis, P.; Van Alphen, J. DMDU into practice: Adaptive Delta Management in the Netherlands. In *Decision Making under Deep Uncertainty: From Theory to Practice*; Marchau, V.A.W.J., Walker, W.E., Bloemen, P.J.T.M., Popper, S.W., Eds.; Springer International Publishing: Cham, Switzerland, 2019; pp. 321–351. [CrossRef]
5. IPCC. *Strategies for Adaptation to sea Level Rise*; Report of the Coastal Zone Management Subgroup, IPCC Response Strategies Working Group; Ministry of Transport and Public Works, Rijkswaterstaat, Tidal Water Division: The Hague, The Netherlands, 1990. Available online: https://puc.overheid.nl/rijkswaterstaat/doc/PUC_26288_31/ (accessed on 9 May 2022).
6. OECD. *Water Governance in the Netherlands, Fit for the Future?* OECD Studies on Water; OECD Publishing: Paris, France, 2014. [CrossRef]
7. Deltacommissaris. *The 2022 Delta Programme, Every New Development Climate-Proof*; Ministry of Infrastructure and Water Management: The Hague, The Netherlands, 2021. Available online: https://english.deltaprogramma.nl/documents/publications/2021/09/21/dp-2022-in-english---print-version (accessed on 20 March 2022).
8. Deltacommissaris. *The 2011 Delta Programme, Working on the Delta*; Ministry of Transport; Public Works and Water Management: The Hague, The Netherlands, 2010.
9. Haasnoot, M.; van't Klooster, S.; Van Alphen, J. Designing a monitoring system to detect signals to adapt to uncertain climate change. *Glob. Environ. Chang.* **2018**, *52*, 273–285. [CrossRef]
10. Haasnoot, M.; Kwadijk, J.; Van Alphen, J.; LeBars, D.; Van den Hurk, B.; Diermanse, F.; Van der Spek, A.; Oude Essink, G.; Delsman, J.; Mens, M. Adaptation to uncertain sea-level rise: How uncertainty in Antarctic mass-loss impacts coastal adaptation strategy of the Netherlands. *Environ. Res. Lett.* **2020**, *15*, 034007. [CrossRef]
11. Deltacommissaris. *The 2015 Delta Programme, Working on the Delta. The Decisions to Keep the Netherlands Safe and Liveable*; Ministry of Infrastructure and the Environment: The Hague, The Netherlands, 2014.
12. Royal Dutch Meteorological Institute KNMI. *KNMI'14 Climate Scenarios for the Netherlands*; A Guide for Professionals in Climate Adaptation; KNMI: De Bilt, The Netherlands, 2015; 34p. Available online: https://www.knmiprojects.nl/projects/climate-scenarios/documents/publications/2015/01/01/brochure-knmi14-climate-scenarios (accessed on 24 December 2021).
13. IPCC. *Climate Change 2014: Synthesis Report. Contribution of Working Groups I, II and III to the Fifth Assessment Report of the Intergovernmental Panel on Climate Change*; IPCC: Geneva, Switzerland, 2014; 151p.

14. Deltacommissaris. *The 2018 Delta Programme, Continuing the Work on a Sustainable and Safe Delta*; Ministry of Infrastructure and the Environment: The Hague, The Netherlands, 2017.
15. Hekman, A.; Booister, N. *Ruimte voor de Toekomst, Flexibel Invullen van Investeringsopgaven om Effecten van Zeespiegelstijging in de Toekomst te Kunnen Opvangen*; White Paper; SWECO Nederland: De Bilt, The Netherlands, 2021. (In Dutch)
16. Broderick, C.; Murphy, C.; Wilby, R.L.; Matthews, T.; Prudhomme, C.; Adamson, M. Using a scenario-neutral framework to avoid potential maladaptation to future flood risk. *Water Resour. Res.* **2019**, *55*, 1079–1104. [CrossRef] [PubMed]
17. Kwadijk, J.; Haasnoot, M.; Mulder, J.P.M.; Hoogvliet, M.M.C.; Jeuken, A.B.M.; Van der Krogt, R.A.A.; Van Oostrom, N.G.C.; Schelfhout, H.A.; Van Velzen, E.H.; Van Waveren, H.; et al. Using adaptation tipping points for climate change and sea level rise: A case study in the Netherlands. *Wiley Interdiscip. Rev. Clim. Chang.* **2010**, *1*, 729–740. [CrossRef]
18. Slangen, A.; Haasnoot, M.; Winter, G. Rethinking Sea-Level Projections Using Families and Timing Differences. *Earth Futures* **2022**, *10*, e2021EF002576. [CrossRef]
19. Zijl, F.; Sumihar, J.; Verlaan, M. Application of data assimilation for improved operational water level forecasting on the northwest European shelf and North Sea. *Ocean Dyn.* **2015**, *65*, 1699–1716. [CrossRef]
20. Oude Essink, G.H.P.; Van Baaren, E.S.; De Louw, P.G.B. Effects of climate change on coastal groundwater systems: A modeling study in the Netherlands. *Water Resour. Res.* **2010**, *46*. [CrossRef]
21. De Lange, W.J.; Prinsen, G.F.; Hoogewoud, J.C.; Veldhuizen, A.A.; Verkaik, J.; Oude Essink, G.H.P.; Van Walsum, P.E.V.; Delsman, J.R.; Hunink, J.C.; Massop, H.T.L.; et al. An operational, multi-scale, multi-model system for consensus-based, integrated water management and policy analysis: The Netherlands Hydrological Instrument. *Environm. Model. Softw.* **2014**, *59*, 98–108. [CrossRef]
22. Deltares. Kustwiki. Available online: https://publicwiki.deltares.nl/display/KWI/Projects (accessed on 28 March 2022). (In Dutch)
23. Van der Spek, A.J.F. The development of the tidal basins in the Dutch Wadden Sea until 2100: The impact of accelerated sea-level rise and subsidence on their sediment budget—A synthesis. *Netherlands J. Geosci.* **2018**, *97*, 71–78. [CrossRef]
24. Haasnoot, M.; Diermanse, F.; Kwadijk, J.; de Winter, R.; Winter, G. *Strategieën voor adaptatie aan hoge en versnelde zeespiegelstijging, een verkenning*; Deltares rapport 11203724-004; Deltares: Delft, The Netherlands, 2019. (In Dutch)
25. Van Alphen, S. Room for the River: Innovation, or Tradition? The Case of the Noordwaard. In *Adaptive Strategies for Water Heritage*; Hein, C., Ed.; Springer: Cham, Switzerland, 2020; pp. 309–323. [CrossRef]
26. Marijnissen, R.J.C.; Kok, M.; Kroeze, C.; Van Loon-Steensma, J.M. Flood risk reduction by parallel flood defences—Case-study of a coastal multifunctional flood protection zone. *Coast. Eng.* **2021**, *167*, 103903. [CrossRef]
27. Haasnoot, M.; Kwakkel, J.H.; Walker, W.E.; Ter Maat, J. Dynamic Adaptive Policy Pathways: A New Method for Crafting Robust Decisions for a Deeply Uncertain World. *Glob. Environ. Chang.* **2013**, *23*, 485–498. [CrossRef]
28. Van Alphen, J. The Delta Programme and updated flood risk management policies in the Netherlands. *J. Flood Risk Manag.* **2016**, *9*, 310–319. [CrossRef]
29. De Neuville, R.; Smet, K.; Cardin, M.A.; Ranjbar-Bourani, M. Dynamic adaptive planning (DAP): The case of intelligent speed adaptation. In *Decision Making under Deep Uncertainty: From Theory to Practice*; Marchau, V.A.W.J., Walker, W.E., Bloemen, P.J.T.M., Popper, S.W., Eds.; Springer International Publishing: Cham, Switzerland, 2019; pp. 223–252. [CrossRef]
30. Available online: Sand Motor—Building with Nature Solution to Improve Coastal Protection along Delfland Coast (The Netherlands). Available online: https://climate-adapt.eea.europa.eu/metadata/case-studies/sand-motor-2013-building-with-nature-solution-to-improve-coastal-protection-along-delfland-coast-the-netherlands (accessed on 20 March 2022).
31. Seven Years after Tsunami, Japanese Live Uneasily with Seawalls. Available online: https://www.reuters.com/article/us-japan-disaster-seawalls/seven-years-after-tsunami-japanese-live-uneasily-with-seawalls-idUSKCN1GL0DK (accessed on 20 March 2022).
32. Saemangeum Seawall. Available online: https://en.wikipedia.org/wiki/Saemangeum_Seawall (accessed on 20 March 2022).
33. Out of the Deep: 7 Massive Land Reclamation Projects. Available online: https://ww3.rics.org/uk/en/modus/natural-environment/land/out-of-the-deep--7-massive-land-reclamation-projects--.html (accessed on 20 March 2022).
34. Sea Level Rise Knowledge Programme. Available online: https://english.deltaprogramma.nl/delta-programme/knowledge-development/sea-level-rise-knowledge-programme (accessed on 30 March 2022).
35. De Bruijn, K.; Diermanse, F.L.M.; Weiler, O.M.; De Jong, J.S.; Haasnoot, M. Protecting the Rhine-Meuse delta against sea level rise: What to do with the river's discharge? *J. Flood Risk Manag.* **2022**, e12782. [CrossRef]

water

MDPI

Article

Linking Urban Floods to Citizen Science and Low Impact Development in Poorly Gauged Basins under Climate Changes for Dynamic Resilience Evaluation

Maria Clara Fava [1,*], Marina Batalini de Macedo [2], Ana Carolina Sarmento Buarque [3], Antonio Mauro Saraiva [4], Alexandre Cláudio Botazzo Delbem [5] and Eduardo Mario Mendiondo [3]

1 Institute of Exact and Technological Sciences, Federal University of Viçosa (UFV), Rio Paranaíba 38810-000, Brazil
2 Institute of Natural Resources, Federal University of Itajubá (UNIFEI), Itajubá 37500-903, Brazil; marinamacedo@unifei.edu.br
3 São Carlos School of Engineering (EESC), University of São Paulo (USP), São Carlos 13566-590, Brazil; acsbuarque@usp.br (A.C.S.B.); emm@sc.usp.br (E.M.M.)
4 Polytechnic School (EP), University of São Paulo (USP), São Paulo 05508-010, Brazil; saraiva@usp.br
5 Institute of Mathematics and Computer Sciences (ICMC), University of São Paulo (USP), São Carlos 13566-590, Brazil; acbd@icmc.usp.br
* Correspondence: maria.fava@ufv.br

Citation: Fava, M.C.; Macedo, M.B.d.; Buarque, A.C.S.; Saraiva, A.M.; Delbem, A.C.B.; Mendiondo, E.M. Linking Urban Floods to Citizen Science and Low Impact Development in Poorly Gauged Basins under Climate Changes for Dynamic Resilience Evaluation. *Water* **2022**, *14*, 1467. https://doi.org/10.3390/w14091467

Academic Editors: Slobodan P. Simonovic, Subhankar Karmakar and Zhang Cheng

Received: 7 March 2022
Accepted: 9 April 2022
Published: 4 May 2022

Publisher's Note: MDPI stays neutral with regard to jurisdictional claims in published maps and institutional affiliations.

Abstract: Cities must develop actions that reduce flood risk in the face of extreme rainfall events. In this study, the dynamic resilience of the Gregorio catchment (São Carlos, Brazil) was assessed. The catchment lacks environmental monitoring and suffers from recurrent floods. The resilience curves were made considering the water depth in the drainage system as the performance index, obtained by simulations with SWMM and HEC-RAS. The calibration of the flood extension was performed using citizen science data. The contribution to increasing the dynamic resilience by implementing decentralized low impact development (LID) practices was also evaluated. For this purpose, bioretention cells were added to the SWMM simulations. The resilience curves were then calculated for the current and future climate scenario, with and without LID, for return periods of 5, 10, 50, and 100 years and duration of 30, 60, and 120 min. Intensity–duration–frequency curves (IDFs) updated by the regional climate model MIROC5 for 2050 and 2100 were used. The results showed a significant improvement in the system's resilience for light storms and the current period due to LID practice interventions. Efficiencies were reduced for moderate and heavy storms with no significant drops in floodwater depth and resilience regardless of the scenario.

Keywords: historical data source; flood mapping; poorly gauged catchments; citizen science; low impact development

1. Introduction

Historically, cities have been affected by extreme rainfall events and their consequences such as floods and landslides [1–4]. Poor planning and management of the urban space contributes to increasing flood risk for the population due to the housing settlement in steep or floodplain areas and excessive impervious areas and consequently higher runoff generation [5–7]. In developing countries, the housing deficit further increases the occupation of hazardous areas by socially vulnerable populations. Additionally, the pattern of urban occupation and creation of cities has been around rivers due to the need for access to water resources; in many cities, the traditional commercial center is located near floodplain areas [8,9].

Climate change aggravates this risk scenario by increasing the probability of extreme events and their intensities [10–12]. Studies in different regions worldwide show a trend of higher occurrence of storm events, even when there is a decrease in the total rainfall volume in the wet season [13–15]. This trend is also observed in Brazil's regional climate models (RCM) developed with climate change [16,17]. A total reduction in the rainfall volume in southeastern Brazil has been observed with an increase in extreme events. These results were reaffirmed, including more pessimistic forecasts, in the IPCC's Sixth Assessment Report, working group 1: the physical science basis (AR6) [18].

As a guideline for disaster risk management caused by extreme events, the United Nations Office on Disaster Risk Reduction (UNDRR) released the Sendai report in 2015 [19], presenting increased resilience as the ultimate goal to be pursued by decision makers. This report presents the importance of taking risk management actions from a science–policy interface for risk-oriented and evidence-based decision making [20], (pp. 3–5, [21]), (pp. 51–68, [22]).

This report defines resilience as "the ability of a system, community or society exposed to hazards to resist, absorb, accommodate, adapt to, transform and recover from the effects of a hazard in a timely and efficient manner, including through the preservation and restoration of its essential basic structures and functions through risk management." Several actions have been proposed to increase the resilience of cities to flood events, considering the different temporal components of risk management actions (anticipation, prevention, mitigation, preparedness, response, and recovery [19,23]). Furthermore, actions can be classified as structural, measures, or non-structural, instruments [24].

Structural prevention and mitigation measures are commonly used by cities with the main application of centralized grey infrastructures such as detention basins and rainwater reservoirs [25–27]. However, in recent years, the application of green infrastructure measures, also known as low impact development (LID) practices, has grown due to their complementary benefits for the hydrological cycle, diffuse pollution control, and the catchment life cycle [28]. According to the 2014 report of the C40 group, drainage solutions with LID ranked third among the most performed actions by the group's cities [29].

Due to the central importance of resilience in the risk-oriented and evidence-based decision-making process, several studies aimed to define its indicators. In [30,31], the authors criticized the application of static resilience indicators since they do not allow a good representation of the event cycle after stress and consequently the evaluation of the temporal components of risk management actions. Thus, the authors proposed a space-time dynamic resilience measure (STDRM). For this measure, resilience is presented as the loss of system performance (e.g., physical, social, economic, health, etc.), which can be calculated as the integral of the curve representing the system performance level between the beginning of the disruptive event and the end of the system's recovery. Over time and space, the different performance metrics can be further normalized and combined into a single resilience curve.

The system's performance can be quantified through computational modelling, allowing the evaluation and comparison of different intervention scenarios [32,33]. In the case of resilience to flood events, a commonly used metric for the physical system is the water depth reached by the flood and its horizontal extent in the affected area. To this end, hydrological, hydraulic, and hydrodynamic models must be used [34,35]. However, limited hydrological data availability can hinder urban flood modelling [36].

In order to simulate both pluvial and fluvial floods happening simultaneously in urban catchments, coupled hydrological and hydrodynamical modelling have been increasingly used. One common configuration is to use HEC-HMS for hydrological modelling and HEC-RAS for the hydrodynamic 2D process, obtaining the flood extensions and flood depths for different scenarios [37–39]. This configuration has been used in ungauged or poorly gauged basins with a high application in India. Nevertheless, there are some limitations in using HEC-HMS for hydrological modelling: it does not allow simulation of the effect of LID practices and it does not account water quality assessment.

To overcome these issues, it is possible to use an alternative configuration such as coupling SWMM and SWAT for hydrological modelling with HEC-RAS. SWMM and SWAT models allow us to incorporate different types of LID practices in the catchment, e.g., rain gardens, green roof, porous pavement, etc. [40,41]. The SWMM model has been widely used for urban catchments [42–44] while the SWAT model has been more applied in large areas [45].

In [46], the authors successfully used SWMM to assess the impact of climate change on an small urban catchment in Athens, Greece by updating IDF curves due to climate change and demonstrated that under the non-stationarity projections of climate, the percent change in future rainfall intensity could vary immensely and showed the potential of a systematic assessment of flood mitigation scenarios by using climate projections, bias correction methods, and temporal downscaling associated with hydraulic models. In the same sense, [47] found that climate change would reduce total rainfall in Tehran but with an increase in heavy storms. Therefore, the authors evaluated these non-stationary effects in the urban hydrology by using SWMM and proposed different intervention scenarios with LID practices to maintain the urban resilience. SWMM was also used in [48] to assess the impact of climate change on urban hydrology and their impacts on combined-sewer overflow occurrences in city of Toledo, Ohio, obtaining an increase of 12–18% for future scenarios. Additionally, the simulation showed that rainwater harvesting implementation in half of the buildings could mitigate the effects of future climates.

Despite the contribution of these studies to evaluate mitigation measures to climate change by LID practices, they only performed hydrological modelling and did not provide information about flood extent and depth. For this purpose, coupling SWMM with HEC-RAS can be an alternative [49].

For a good representation of the area by coupled hydrological and hydrodynamic modelling, it is necessary to calibrate and validate the models with observed data with different severities and probabilities of occurrence. However, in the case of developing countries, it is common for urban catchments to be poorly gauged, lacking observed data with suitable temporal and spatial discretization [39,50–52]. To overcome this problem, new studies have used citizen science data—the involvement of citizens in collecting data and knowledge for scientific research [53] and historical records—and social data such as photos, footage from security cameras, newspapers, and population memory to extract records of depths and flows of flood events [54–56], showing a clear potential for using data collected by citizens to complement traditional monitoring data for flood model development and validation, especially at ungauged or poorly gauged catchments. Data reported from citizens also benefit from being spatial and can be compared to other spatial datasets [57]. Case studies have shown that even when the amount of data is not so expressive, they can still provide an effective form of model validation [58]. Additionally, [59] showed that regionalization methods and optimization techniques [60] were helpful for LID simulation to overcome the lack of hydrological and stormwater quality data.

The state of São Paulo (SP) in the city of São Carlos in Brazil has had recurrent cases of flooding over the years, with critical flooded areas in its main commercial center [61,62]. However, due to traditional cultural aspects, there is a permanence of commercial establishments in the risk area, making an eviction of the area unfeasible [62]. Therefore, it is necessary to develop risk management actions to reduce peak flows and flooding water depths in critical areas, reducing the damage generated.

This study aimed to explore the effect of different climate scenarios, combined with the implementation of LID practices as a structural measure of risk management, on the extent and depth of flooding in the center of São Carlos. To this end, SWMM and HEC-RAS software were used for hydrological and hydrodynamic modelling of the catchment. Due to the absence of observed monitored data as the catchment is poorly gauged, records extracted from population memory and register (citizen science) were used to calibrate and validate the models. Finally, the different scenarios were compared for their effect on the catchment resilience using the STDRM approach.

2. Case Study and Datasets

2.1. Gregorio Creek and Its Flood History

The case study was the Gregorio catchment located in the city of São Carlos, state of São Paulo, Brazil (Figure 1, zoom map on the left). The total area of the municipality of São Carlos is about 1136.91 km^2. The city's population density is about 194.53 inhabitants/km^2, and the entire population is estimated as 249,415 inhabitants [63]. The average altitude is 856 m a.s.l. and the soil is highly permeable. As for its climatological characteristics, São Carlos is classified as Cfa (humid summer subtropical climate) according to the Köppen-Geiger climate classification, with an annual average rainfall of 1558.3 mm and average daily temperature of 20.6 °C [64]. The rainy season extends from October to March and accounts for almost 80% of the total annual rainfall, with January having the highest records of rainfall, totaling 303.8 mm [64].

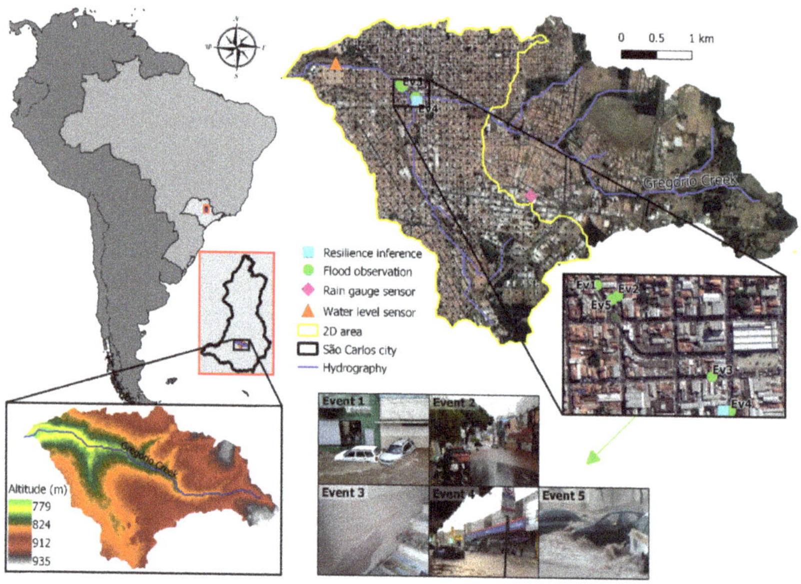

Figure 1. Gregório Creek Catchment in the city of São Carlos, São Paulo State, Brazil. The photographs show citizen science data collected during flood events that occurred in the area.

The Gregorio Creek has its source in the southeast of São Carlos, in the city's rural area, with an approximate altitude of 900 m a.s.l. In its course, the creek crosses a highly urbanized area from the catchment middle to the outlet. The stream has natural coverage until the beginning of the urbanized area, where the open channel is then coated with cement and its natural course is straightened. The total area of the Gregorio catchment is around 19.14 km^2. The catchment has a long history of urban flooding. A critical area is the Municipal Market region, where the city's commercial center is concentrated. The urban flooding is due to the insufficiency of the drainage system both on a micro-drainage scale and in the Gregório Creek channel overflow. In this area, there are reports of flooding events occurring since 1970, with an intensification of hazard and damages to the population since 2004 due to the accelerated increase in occupation density at the site [62]. According to local shopkeepers, there is an interval of up to 15 min between when the rain starts and flood occurrence [61]. In addition, traders reported losses associated with floods ranging from USD 100–USD 56,000 in 2015 and 2018 [61].

Ref. [65] estimated, based on historical data from 1940 to 2004, that the average return period of flood occurrence is 0.65 events per year with a standard deviation of 0.84. Their study also showed that the cumulative number of flood occurrences increased with catchment urbanization. This historical data suggests a clear need to adopt structural and non-structural means to contain floods. Still, the proposals to solve the problem recorded in official documents are onerous and have not yet been adopted [54,66].

2.2. Datasets

2.2.1. Water Level Data

Water level data registered with a time-step of 5 min was used. The sensor was installed inside the Gregorio Creek canal downstream of the area most affected by floods (Figure 1). The canal has cement coating and has no margin conservation at this location. The riverbank has a 2 m wide strip of grass coverage. The water levels were converted to flow data through a rating curve (Equation (1)) proposed by [67].

$$Q = 35.466637y^{1.548284} \quad (1)$$

where Q is the estimated flow (m^3/s); y is the observed water level (m).

2.2.2. Citizen Science Data

Buarque et al. [56] extracted records of water depth in the streets from pictures taken by people who witnessed the events (Figure 1, Events 1–4) and *in loco* after the events using the mark left on the walls as evidence (Figure 1, Event 5). We applied these data for the urban flood modelling performed in this study.

2.2.3. Rainfall Data

Rainfall data was registered at 10 min intervals between September 2013 and November 2020. The input data come from a rain gauge installed inside the Gregório Creek catchment. The gauge location was close to the flooding area and approximately in the boundary of the area where hydrodynamic simulation was performed (2D area, Figure 1).

3. Methodology

In this study, the SWMM model was selected as a computational engine to simulate the resulting flow from several design storms at the actual scenario of Gregório Creek catchment and for future conditions considering climate changes. After this, HEC-RAS was chosen to perform a 2D rain-on-grid simulation using SWMM inputs as a boundary condition and citizen science data (CS data) for model validation. After model validation, we used it to achieve the flood scenarios for the current and future situation of the catchment due to climate changes—Figure 2 details the main steps performed towards generating flood risk scenarios. The purpose is to assess the dynamic resilience of the system and flood risk by obtaining flooded area scenarios, representing the risk component related to the hazard and exposure.

3.1. Hydrological and Hydraulic Modelling

The USEPA Storm Water Management Model (SWMM) is a computational program capable of simulating the rainfall runoff transformation over a catchment [40]. The model has a hydrological and a hydraulic modelling module capable of simulating single events or long-term simulations of runoff quantity and quality in micro- and macro-drainage of urban catchments. The runoff component of SWMM is lumped and conceptual at the subcatchment scale and is responsible for receiving rainfall and generating runoff and pollutant loads. The routing module transports the surface runoff from the hydrological model along rivers, canals, and other conduits [68]. SWMM is a semi-distributed model, as the user can define the number of subcatchments according to the level of discretization desired [69].

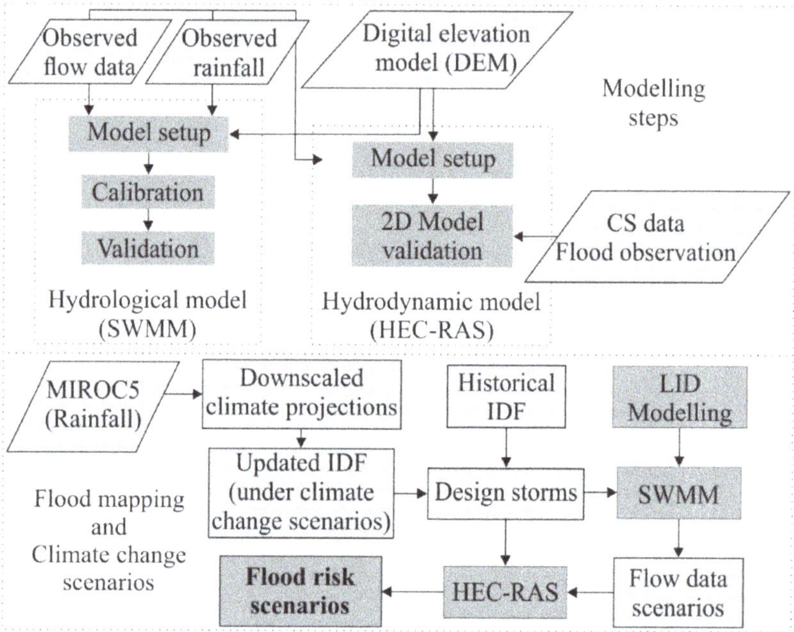

Figure 2. Methodological activity diagram. The light gray rectangles represent the input data (primary and secondary) for each activity and the dark gray rectangles represent the main methodological steps.

Gregório Creek catchment was modelled using SWMM, setting the dynamic wave equations and the SCS curve number (CN) method for the model routing and infiltration module. The inputs used for model set-up were width, average slope, infiltration, Manning's coefficient for the pervious and impervious area (N-perv and N-imperv, respectively), percentage of impervious area with no depression storage, depth of depression storage on pervious and impervious areas, the percentage of impervious area (%Imperv), and canal roughness. Aiming to keep the physical meaning of the model, two groups of model features with a different range of parameters value were defined, one for highly urbanized areas and other for green areas. The curve number and percent of impervious area parameter groups were defined based on the map of land use of São Carlos city [70].

As a multi-parameter model, many methodologies to calibrate SWMM have been developed [60,71]. We performed model calibration using an automatic calibrator that uses genetic algorithms for optimization developed by [72]. The stop criteria is based on the deviation between the evaluation function of the best set of decision variables found for the current generation $(n)[f(x_n)]$ and the one of the previous generation $[f(x_{n-1})]$. A threshold of 0.0001 was defined; when the deviation is lower than that value during three generations in sequence, the optimization process stops. Table 1 shows the calibration settings.

Seven rainfall events between September 2013 and January 2014 were divided into two groups according to the average intensity of the rainfall: moderate and heavy—according to the American Meteorological Society [73] classification. Then, the events were chosen at random for two groups: calibration and validation (Table 2). The Nash–Sutcliffe efficiency (NSE) was used as OF and to assesses model accuracy. Model calibration resulted in an NSE value of 0.63 for the calibration period and 0.51 for the validation period. Table 2 presents the events selected.

Table 1. Calibration settings.

Genetic Algorithm	Settings
Two-point crossover probability	0.8
Flip bit mutation probability	0.05
Individuals' selection	Tournament selection
Decision variables (parameters)	CN, %Imperv, Roughness, N-perv, N-imperv
Population size	50
Generations	100
Objective function (OF)	NSE
Stop criteria	$\lvert f(x_n) - \lvert f(x_{n-1}) / f(x_n) \leq 0.0001$

Table 2. Rainfall events and the resulting water depth peak used for calibration and validation of the hydrological and hydrodynamic models.

Hydrological Model (SWMM)				
Event	Date (Day Month Year)	Rainfall intensity (mm/h)	Total rainfall (mm)	Peak water depth (m)
Calibration				
Event 1	3 September 2013	6.13	63.2	1.20
Event 2	25 March 2013	16.07	46.6	1.35
Event 3	28 May 2013	4.58	75.8	1.14
Event 4	29 November 2013	13.8	35.2	1.26
Validation				
Event 1	4 October 2013	4.41	58.4	1.23
Event 2	20 December 2013	14.63	17.8	0.52
Event 3	14 January 2014	3.19	42.6	0.95
Hydrodynamic model (HEC-RAS)				
Validation				
Event 1	23 November 2015	2.21	16.6	0.85 *
Event 2	20 March 2018	37.42	68.6	0.75 *
Event 3	12 January 2020	21.47	94.2	0.72 *
Event 4	13 November 2020	8.07	39	0.28 *
Event 5	26 November 2020	8.77	38	1.65 *

* Citizen science data collected during the flood events may not represent the maximum water depth that occurred.

The rainfall runoff transformation was made on SWMM to feed the hydrodynamic model (HEC-RAS) with different flow values to compare water depth reached when considering LIDs insertion and climate changes. The flow data used for the model's accuracy evaluation came from a sensor installed near the catchment outlet (Figure 1); then, the entire catchment was modeled and calibrated on SWMM based on the inference made at this location. Though the whole catchment was modeled, the flow outputs that fed the HEC-RAS model were accessed in an intermediary location of the catchment near the 2D area border and at the Gregório Creek.

3.2. LID Modelling

SWMM can explicitly model eight types of LID controls: rain gardens, bioretention cells, infiltration trenches, porous pavements, rain barrels, vegetative swales, rooftop disconnection, and green roofs. The bioretention cells modelled in SWMM are depressions containing vegetation grown in soil mixture placed above a gravel drainage bed, which provides storage and infiltration of the runoff captured from surrounding areas [50] as a vertical layer. The flux and storage of water resulting from each layer are tracked at each model time-step. The LID practice's flow is combined with the subcatchments' runoff to establish the aggregate values for each process at the end of every time step [74]. We choose the bioretention cells as LID control to simulate its impacts on water depth peaks and flood spots at the Gregório Creek.

The input data used for the bioretention cells were calibrated by a previous study developed by [75] for the city of São Carlos in an adjacent catchment of Gregório Creek. The study built a prototype and performed field experiments with observed rainfall events to determine all the coefficients. Each cell had an area of 12 m^2.

In the SWMM model, the rainfall runoff component was lumped and conceptual at the sub-basin scale. The Gregório creek catchment was discretized into 13 sub-basins, with areas ranging from 0.67 km^2 up to 4.71 km^2, according to their main natural sources. The sub-catchments had their area divided into pervious and impervious portions. The LIDs were designed to capture the runoff coming 100% from the catchment's impervious areas. From this, we modeled the LIDs on the sub-catchments until they corresponded to 1% of each sub-area. The outflow from the LIDs was sent to the catchment's pervious portion. Table 3 presents the main parameters used for the bioretention cells set-up.

Table 3. Input data * for the bioretention cells modelled on SWMM.

Surface	**Value Adopted**	**Typical Range ****	**Soil**	**Value Adopted**	**Typical Range ****
Storage depth (mm)	600	-	Thickness (mm)	1000	450–900
Vegetative volume fraction	0.1	0.1–0.2	Porosity	0.32	0.45–0.60
Storage			Field Capacity	0.43	0.15–0.25
Thickness (mm)	500	100–150	Wilting Point	0.12	0.05–0.15
Void ratio	0.4	0.12–0.21	Conductivity (mm/h)	195.48	50.80–1397
Seepage rate (mm/h)	5.83	-	Conductivity slope	46.38	30–60
Clogging factor	0	-	Suction Head	66.98	50.80–101.60

* Values calibrated by [55]; ** Typical range from [40].

To evaluate the performance of the LID practices, runoff retention efficiency and peak flow attenuation efficiency were calculated according to Equations (2) and (3).

$$Eff_{rr} = \frac{V_{in} - V_{over}}{V_{in}} \quad (2)$$

$$Eff_{peak} = \frac{Q_{peak,in} - Q_{peak,over}}{Q_{peak,in}} \quad (3)$$

where: Eff_{rr} is the runoff retention efficiency; V_{in} (m^3) is the total inflow volume; V_{over} (m^3) is the total overflow volume; Eff_{peak} is the peak attenuation efficiency; $Q_{peak,in}$ (m^3/s) is the maximum inflow value; $Q_{peak,over}$ (m^3/s) is the maximum overflow value.

3.3. Hydrodynamic Modelling

The Hydrologic Engineering Center's (HEC) River Analysis System (HEC-RAS) software is capable of performing steady flow simulations and one-dimensional (1D) and two-dimensional (2D) unsteady flow analysis. The HEC-RAS system comprises a graphical user interface, separate hydraulic analysis, and mapping facilities (HEC-RAS Mapper) [76]. HEC-RAS can perform coupled 1D and 2D modelling and 2D modelling with no 1D elements. We used the second option to reproduce the hydrodynamic modelling of the Gregório flooding area.

To perform the 2D HEC-RAS rain-on-grid simulations, the Advanced Land Observing Satellite Digital Elevation Model (ALOS DEM) with a spatial resolution of 12.5 m was used to represent surface terrain. We used this input due to a lack of more accurate terrain information for the area. Several hydrological studies have already been carried out in the region [77–80]. Still, none of them performed measurements or purchased more accurate images, which is a reality in many river basins in developing countries.

The HEC-RAS model geometry was set up in the HEC-RAS mapper. The 2D area (Figure 1, map on the top right) is defined as the boundary for which 2D computations occur, a computational mesh based on the DEM is created within the 2D flow area. The delineation of the 2D area was made as a sub-catchment delineation, where the inlet point was defined as the location at Gregório Creek immediately upstream of the main flooding area (the Municipal Market region).

An unsteady flow plan (2D only) including rain-on-grid rainfall time-series and a flow input as upstream boundary condition was created in HEC-RAS. This flow input came from the SWMM simulations at the location in Grregório Creek near the 2D area boundary. The unsteady flow routing was performed using diffuse wave momentum equations. Rainfall inputs were at a 10 min time-step distributed uniformly throughout the catchment area and flow data at a 5 min time-step. The value adopted for the global roughness was 0.01 [81]. Computational interval is an end-user-defined parameter that defines the temporal resolution of hydrodynamic calculations [82]. The hydrograph output interval was set to every 5 min and the mapping output interval each hour. A 15 s computational interval was determined to achieve satisfactory accuracy at HEC-RAS output, considering the rapid response from urban floods and the spatial scale and resolution of the project.

There are no flood monitoring data from sensors for the Gregório catchment. Buarque et al. [56] developed a socio-hydrological study and validated an urban flood model as one of its methodological steps. In order to validate the model, they extracted records of water depths in the streets from pictures provided by witnesses of the flood events. Furthermore, they collected the water depth *in loco* after the events using the mark left on the walls. They used the CAFLOOD application to develop the flood model, a low-complexity model capable of providing fast and accurate urban flood simulations based on a cellular automata approach [83]. The model uses the DEM, rainfall intensity, and global roughness as primary inputs.

The main purpose of their study in the Gregório Creek was to replicate the socio-hydrological mechanisms and design possible scenarios resulting from the interaction between water and human variables instead of seeking to best capture the catchment hydrodynamic behavior [56]. Even though the model reached satisfactory results (RMSE = 0.62), the same model approach could not be applied for our study as the CAFLOOD model is unsuitable for using flow time series as input data. To capture flow changes at the catchment due to LID controls set-up and its influence on flood spots, we needed to couple the hydrological simulations on SWMM with the 2D unsteady flow analysis at HEC-RAS.

We followed a similar approach to validate the 2D HEC-RAS for Gregório Creek and used a dataset provided by [56] (Events 1–5, Table 2). Model calibration was not performed due to the lack of data. The five flood events obtained from the cited study were used for the 2D model validation (Figure 1, Event 1–5). The model performance was inferred at the pixel closest to the water depth observation location (RMSE = 0.52). Figure 3 demonstrates the model outputs used for accuracy assessment during the validation events.

3.4. Design Storms

Design storms were used as input for scenarios simulation of current and future perio5ds with climate change. For the construction of the storms, the alternating block method [84–86] with different configurations of duration, return period, and temporal distribution was adopted to obtain ranges of variability. Variability assessment was proposed by [87] as an alternative to uncertainty assessment for climate change scenarios due to the difficulty of accessing statistical uncertainty of GCM and RCM itself, in addition to downscaling and bias correction methods.

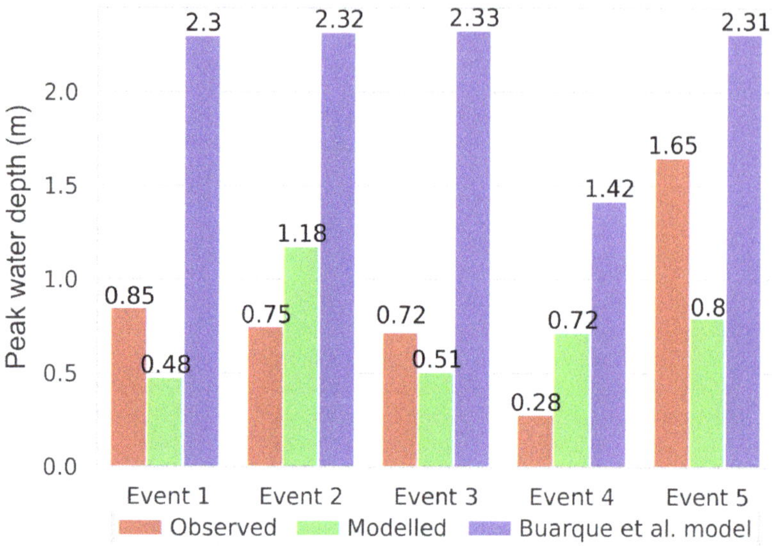

Figure 3. Comparison between observed water depth with maximum water depth reached at the closest pixel to the observation using HEC-RAS and CAFLOOD (Buarque et al., 2021).

As for duration, three different intervals were adopted: 30 min, 1 h, and 2 h, representing more intense and more recurrent rainfalls in the city of São Carlos. As for the return period (RP), 5, 10, 50, and 100 years were adopted to obtain a variability between higher recurrence and lower intensity and lower recurrence and extreme intensity. Finally, two patterns of temporal distribution were adopted: centralized (Ce) and delayed (At) (similar to Huff's third quartile [88,89]). The delayed pattern was included to represent scenarios with higher runoff generation. In this temporal pattern, the initial and less intense rainfall is responsible for saturating the soil. At peak intensity, the infiltration of water is lower, leading to more significant runoff and higher flow peaks.

For the current period, the design storms were constructed using the standard intensity–duration–frequency curve (IDF) for the city of São Carlos [90] (Appendix A). Regarding future periods, the updated IDFs with climate change developed by [75] were adopted for the periods from 2015 to 2050 and 2050 to 2100 (Appendix A). The future IDFs were updated using the daily rainfall data obtained by the RCM Eta-MIROC5 [16,17] for the scenarios, RCP 4.5 and 8.5 (representing optimistic and pessimistic variabilities) bias-corrected by the methods power transformation (PT), and distribution mapping (MD).

The combination of all design storm configurations and evaluation scenarios resulted in 216 input rainfalls for hydrological and hydrodynamic simulation. Storms were selected to better visualize the results, representing weak, moderate, strong, and extreme intensities for the current and future periods defined according to extremes of the return periods and rainfall duration, presented in Table 4.

Table 4. Representative design storms.

	Duration (min)	RP (years)	Temporal Distribution *	Period	RCM	Bias Correction **	RCP
Weak	30	10	Ce and At	Current	-	-	-
Moderate	60	10	Ce and At	Current	-	-	-
Strong	60	50	Ce and At	Current	-	-	-
Weak	30	10	Ce and At	2050	MIROC5	PT and MD	4.5 and 8.5
Moderate	60	10	Ce and At	2050	MIROC5	PT and MD	4.5 and 8.5
Strong	60	50	Ce and At	2050	MIROC5	PT and MD	4.5 and 8.5
Extreme	120	50	At	2100	MIROC5	PT and MD	4.5 and 8.5

* Rainfall hydrograph centralized (Ce) and delayed (At); ** bias-correction by power transformation (PT) and distribution mapping (MD) methods.

3.5. Dynamic Resilience

To assess resilience, the STDRM approach [30] was used, here called dynamic resilience. In this approach, resilience was presented as a unitary variable representing the system's performance losses during a disruptive event and its subsequent recovery.

As a first step to evaluate dynamic resilience, it is necessary to establish the system performance evaluation metrics. In this study, the peak water depth at the most critical point of the catchment along with the event—$h_c(t)$—was considered a performance indicator. The most critical point (Figure 1) was defined from physical and social characteristics, i.e., water depth and associated economic losses.

Finally, the resilience—$r(t)$—was calculated according to Equation (4) as a function of the normalized performance, i.e., dividing it by the system's maximum performance considering all the events evaluated—$h_{c,\ max}$ (m). Resilience was deemed to be maximum if the floodwater depth was lower than the minimum water depth for houses and stores flooding—$h_{c,min}$ (m).

$$r(t) = \begin{cases} if\ h_c(t) \leq h_{c,min}, & 1 \\ if\ h_c(t) > h_{c,min}, & 1 - \frac{h_c(t)}{h_{c,\ max}} \end{cases} \quad (4)$$

4. Results

4.1. Hydrological Scenarios

After the SWMM calibration and validation, different climate scenarios and structural intervention for risk management were simulated for the current and future periods, considering climate change. Figure 4 shows the simulated hydrographs at the midpoint of the Gregório basin (2D model input) with the input of representative design storms considered weak, moderate, strong, and extreme for the different scenarios, presented in Table 4.

It is possible to notice from Figure 4 that peak flows for current periods varied between maximum values of 246.7 m^3/s to 398.6 m^3/s considering no risk management intervention. In intervention scenarios (Table 4) with LID practices, these values were reduced between 78 m^3/s and 97 m^3/s, representing an attenuation of approximately 40% and 30% for weak and moderate design storms, respectively. However, for heavy storms and maximum peaks in the most extreme scenario, the peak flows were reduced by only 35 m^3/s, representing a 9% reduction.

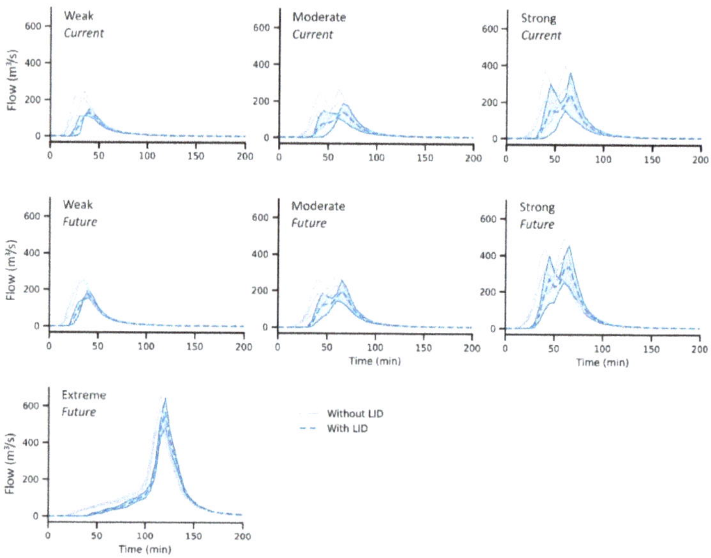

Figure 4. Hydrographs for representative climate scenarios for the current and future period, with and without applying LID practices. Lines represent the average value of the hydrographs, and the shaded area is the confidence interval, considering the variation evaluation for the representative design storms.

In the future scenarios presented in Figure 4, there was an increase in the maximum peak values between 72 m^3/s and 235 m^3/s when compared with the current period, considering scenarios of weak, moderate, and heavy design storms and without the intervention of LID practices. The application of bioretention cells collaborated to reduce the peak increase to a range of 78 m^3/s to 221 m^3/s for scenarios with LID intervention.

The intensification of rainfalls in future periods due to climate changes reflects on the efficiency of bioretention to mitigate peak flows. This drop was more accentuated for the moderate rainfall scenario (from 78 m^3/s to 35 m^3/s, equivalent to an efficiency drop from 30% to 10%). For heavy storms, the peaks were reduced by 35 m^3/s in the current period and 50 m^3/s in the future period, equivalent to an efficiency drop of 9% to 8%. Even with higher peak flow reductions for future periods, there was a reduction in the system's overall efficiency. In the case of an extreme storm for the final year of 2100, the flow reductions were even less significant when evaluating the implementation of LID practices. There was a reduction in maximum peak by 31 m^3/s, representing a 4% efficiency of peak attenuation. From the hydrographs, it is also possible to observe that bioretention cells contributed to a delay in the occurrence of peak flow between 5 and 10 min.

In addition to the hydrographs evaluation for the representative storms, the reduction in volumes and peak flow for all scenario combinations and the current and future periods were also evaluated (Figure 5). There was a general increase in runoff volumes and peak flows for future periods with climate change. The increase was significant and more pronounced for higher RPs (50 and 100 years) than smaller RPs.

Considering the scenarios without LID practices, there was an increase of 25.7 m^3 and 56.9 m^3 in the medians for RP 5 years (representing a relative increase of 1.2× and 1.4) for periods 2050 and 2100, respectively. For 100 year RP in the same periods, there was an increase of 98.6 m^3 and 111.3 m^3, representing a relative increase of 1.8× and 1.9×, respectively.

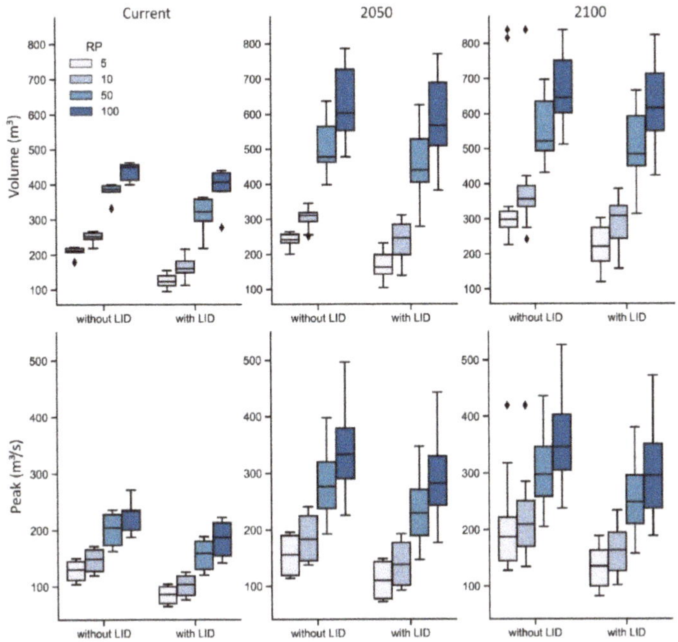

Figure 5. Comparison of hydrological variables for current and future scenarios with and without applying LID practices. The black rhombuses are the data outliers.

Furthermore, there is an increase in the variability of volumes for the larger RPs, both for the current period and for future periods (Figure 5). In future periods, the variability was more accentuated with a greater tendency to higher values [87]. As for peak flows, the variability was more constant among all RPs but increased when comparing the future periods with current periods.

From Figure 5, it is also noted that the LID practices could reduce the runoff volumes in values within a fixed range of approximately 45 m^3 to 55 m^3. This interval represents the volume retention capacity of all bioretention cells applied together in the catchment. The variations are relative to the soil saturation level and therefore its infiltration capacity. Similar results were observed in [75]. This reflects on the peak flow attenuation, which also presented limitations. In Figure 4, it is possible to observe that the maximum peaks were reduced to values approximately around 30 m^3/s. Therefore, the drop inefficiencies over time (Figure 6) are related to an increase in rainfall volumes and peaks while keeping the maximum capacity of bioretention constant.

4.2. Hydrodynamic Scenarios

The HEC-RAS model was validated for 2D rain-on-grid simulations. The flow hydrographs simulated using SWMM for all the climate scenarios, with and without LID interventions, and the rainfall for the different climate scenarios (Table 4) were used as inputs to HEC-RAS to assess its effects over the flooding area. Figure 7 shows these results by comparing the changes in peak water depth in the flooding area for weak, moderate, and strong scenarios of rainfall for the current situation of the catchment when adding LID interventions and without LID intervention.

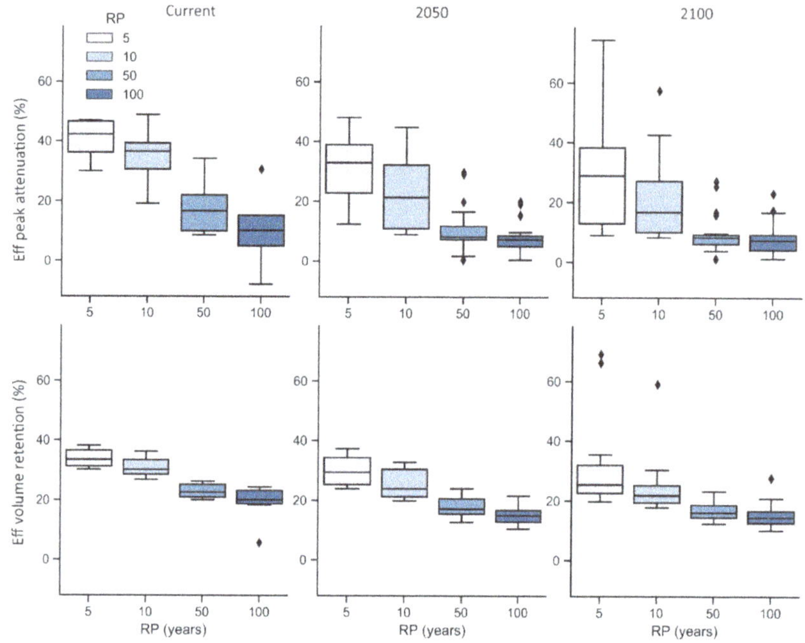

Figure 6. LID practices efficiency in peak attenuation and volume reduction for the current and future scenarios. The black rhombuses are the data outliers.

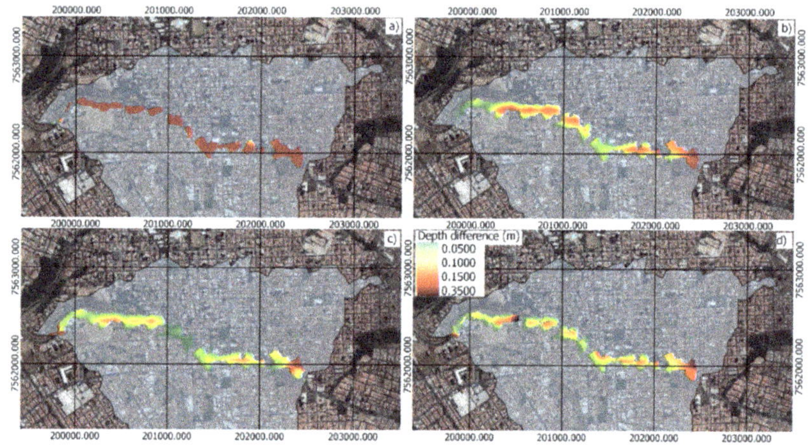

Figure 7. Maximum water depth difference between catchment conditions with and without LID practices for the current climate conditions. Hydrodynamic simulations with: (**a**) a weak rainfall input, (**b**) a moderate rainfall input, (**c**) a strong rainfall input, and (**d**) the average of all rainfall inputs for the current scenario.

The results obtained in the spatial distribution of events reinforced the results of Section 4.2 (simulated scenarios using SWMM). The reduction in the flood spot peaks was strongly noticed for weak rain with a predominance of about 30 cm reduction in the maximum depths for this scenario (Figure 7a) whereas for a heavy rain scenario, the decline of the peak between 5 and 10 cm was more noticeable (Figure 7c). LID interventions mitigated the risk of flooding but were more notable for less extreme events.

Figure 7d presents the difference between the average peak water depths for all scenarios with and without adding LID practices to evaluate the average performance of flooding area peaks reduction by introducing bioretention cells over the catchment, when considering the current climate scenario. The maximum average difference in peak water depth was about 0.35 m.

Figure 8 shows the efficiency of bioretention to mitigate peak flows in future periods due to climate changes. Compared to the current climate condition, the LIDs efficiency decreasing for climate change scenarios was remarkable, especially for strong and extreme rainfall events (Figure 8c,d). Even when focusing on weak rainfall, water peak reduction due to adding bioretention cells dropped from a maximum decrease of 0.35 m for the current scenario to 0.15 m for future climate conditions.

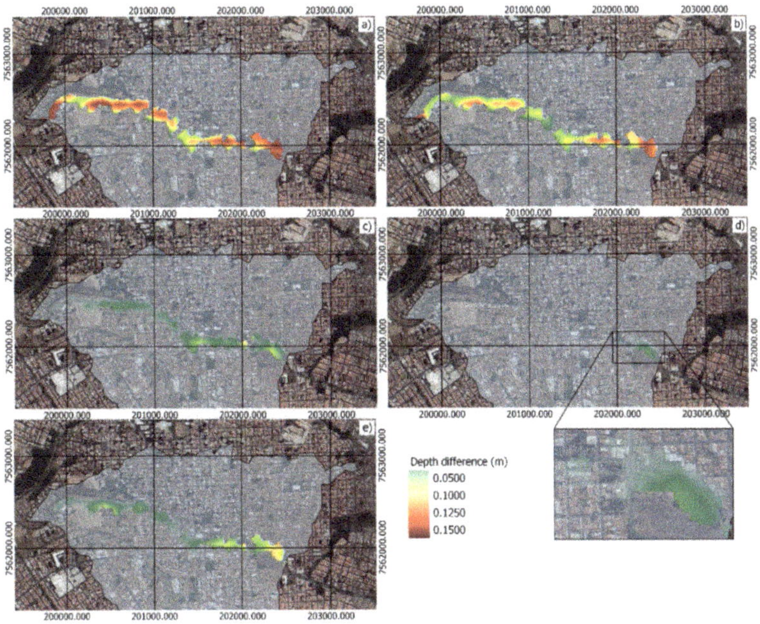

Figure 8. Maximum water depth difference between catchment conditions with and without LID practices for the future climate conditions. Hydrodynamic simulations with: (**a**) a weak rainfall input, (**b**) a moderate rainfall input, (**c**) a strong rainfall input, (**d**) an extreme rainfall input, and (**e**) the average of all rainfall inputs for the future scenario.

A slight reduction in peak water depth over the flooding area could be noticed for heavy storms, with a maximum reduction of about 0.1 m (Figure 8c). In the case of an extreme storm for the year 2100, the peak water depth reductions over the flooded spot were even less significant when evaluating the implementation of bioretention cells (Figure 8d); the only effect on water depth peak reduction seen was about 0.05 m at the most upstream area of the 2D mesh.

Figure 8e shows the difference between the average peak water depths for all scenarios with and without adding LID practices to evaluate the average performance of flooding area peaks reduction by introducing bioretention cells over the catchment when considering climate changes. The maximum average difference in peak water depth was about 0.15 m. When comparing the average drop of water depth peaks for the climate change scenario (a maximum average reduction of about 0.13 m, Figure 8e) with the current climate scenario (a maximum average drops of about 0.35 m, Figure 7d), the efficiency decreasing of the LID cell's introduction due to climate changes became evident.

4.3. Dynamic Resilience

For the evaluation of dynamic resilience, the system's performance was first obtained from the assessment of water depths over time at the critical point of the catchment (Figure 1—resilience inference) for the representative storm scenarios and intervention measures (Appendix B). The performances were normalized according to Equation (4) to obtain the unitary resilience during the event (Figure 9).

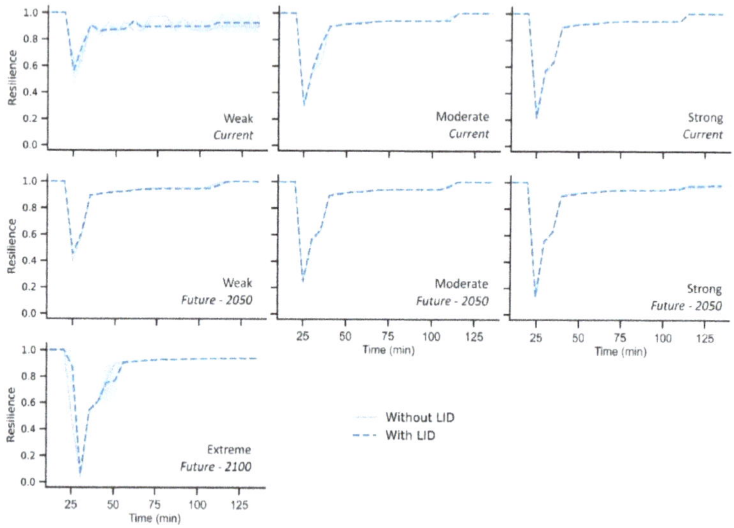

Figure 9. Resilience curves for current and future scenarios with and without applying LID practices.

The scenarios with the greatest resilience were those of weak storms in which the resilience reached its worst values around 0.6 and 0.4 for both the current and future periods. Regarding the minimum extreme of the variation range, without LIDs, the resilience values reached minimums of 0.39 and 0.37 for the current and future periods, respectively. With LID practices, the resilience reached minimum values around 0.45 and 0.6. For the moderate to extreme scenarios, the increase in resilience with LIDs was slight, around 0.01, which was not significant.

From Figure 9, it is possible to observe that there was a more significant time until the resilience began to drop in scenarios with LID practices. In the case of an extreme storm, the resilience reached minimum values of 0 and 0.023 with and without LID, respectively. However, it reached worse values faster (greater slope of the recession curve), i.e., bioretention cells contributed to a more abrupt reduction in resilience. In addition, a longer delay for full recovery was also observed (smaller slope in the recovery curve). The combination of two factors could explain this behavior: (1) the considered scenarios presented rainfall of 120 min, and (2) the LID practices were able to delay the peak flow occurrence by up to 10 min. Therefore, the maximum values and their recession occurred after the peak of hydrographs without LID.

5. Discussion

5.1. Coupled 1D and 2D Model Accuracy

Studies [91,92] discussed that optimal values for flood extent and flood depth are obtained for hydrodynamic models not calibrated when using DEM or Digital Terrain

Models (DTM) with resolution of 2 m. This data resolution can be obtained by airborne light detection (LiDAR), aerial photography, and topographical maps. However, these types of data are frequently not available in poorly gauged and ungauged basin or are not open source and presents high associated costs, especially in developing countries. Using terrain data with less resolution that can lead to less accuracy in the hydrodynamic models is a common reality in those locations.

Examples of studies with coupled hydrological and hydrodynamic modelling approach performed with DEM data with 10 m resolution for poorly gauged basins could be found in [37] for Toronto, Canada, [39,93] in India, and [45] in Missouri, USA. In those studies, the basins evaluated ranged from 6.8 km^2 to 658.9 km^2. Therefore, application purposes of flood maps obtained by hydrodynamic models with less accuracy must be discussed.

Additionally, the majority of these studies have not performed a calibration and validation for the hydrodynamic modelling. An exception is the study of Jha and Afreen (2020) that calibrated both SWAT and HEC-RAS models independently; however, the calibration of HEC-RAS was made for peak flows and not for flood extent and flood depth. In the same sense, [92] also performed a calibration and validation for HEC-RAS considering discharges and peak flow. In their study, they discussed that calibrating the hydrodynamic model was difficult due to the uncertainties in the parameters and variables, especially rainfall depth and discharge data synchronization.

Ref. [49] also discussed that calibrating flood extension and flood depth was a difficult process because it requires spatial data. Spatial data can be obtained from remote sensing, but during rainfall events it is difficult to obtain cloud-free images, impairing the water depth classification. Additionally, the images may also lack synchronization with the flood event.

To overcome this problem, in this study, we proposed an alternative methodology to acquire spatial data using citizen science that can be used for model calibration and validation. Here, we opted for the Manning coefficient from the literature (n = 0.01) [81] to validate the model due to few monitored events. From Figure 3 and the obtained NSE = 0.52, the model accuracy was considered average and could be improved by calibration of Manning coefficient and infiltration process after obtaining more citizen science data for other flood events.

In general terms, the hydrodynamic model tended to overestimate the flood depths (Figure 3). Other studies [39,93] with coupled hydrological models with HEC-RAS and without calibration have adopted roughness values of 0.025 and 0.03. The higher value of global roughness adopted in this study may be one of the reasons to obtain higher flood depths. Additionally, the coarse resolution of DEM used generated higher uncertainties in the model.

Finally, as mentioned before, the purpose of applying hydrodynamic models to obtain flood maps must be discussed, particularly in partially gauged or ungauged basins. Study [37] stated that the major purpose of evaluating the flood maps for different future scenarios (climate, land use, or mitigation intervention) is to evaluate the difference between them rather than to obtain flood precision. A similar discussion was made by [56] in their study for the Gregório Creek where they highlighted that the best capture of the catchment hydrology was not their main purpose.

Therefore, it is important to highlight that this approach of coupled hydrological and hydrodynamic model, calibrated or validated with citizen science data, has the main purpose to serve as a tool to contribute to a risk-informed and evidence-based decision-making process in poorly gauged or ungauged catchments. For example, the results obtained in this study supports the decision that LID practices can be used as mitigation measures to more recurrent events (RP less than 10 years) and it is necessary to guarantee more than 1% cover with LID practices to mitigate the non-stationary effects of future climate projections.

5.2. Urban Flood Mitigation Scenarios

For weak and moderate design storms in the current scenario, the intervention scenario with LID practices was able to mitigate peak flows by approximately 40% and 30%, respectively. These results are similar to those obtained by [49] for rainwater harvesting applied, decentralized, and distributed in the basin using the same SWMM and HEC-RAS coupled methodology. Additionally, in agreement with these results, several studies showed that LID practices, especially bioretention, are more efficient and (therefore) recommended for more recurrent and less intense rainfalls [94–96].

However, when considering the effects of climate change in the design rainfalls, there was a drop in peak flow attenuations. For the moderate rainfall scenario, the peak mitigation was reduced to only 10%. The drop in efficiencies over time (Figure 6) was related to an increase in rainfall volumes and peaks while keeping the maximum capacity of bioretention constant. Therefore, there is a need to increase the practice's useful volumes. This can be done by:

(1) Increasing the bioretention volume of each practice over time. In this sense, [28,75] discussed the importance of incorporating future climate change scenarios in the conception and design of bioretention structures based on a modular design that allows the expansion of their areas and volumes over pre-defined periods.
(2) Increasing the total bioretention volume in the entire catchment, i.e., increasing the number of practices applied in the catchment.

It should be noted that this study evaluated a scenario with the implementation of bioretention cells corresponding to 1% of the catchment area (due to the availability of space-related to the current occupation of the soil in the city of São Carlos).

LID practice manuals recommend coverage of 1% to 5% of the total area of the catchment as optimal values for flood control [97–99]. Additionally, [100] recommended rates of 10% to 20% between the bioretention area and the impermeable area to obtain optimal exfiltration and groundwater recharge. However, these metrics do not consider the changes in climate patterns in future scenarios and may lead to areas that are smaller than necessary for good performances, as observed in the results from this study.

One of the complementary purposes of this study was to assess whether LID practices can at least mitigate the surplus generated by climate change on river discharges and consequently flood depths, keeping the volumes and peak flows constant and equal to the values of the current period. From Figure 5, it is possible to notice that the proposed level of coverage by bioretention was not sufficient to guarantee this mitigation. However, as shown in Figure 6, the efficiency of these techniques with only 1% coverage of the total catchment area (a viable scenario in terms of spatial availability) could maintain average efficiencies of around 35% for more recurring rainfall depths.

Finally, for resilience evaluation, it was possible to observe a more significant variability in the resilience curve for the weak storms and the current period. This is because the LID practices have higher volume retention and peak attenuation efficiencies for weak storms. The effects varied depending on the rainfall pattern, presenting different final reductions in the peak water level. The efficiencies were reduced for moderate and heavy storms, with no significant decreases in the flooded water level (as discussed in Figure 5) and resilience, regardless of the scenario.

6. Conclusions

This research proposed a novel approach of coupling the hydrological model SWMM and the hydrodynamic model HEC-RAS calibrated and validated with citizen science data for modelling a partially gauged urban basin. This approach was applied to simulate the effects of adding bioretention cells on the system's resilience for several rainfall scenarios that considered different return periods and climate changes. The methodology was applied at the Gregório Creek catchment. The effects on the system's resilience by adding LID practices were evaluated by comparing the model's outputs for all the rainfall scenarios with and without the bioretention cells modelled on SWMM.

The results showed that LID practices can reduce peak flows by around 30% to 40% for moderate and weak rainfall, respectively, in the current climate scenario. However, when considering the effects of climate change, the efficiency for moderate rainfall dropped to only 10%. As for heavy rainfall, in the current climate scenario, the peak retention efficiency was significantly smaller than compared with moderate and weak rainfalls, with values around only 9%. In the future scenario, this efficiency dropped to 8%, representing a less accentuated drop than for less intense rainfall. This happens because bioretention has a maximum storage capacity, which is responsible for its mitigation effect. This maximum capacity was already reached in the current extreme rainfall scenarios; thus, there were no significant reductions in future scenarios with climate change.

These effects on main channel flows were reflected in the flood map generated by the hydrodynamic model and consequently in the resilience assessment. For more extreme rainfall and future climate change scenarios, the reductions in flood depth and increases in resilience were minimal. This demonstrates that the total volume retention capacity of LID practices was undersized, requiring an increase in their individual volumes (modular expansion) or an increase in their coverage in the basin (>1%).

The hydrological simulation coupled with hydrodynamics allowed establishment of maximum peak water depths that represent acceptable risks and desired resilience. From these goals, the optimal intervention scenario could be selected (e.g., minimum coverage of LID practices over time). The modular design of LID practices or the expansion of their coverture in the catchment over time can be a viable alternative for long-term planning, both in terms of risk management actions and economic resource availability. Planning to increase the coverage of LID practices through policies to encourage their individual and decentralized adoption as well as their implementation through the local government can contribute to at least keep the current flood risk constant for future climate change scenarios (which was not achieved with the intervention scenario proposed in this study).

We strongly suggest performing uncertainty analysis for future studies, as data coming from citizen observatories are subject to significant uncertainty beyond the complexities in both models used (SWMM and HEC-RAS). Moreover, the methodology was only tested as a case study on the Gregório Creek catchment applying bioretention cells. Further comparison using different green infrastructure measures such as porous pavements, rain barrels, vegetative swales, and green roofs as well as in other catchments may foster sustainable approaches towards systems resilience improvement and mitigation of climate changes effects. Furthermore, adopting criteria of selecting check-points decentralized across the watershed to assess runoff retention efficiency and peak flow attenuation efficiency spatially distributed may provide a better understanding of measures to take and find the optimal distribution of LID practices over the area.

Author Contributions: Conceptualization, M.C.F., M.B.d.M., A.C.S.B., E.M.M. and A.C.B.D.; methodology, M.C.F., M.B.d.M. and A.C.S.B.; software, M.C.F.; validation, M.C.F. and A.C.S.B.; formal analysis, M.C.F. and M.B.d.M.; investigation, M.C.F. and M.B.d.M.; data curation, M.C.F. and M.B.d.M.; writing—original draft preparation, M.C.F. and M.B.d.M.; writing—review and editing, A.M.S., A.C.B.D. and E.M.M.; supervision, A.M.S., A.C.B.D. and E.M.M.; funding acquisition, A.M.S., A.C.B.D. and E.M.M. All authors have read and agreed to the published version of the manuscript.

Funding: The authors of this work would like to thank the Center for Artificial Intelligence (C4AI-USP) and the support from the São Paulo Research Foundation (FAPESP grant #2019/07665-4) and from the IBM Corporation. We would like also to thank the Artificial Intelligence group of the Institute of Mathematics and Computer Science of the University of São Paulo (CEMEAI).

Informed Consent Statement: Not applicable.

Data Availability Statement: Not applicable.

Acknowledgments: The authors of this work would like to thank the Fundação de Amparo à Pesquisa do Estado de São Paulo (FAPESP grants #2014/50848-9, #2020/03029-3) throughout project n. 2019/23393-4 "TOCO_DR Theory of Change Observatory in Disaster Resilience"; the Coordenação de Aperfeiçoamento de Pessoal de Nível Superior do Brasil (CAPES)—finance Code 001—and regular funding to post-graduate program in Hydraulics and Sanitation of University of São Paulo, São Carlos School of Engineering by the Brazilian National Council for Scientific and Technological Development (CNPq); the National Institute of Science and Technology for Climate Change Phase 2 (INCT-II) under the CNPq Grant 465501/2014-1; and CAPES Grant 16/2014.

Conflicts of Interest: The authors declare no conflict of interest.

Appendix A

Table A1. Parameters of the IDF curves type Sherman *, for the current period and updated with climate change patterns (Macedo, 2020).

	Current	MIROC5 4.5 PT **	MIROC5 4.5 MD ***	MIROC5 8.5 PT	MIROC5 8.5 MD
			2015–2050		
K	819.67	772.4	764.56	899.82	890.51
m	0.138	0.311	0.2956	0.2182	0.2176
t_0 (min)	10.77	12	12	12	12
n	0.75	0.764	0.764	0.764	0.764
			2050–2100		
K	819.67	1007.77	965.93	1036.49	1034.01
m	0.138	0.2645	0.2113	0.2356	0.2007
t_0 (min)	10.77	12	12	12	12
n	0.75	0.764	0.764	0.764	0.764

* $I = \frac{K \,.\, TR^m}{(t+t_0)^m}$ ** Power transformation and *** distribution mapping (MD) bias correction method.

Appendix B

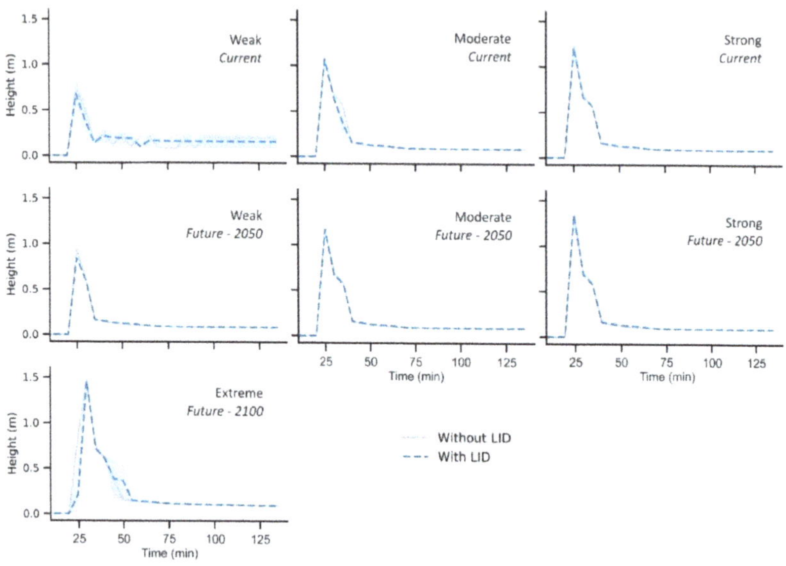

Figure A1. Performance curves for representative design storms and intervention scenarios.

References

1. Hederra, R. Environmental sanitation and water supply during floods in Ecuador (1982–1983). *Disasters* **1987**, *11*, 297–309. [CrossRef]
2. Ahmed, M.F.; Ashfaque, K.N. Sanitation and Solid Waste Management in Dhaka City During the 1998 Flood. In *Engineering Concerns of Flood*, 1st ed.; Ali, M.A., Seraj, S.M., Ahmad, S., Eds.; Directorate of Advisory, Extension and Research Services Bangladesh University of Engineering and Technology: Dhaka, Bangladesh, 2002; Volume 1, pp. 1–13. ISBN 984-823-002-5.
3. Baig, S.A.; Xu, X.; Navedullah, M.N.; Khan, Z.U. Pakistan's drinking water and environmental sanitation status in post 2010 flood scenario: Humanitarian response and community needs. *J. Appl. Sci. Environ. Sanit.* **2012**, *7*, 49–54.
4. Bastawesy, M.E.; Ella, E.M.A.E. Quantitative estimates of flash flood discharge into wastewater disposal sites in Wadi Al Saaf, the Eastern Desert of Egypt. *J. Afr. Earth Sci.* **2017**, *136*, 312–318. [CrossRef]
5. Leopold, L.B. *Hydrology for Urban Land Planning: A Guidebook on the Hydrological Effects of Urban Land Use*; USA Geological Survey: Washington, DC, USA, 1968; Volume 554.
6. Wong, T.H.F.; Eadie, M.L. Water Sensitive Urban Design—A Paradigm Shift in Urban Design. In Proceedings of the 10th World Water Congress, Copenhagen, Denmark, 16 December 2000. Available online: http://gabeira.locaweb.com.br/cidadesustentavel/biblioteca/%7B30788FE6-98A8-44E5-861E-996D286A78B3%7D_Wong1.pdf (accessed on 1 April 2018).
7. Konrad, C.P.; Booth, D.B. Hydrologic changes in urban streams and their Ecological significance. *Am. Fish. Soc. Symp.* **2005**, *47*, 157–177.
8. Stark, B.; Ossa, A. Ancient Settlement, Urban Gardening, and Environment in the Gulf Lowlands of Mexico. *Lat. Am. Antiq.* **2007**, *18*, 385–406. [CrossRef]
9. Nunes, M.F.; Figueiredo, J.A.S.; Rocha, A.L.C.D. Sinos River Hydrographic Basin: Urban occupation, industrialization and environmental memory. *Braz. J. Biol.* **2015**, *75*, 3–9. [CrossRef]
10. IPCC. *Contribution of Working Groups I, II and III to the Fourth Assessment Report of the Intergovernmental Panel on Climate Change*; IPCC: Geneva, Switzerland, 2007; p. 104.
11. Marengo, J.A.; Schaeffer, R.; Zee, D.; Pinto, H.S. Mudanças Climáticas e Eventos Extremos no Brasil. 2010. Available online: http://www.fbds.org.br/cop15/FBDS_MudancasClimaticas.pdf (accessed on 1 October 2010).
12. Debortoli, N.S.; Camarinha, P.I.M.; Marengo, J.A.; Rodrigues, R.R. An index of Brazil's vulnerability to expected increases in natural flash flooding and landslide disasters in the context of climate change. *Nat. Hazards* **2017**, *86*, 557–582. [CrossRef]
13. Semadeni-Davies, A.; Hernebring, C.; Svensson, G.; Gustafsson, L.G. The impacts of climate change and urbanisation on drainage in Helsingborg, Sweden: Suburban stormwater. *J. Hydrol.* **2008**, *350*, 114–125. [CrossRef]
14. Houston, D.; Werritty, A.; Bassett, D. *Pluvial (Rain-Related) Flooding in Urban Areas: The Invisible Hazard*; Joseph Rowntree Foundation: York, UK, 2011.
15. Gersonius, B.; Nasruddin, F.; Ashley, R.; Jeuken, A.; Pathirana, A.; Zevenbergen, C. Developing the evidence base for mainstreaming adaptation of stormwater systems to climate change. *Water Res.* **2012**, *46*, 6824–6835. [CrossRef]
16. Chou, S.C.; Lyra, A.; Mourão, C.; Dereczynski, C.; Pilotto, I.; Gomes, J.; Bustamante, J.; Tavares, P.; Silva, A.; Rodrigues, D.; et al. Assessment of climate change over South America under RCP 4.5 and 8.5 downscaling scenarios. *Am. J. Clim. Chang.* **2014**, *3*, 512–525. [CrossRef]
17. Lyra, A.; Tavares, P.; Chou, S.C.; Sueiro, G.; Dereczynski, C.; Sondermann, M.; Silva, A.; Marengo, J.; Giarolla, A. Climate change projections over three metropolitan regions in Southeast Brazil using the non-hydrostatic Eta regional climate model at 5-km resolution. *Theor. Appl. Climatol.* **2018**, *132*, 663–682. [CrossRef]
18. IPCC. *Climate Change 2021–The Physical Science Basis. Working Group I Contribution to the Sixth Assessment Report of the Intergovernmental Panel of Climate Change*; IPCC: Geneva, Switzerland, 2021.
19. UNDRR–United Nations Office for Disaster Risk Reduction. *Sendai Framework for Disaster Risk Reduction 2015–2030*; UNDRR: Geneva, Switzerland, 2015.
20. Urban Institute. *Beyond Ideology, Politics and Guesswork: The Case for Evidence-Based Policy*; Urban Institute: Washington, DC, USA, 2008.
21. Zio, E.; Pedroni, N. *Overview of Risk-Informed Decision-Making Processes*; Number 2012-10 of the Cahiers de la Sécurité Industrielle; Foundation for an Industrial Safety Culture: Toulouse, France, 2012. Available online: http://www.FonCSI.org/en/ (accessed on 1 October 2021).
22. Kurian, M.; Ardakanian, R.; Veiga, L.G.; Meyer, K. *Resources, Services and Risks: How Can Data Observatories Bridge the Science-Policy Divide in Environmental Governance?* Springer: Berlin/Heidelberg, Germany, 2016.
23. Toklu, A. Improving Organisational Performance with Balanced Scorecard in Humanitarian Logistics: A Proposal for Key Performance Indicators. *Int. J. Acad. Res. Account. Financ. Manag. Sci.* **2017**, *7*, 131–137. [CrossRef]
24. Schanze, J. Flood Risk Management–A Basic Framework. In *Flood Risk Management: Hazards, Vulnerability and Mitigation Measures*; Springer: Dordrecht, The Netherlands, 2006; pp. 1–20.
25. Vieira, I.; Barreto, V.; Figueira, C.; Lousada, S.; Prada, S. The use of detention basins to reduce flash flood hazard in small and steep volcanic watersheds–a simulation from Madeira Island. *J. Flood Risk Manag.* **2018**, *11*, S930–S942. [CrossRef]
26. Jacob, A.C.P.; Rezende, O.M.; de Sousa, M.M.; de França Ribeiro, L.B.; de Oliveira, A.K.B.; Arrais, C.M.; Miguez, M.G. Use of detention basin for flood mitigation and urban requalification in Mesquita, Brazil. *Water Sci. Technol.* **2019**, *79*, 2135–2144. [CrossRef]

27. Manfreda, S.; Miglino, D.; Albertini, C. Impact of detention dams on the probability distribution of floods. *Hydrol. Earth Syst. Sci.* **2021**, *25*, 4231–4242. [CrossRef]
28. Macedo, M.B.; Gomes Junior, M.N.; Oliveira, T.R.P.; Giacomoni, H.M.; Imani, M.; Zhang, K.; Ambrogi Ferreira do Lago, C.; Mendiondo, E.M. Low Impact Development practices in the context of United Nations Sustainable Development Goals: A new concept, lessons learned and challenges. *Crit. Rev. Environ. Sci. Technol.* **2021**, 1–44. [CrossRef]
29. C40. Climate Action in Megacities: C40 Cities Baseline and Opportunities. *Volume 2.0. February 2014.* Available online: http://issuu.com/c40cities/docs/c40_climate_action_in_megacities/149?e=10643095/6541335 (accessed on 1 October 2016).
30. Simonovic, S.; Peck, A. Dynamic resilience to climate change caused natural disasters in coastal megacities quantification framework. *Int. J. Environ. Clim. Change* **2013**, *3*, 378–401. [CrossRef]
31. Simonovic, S. Adapting to Climate Change: A Web Based Intensity-Duration-Frequency (IDF) Tool. *Geotech. News* **2017**, *35*, 40–42.
32. Chen, J.; Hill, A.A.; Urbano, L.D. A GIS-based model for urban flood inundation. *J. Hydrol.* **2009**, *373*, 184–192. [CrossRef]
33. Wu, X.; Wang, Z.; Guo, S.; Liao, W.; Zeng, Z.; Chen, X. Scenario-based projections of future urban inundation within a coupled hydrodynamic model framework: A case study in Dongguan City, China. *J. Hydrol.* **2017**, *547*, 428–442. [CrossRef]
34. Leandro, J.; Martins, R. A methodology for linking 2D overland flow models with the sewer network model SWMM 5.1 based on dynamic link libraries. *Water Sci. Technol.* **2016**, *73*, 3017–3026. [CrossRef] [PubMed]
35. Bisht, D.S.; Chatterjee, C.; Kalakoti, S.; Upadhyay, P.; Sahoo, M.; Panda, A. Modeling urban floods and drainage using SWMM and MIKE URBAN: A case study. *Nat. Hazards* **2016**, *84*, 749–776. [CrossRef]
36. Pati, A.; Sahoo, B. Effect of Low-Impact Development Scenarios on Pluvial Flood Susceptibility in a Scantily Gauged Urban–Peri-Urban Catchment. *J. Hydrol. Eng.* **2022**, *27*, 05021034. [CrossRef]
37. Feng, B.; Zhang, Y.; Bourke, R. Urbanization impacts on flood risks based on urban growth data and coupled flood models. *Nat. Hazards* **2021**, *106*, 613–627. [CrossRef]
38. Natarajan, S.; Radhakrishnan, N. An integrated hydrologic and hydraulic flood modeling study for a medium-sized ungauged urban catchment area: A case study of Tiruchirappalli City Using HEC-HMS and HEC-RAS. *J. Inst. Eng.* **2020**, *101*, 381–398. [CrossRef]
39. Rangari, V.A.; Sridhar, V.; Umamahesh, N.V.; Patel, A.K. Floodplain mapping and management of urban catchment using HEC-RAS: A case study of Hyderabad city. *J. Inst. Eng.* **2019**, *100*, 49–63. [CrossRef]
40. Rossman, L.A. *Storm Water Management Model User's Manual*; Version 5.0. U.S.; Environmental Protection Agency: Cincinatti, OH, USA, 2004.
41. Jeong, J.; Her, Y.; Arnold, J.; Gosselink, L.; Glick, R.; Jaber, F. SWAT LID Module. In Proceedings of the 2015 SWAT Conference, Pula, Italy, 18–26 June 2015.
42. Jiang, L.; Chen, Y.; Wang, H. Urban flood simulation based on the SWMM model. *Proc. Int. Assoc. Hydrol. Sci.* **2015**, *368*, 186–191. [CrossRef]
43. Pina, R.; Ochoa-Rodriguez, S.Ç.; Simões, N.; Mijic, A.; Marques, A.; Maksimović, Č. Semi-vs. *fully-distributed urban stormwater models: Model set up and comparison with two real case studies. Water* **2016**, *8*, 58.
44. Uchiyama, S.; Bhattacharya, Y.; Nakamura, H. Efficacy Analysis of Urban Planning Scenarios for Flood Mitigation with Low Impact Development Technologies Using SWMM: A Case Study in Saitama City, Japan. In *IOP Conference Series: Earth and Environmental Science*; IOP Publishing: Bristol, UK, 2022; Volume 973, p. 012012.
45. Jha, M.K.; Afreen, S. Flooding urban landscapes: Analysis using combined hydrodynamic and hydrologic modeling approaches. *Water* **2020**, *12*, 1986. [CrossRef]
46. Kourtis, I.M.; Bellos, V.; Kopsiaftis, G.; Psiloglou, B.; Tsihrintzis, V.A. Methodology for holistic assessment of grey-green flood mitigation measures for climate change adaptation in urban basins. *J. Hydrol.* **2021**, *603*, 126885. [CrossRef]
47. Ghazal, R.; Ardeshir, A.; Rad, I.Z. Climate change and stormwater management strategies in Tehran. *Procedia Eng.* **2014**, *89*, 780–787. [CrossRef]
48. Tavakol-Davani, H.; Goharian, E.; Hansen, C.H.; Tavakol-Davani, H.; Apul, D.; Burian, S.J. How does climate change affect combined sewer overflow in a system benefiting from rainwater harvesting systems? *Sustain. Cities Soc.* **2016**, *27*, 430–438. [CrossRef]
49. Akter, A.; Tanim, A.H.; Islam, M.K. Possibilities of urban flood reduction through distributed-scale rainwater harvesting. *Water Sci. Eng.* **2020**, *13*, 95–105. [CrossRef]
50. Ogie, R.I.; Perez, P.; Win, K.T.; Michael, K. Managing hydrological infrastructure assets for improved flood control in coastal mega-cities of developing nations. *Urban Clim.* **2018**, *24*, 763–777. [CrossRef]
51. Fava, M.C.; Mazzoleni, M.; Abe, N.; Mendiondo, E.M.; Solomatine, D.P. Improving flood forecasting using an input correction method in urban models in poorly gauged areas. *Hydrol. Sci. J.* **2020**, *65*, 1096–1111. [CrossRef]
52. Nkwunonwo, U.C.; Whitworth, M.; Baily, B. A review of the current status of flood modelling for urban flood risk management in the developing countries. *Sci. Afr.* **2020**, *7*, e00269. [CrossRef]
53. Roy, H.E.; Pocock, M.J.O.; Preston, C.D.; Roy, D.B.; Savage, J.; Tweddle, J.C.; Robinson, L.D. *Understanding Citizen Science and Environmental Monitoring: Final Report on Behalf of UK-Environmental Observation Framework*; NERC Centre for Ecology & Hydrology and Naturel History Museum: Wallingford, UK, 2012.
54. Fava, M.C.; Abe, N.; Restrepo-Estrada, C.; Kimura, B.Y.; Mendiondo, E.M. Flood modelling using synthesised citizen science urban streamflow observations. *J. Flood Risk Manag.* **2019**, *12*, e12498. [CrossRef]

55. Nardi, F.; Cudennec, C.; Abrate, T.; Allouch, C.; Annis, A.; Assumpção, T.; Aubert, A.H.; Bérod, D.; Braccini, A.M.; Buytaert, W.; et al. Citizens AND HYdrology (CANDHY): Conceptualizing a transdisciplinary framework for citizen science addressing hydrological challenges. *Hydrol. Sci. J.* **2021**, 1–18. [CrossRef]
56. Buarque, A.C.S.; Souza, C.F.; Souza, F.A.A.; Mendiondo, E.M. Urban flood risk under global changes: A socio-hydrological and cellular automata approach in a Brazilian catchment. *Hydrol. Sci. J.* **2021**, *66*, 2011–2021. [CrossRef]
57. Smith, B.; Rodriguez, S. Spatial analysis of high-resolution radar rainfall and citizen-reported flash flood data in ultra-urban New York City. *Water* **2017**, *9*, 736. [CrossRef]
58. See, L. A review of citizen science and crowdsourcing in applications of pluvial flooding. *Front. Earth Sci.* **2019**, *7*, 44. [CrossRef]
59. Saadatpour, M.; Delkhosh, F.; Afshar, A.; Solis, S.S. Developing a simulation-optimization approach to allocate low impact development practices for managing hydrological alterations in urban watershed. *Sustain. Cities Soc.* **2020**, *61*, 102334. [CrossRef]
60. De Paola, F.; Giugni, M.; Pugliese, F. A harmony-based calibration tool for urban drainage systems. Proceedings of the Institution of Civil Engineers. *Water Manag.* **2018**, *171*, 30–41. [CrossRef]
61. Abreu, F.G. Quantificação dos Prejuízos Econômicos à Atividade Comercial Derivados de Inundações Urbanas [Quantification of Economic Damage to Commercial Activity from Urban Flooding]. Doctoral Dissertation, Escola de Engenharia de São Carlos, Universidade de São Paulo, São Carlos, Brazil, 2019. (In Portuguese).
62. Sarmento Buarque, A.C.; Bhattacharya-Mis, N.; Fava, M.C.; Souza, F.A.A.D.; Mendiondo, E.M. Using historical source data to understand urban flood risk: A socio-hydrological modelling application at Gregório Creek, Brazil. *Hydrol. Sci. J.* **2020**, *65*, 1075–1083. [CrossRef]
63. IBGE-Instituto Brasileiro de Geografia e Estatística. Estimativas da População Residente no BRASIL e Unidades da Federação Com Data de Referência em 1° de Julho de 2018. 2015. Available online: https://www.ibge.gov.br/estatisticas/sociais/populacao (accessed on 1 October 2021). (In Portuguese)
64. INMET–Instituto Nacional de Meteorologia. Normais Climatológicas do Brasil [Brazilian Climatological Norms]. 2016. Available online: https://portal.inmet.gov.br/normais (accessed on 1 October 2021). (In Portuguese)
65. Mendes, H.C.; Mendiondo, E.M. Histórico da expansão urbana e incidência de inundações: O Caso da Bacia do Gregório, São Carlos–SP. *Rev. Bras. De Recur. Hídricos* **2007**, *12*, 17–27.
66. Barros, R.M.; Mendiondo, E.M.; Wendland, E. Cálculo de áreas inundáveis devido a enchentes para o plano diretor de drenagem urbana de Sao Carlos (PDDUSC) na bacia escola do córrego do Gregório. *Rev. Bras. De Recur. Hídricos* **2007**, *12*, 5–17.
67. Negrão, A.C. One-Dimensional Hydrodynamic Modeling of Flood Wave Passage in an Urban Stream Considering Transcritical Flow. Master's Thesis, Escola de Engenharia de São Carlos, Universidade de São Paulo, São Carlos, Brazil, 2015.
68. James, W.; Rossman, L.E.; James, W.R. *User's Guide to SWMM5*, 13th ed.; CHI: Guelph, ON, Canada, 2010.
69. Perin, R.; Trigatti, M.; Nicolini, M.; Campolo, M.; Goi, D. Automated calibration of the EPA-SWMM model for a small suburban catchment using PEST: A case study. *Environ. Monit. Assess.* **2020**, *192*, 1–17. [CrossRef] [PubMed]
70. Costa, C.W.; Dupas, F.A.; Pons, N.A.D. Regulamentos de uso do solo e impactos ambientais: Avaliação crítica do plano diretor participativo do município de São Carlos, SP. *Geociências* **2012**, *31*, 143–157.
71. Behrouz, M.S.; Zhu, Z.; Matott, L.S.; Rabideau, A.J. A new tool for automatic calibration of the Storm Water Management Model (SWMM). *J. Hydrol.* **2020**, *581*, 124436. [CrossRef]
72. Fava, M.C. Improving Flood Forecasting Using Real-Time Data To Update Urban Models in Poorly Gauged Areas. Doctoral Dissertation, Escola de Engenharia de São Carlos, Universidade de São Paulo, São Carlos, Brazil, 2019.
73. American Meteorological Society (2022). Rain. *Glossary of Meteorology.* Available online: http://glossary.ametsoc.org/wiki/Rain (accessed on 1 February 2022).
74. Di Vittorio, D. Spatial Translation And Scaling Up Of Lid Practices in Deer Creek Watershed in East Missouri. Doctoral Dissertation, Southern Illinois University at Edwardsville, Edwardsville, IL, USA, 2014.
75. Macedo, M.B. Decentralized Urban Runoff Recycling Facility Addressing the Security of the Water-Energy-Food Nexus. Doctoral Dissertation, Escola de Engenharia de São Carlos, Universidade de São Paulo, São Carlos, Brazil, 2020.
76. HEC-Hydrologic Engineering Center. HEC-RAS 5.0. *Hydraulic Reference Manual [Online].* 2018. Available online: http://www.hec.usace.army.mil/software/hec-ras/documentation.aspx (accessed on 1 May 2018).
77. Oliveira, R.T.; Machado, S.A. Quantificação do pesticida diclorvos por voltametria de onda quadrada em águas puras e naturais. *Química Nova* **2004**, *27*, 911–915. [CrossRef]
78. Marotti, A.C.B.; Santos, K.E.L.; Macera, L.G.; de Lima Neves, L.; Gonçalves, J.C.; Pugliesi, É. Levantamento histórico e relatos de inundações do córrego do Gregório na região central do município de São Carlos-SP. *Rev. Eixo* **2014**, *3*, 25–37. [CrossRef]
79. Stangonini, F.N.; de Lollo, J.A. The growth of the urban area of São Carlos/SP between the 2010 and 2015: The advancement of environmental degradation. *Urbe* **2018**, *10*, 118–128.
80. Fialho, H.C.P.; Abreu, F.G.; Sousa, B.J.D.O.; Souza, F.A.A.; Bhattacharya-Mis, N.; Mendiondo, E.M.; Oliveira, P.T.S.D. Anticipated Memories and Adaptation from Past Flood Events in Gregório Creek Basin, Brazil. *Water* **2021**, *13*, 3394. [CrossRef]
81. Chow, V.T. *Open-Channel Hydraulics*; McGraw-Hill: New York, NY, USA, 1959.
82. Zeiger, S.J.; Hubbart, J.A. Measuring and modeling event-based environmental flows: An assessment of HEC-RAS 2D rain-on-grid simulations. *J. Environ. Manag.* **2021**, *285*, 112125. [CrossRef] [PubMed]
83. Guidolin, M.; Chen, A.S.; Ghimire, B.; Keedwell, E.C.; Djordjević, S.; Savić, D.A. A weighted cellular automata 2D inundation model for rapid flood analysis. *Environ. Model. Softw.* **2016**, *84*, 378–394. [CrossRef]

84. Chow, V.T.; Maidment, D.R.; Mays, L.W. *Applied Hydrology*; McGraw-Hill: New York, NY, USA, 1988.
85. McCuen, R.H. *Hydrologic Analysis and Design*, 3rd ed.; Pearson Prentice Hall: Hoboken, NJ, USA, 2005.
86. Balbastre-Soldevila, R.; García-Bartual, R.; Andrés-Doménech, I. A comparison of design storms for urban drainage system applications. *Water* **2019**, *11*, 757. [CrossRef]
87. Willems, P.; Vrac, M. Statistical precipitation downscaling for small-scale hydrological impact investigations of climate change. *J. Hydrol.* **2011**, *402*, 193–205. [CrossRef]
88. Vandenberghe, S.; Verhoest, N.E.C.; Buyse, E.; De Baets, B. A stochastic design rainfall generator based on copulas and mass curves. *Hydrol. Earth Syst. Sci.* **2010**, *14*, 2429–2442. [CrossRef]
89. Vieux, B.E.; Vieux, J.E. Development of regional design storms for sewer system modeling. *Proc. Water Environ. Fed.* **2010**, *2010*, 6248–6263. [CrossRef]
90. Barbassa, A.P. Simulação Do Efeito Da Urbanização Sobre A Drenagem Pluvial Da Cidade De São Carlos–Sp. [Simulation of the Effect of Urbanization on the Storm Drainage of the City of São Carlos-Sp.]. Ph.D. Thesis, Escola de Engenharia de São Carlos, Universidade de São Paulo, São Carlos, Brazil, 1991. (In Portuguese).
91. Yalcin, E. Assessing the impact of topography and land cover data resolutions on two-dimensional HEC-RAS hydrodynamic model simulations for urban flood hazard analysis. *Nat. Hazards* **2020**, *101*, 995–1017. [CrossRef]
92. Shrestha, A.; Bhattacharjee, L.; Baral, S.; Thakur, B.; Joshi, N.; Kalra, A.; Gupta, R. Understanding suitability of MIKE 21 and HEC-RAS for 2D floodplain modeling. In *World Environmental and Water Resources Congress 2020: Hydraulics, Waterways, and Water Distribution Systems Analysis*; American Society of Civil Engineers: Reston, VA, USA, 2020; pp. 237–253.
93. Rangari, V.A.; Umamahesh, N.V.; Bhatt, C.M. Assessment of inundation risk in urban floods using HEC RAS 2D. *Model Earth Syst Env.* **2019**, *5*, 1839–1851. [CrossRef]
94. Zahmatkesh, Z.; Burian, S.; Karamouz, M.; Tavakol-Davani, H.; Goharian, E. Low-impact development practices to mitigate climate change effects on urban stormwater runoff: Case study of New York City. *J. Irrig. Drain. Eng.* **2014**, *141*, 04014043. [CrossRef]
95. Macedo, M.; Lago, C.; Mendiondo, E. Stormwater volume reduction and water quality improvement by bioretention: Potentials and challenges for water security in a subtropical catchment. *Sci. Total Environ.* **2019**, *647*, 923–931. [CrossRef] [PubMed]
96. Zhang, K.; Manuelpillai, D.; Raut, B.; Deletic, A.; Bach, P. Evaluating the reliability of stormwater treatment systems under various future climate conditions. *J. Hydrol.* **2019**, *568*, 57–66. [CrossRef]
97. The Prince George's County. *Bioretention Manual*; Environmental Services Division, Department of Environmental Resources: Beltsville, MD, USA, 2007.
98. Waterways, M.B. *Water Sensitive Urban Design: Technical Design Guidelines for South East Queensland*; Moreton Bay Waterways and Catchment Partnership; Australian Government: Brisbane, Australia, 2006.
99. Council, G.C. *Water Sensitive Urban Design Guidelines*; Gold Coast City Council: City of Gold Coast, Australia, 2007.
100. Dussaillant, A.R.; Wu, C.H.; Potter, K.W. Richards equation model of a rain garden. *J. Hydrol. Eng.* **2004**, *9*, 219–225. [CrossRef]

water

Article

The Dutch Flood Protection Programme: Taking Innovations to the Next Level

Ellen Tromp [1,2,*], Anouk te Nijenhuis [2] and Han Knoeff [1,2]

1 Deltares, 2600 MH Delft, The Netherlands; han.knoeff@deltares.nl
2 Dutch Flood Protection Programme, Location B4, 3500 GE Utrecht, The Netherlands; anouk.te.nijenhuis@rws.nl
* Correspondence: ellen.tromp@deltares.nl

Abstract: The Dutch regional water authorities face an enormous task: the strengthening of about 1500 km of dikes and 500 civil-engineering structures before 2050. This immense operation is being funded, prioritised and supported by the Dutch Flood Protection Programme (DFPP), an alliance of regional water authorities and the Ministry of Infrastructure and Water Management. The work will be executed in nearly 300 projects located throughout the country on the coast, lakes and major rivers. To complete this task on time and within budget, innovation (a better insight into the behaviour of flood defences, new techniques and processes) is believed to be the way forward. In this paper, we look at how the DFPP has encouraged innovations between 2012 and the present. We stress the importance of using a sender–receiver approach to further knowledge transfer and uptake, and we describe how, by using an action research approach, the Dutch Flood Protection Programme is currently adapting its innovation strategy on the basis of lessons learned to improve knowledge uptake. We will address some of the innovations that have been developed over the years and how monitoring knowledge uptake helps to further improve the learning-by-doing approach.

Keywords: flood risk management; innovations; dikes; flood decision-making; knowledge uptake

Citation: Tromp, E.; te Nijenhuis, A.; Knoeff, H. The Dutch Flood Protection Programme: Taking Innovations to the Next Level. *Water* **2022**, *14*, 1460. https://doi.org/10.3390/w14091460

Academic Editors: Slobodan P. Simonovic, Subhankar Karmakar and Zhang Cheng

Received: 31 January 2022
Accepted: 27 April 2022
Published: 3 May 2022

Publisher's Note: MDPI stays neutral with regard to jurisdictional claims in published maps and institutional affiliations.

1. Introduction

Two-thirds of the Netherlands is below the current sea level. In the distant past, the Dutch responded reactively to floods. However, after the catastrophic floods of 1953, the Dutch took measures to prevent a similar disaster: a system of dike rings with flood protection levels was developed for flood risk management [1], and statutory standards were set out in the Flood Protection Act of 1996 to maintain this system. Those standards related to the primary flood defences: those on the coast, major rivers and lakes. In 2008, the second Delta Commission advised changes to take future uncertainties, such as climate change and land subsidence, into account in order to protect the Netherlands.

The Dutch national government recently adopted a risk-based approach [1] for the Dutch flood risk management policy based on new knowledge about the safety of dikes and the impact of serious floods as a result of the failure of dikes and/or other structures. This is a proactive approach, in which protection standards are based on both the probability and the impact of flooding in 2050, considering climate change and socio-economic developments. The Water Act (soon to be the Environment Act) sets out the statutory standards for the flood protection structures, such as dikes, dams, and other civil-engineering structures. Every 12 years, the regional water authorities are required to conduct assessments to ensure that the flood defences still meet the statutory standards. Where the standards are not met, the responsible body can apply for funding from the Dutch Flood Protection Programme (DFPP) [2]. Over 1500 km of dikes and 500 civil-engineering structures will have to be upgraded between now and 2050. The challenge formulated in the overall programme goals for the DFPP is to:

- Increase the production rate (*effectiveness*) of flood defence projects; In programmes in the past, approximately 25 km was upgraded annually. In order to complete the objective of around 1500 km by 2050, this rate will have to rise to an average of ~50 km/year [2].
- Improve *efficiency* by reducing the costs per kilometre; In recent decades, the average cost per kilometre in the Netherlands has increased to more than €10 million (in the 2nd Dutch Flood Protection Programme), and in the early years of the DFPP, the average cost rose towards €18 million per kilometre. Given the annual budget of € 360 million, this figure will have to be reduced again to ~€7 million per kilometre [3].

Although flood risk management is rooted in technical knowledge, the current problems are particularly thorny due to high levels of complexity, uncertainty and conflict [4,5]. The field is complex because flood risk issues are entwined with other local problems involving diverse stakeholders, while the sector is institutionally fragmented, and resources are distributed in a non-hierarchical way [6]. The ever-increasing complexity is producing new challenges and demands.

As regional stakeholders, and society in general, seek to realise additional value (notably, housing and biodiversity) in dikes, the spatial integration of dikes in the local surroundings is becoming more and more important. The Dutch, therefore, changed their approach: from one of coping with water levels associated with specific historical weather events to working with a minimum protection standard based on the probability of flooding and weather events expected in the future, where flood risk management is also an integral part of the living environment [7].

Multiple parties, with their own individual interests and responsibilities, play a role, and they seek to minimise trade-offs. This complexity further emphasises the importance of knowledge management and continuous learning. To achieve its goals, the DFPP recognised that a different approach using new knowledge and innovations was required. The programme has therefore developed a learning-by-doing knowledge and innovation strategy. To understand better why some elements of the knowledge and innovation strategy work and others fail, the authors of the present paper initiated an action-research project to observe, diagnose and intervene in multiple meetings in order to encourage knowledge uptake in dike redesign projects and make recommendations to change the knowledge and innovation strategy.

The paper is structured as follows: Section 2 presents the materials and methods, looking at the Dutch Flood Protection Programme and its knowledge and innovation strategy, followed by a description of the methodology used in our study; in Section 3, we describe how the framework helped to diagnose and remedy problems with knowledge transfer and uptake and the ultimate adoption of the new approach, which is now in use and being monitored. Finally, Section 4 presents the discussion and conclusions of this paper.

2. Materials and Methods

2.1. Dutch Flood Protection Programme: Context

The present paper looks at the knowledge and innovation strategy of the Dutch Flood Protection Programme (DFPP). In 2012, the DFPP started as an alliance between the national government and regional water authorities. It was the successor to the previous Flood Protection Programme, which was 100% funded by the then Ministry of Transport, Public Works and Water Management. After an evaluation of this programme [8], the funding and organisation became a 50–50 responsibility of the current Ministry of Infrastructure and Water Management and the 21 regional water authorities. As mentioned above, the task facing the DFPP is to upgrade about 1500 km of dikes and about 500 civil-engineering structures before 2050 (see Figure 1). The exact numbers depend on the results of the statutory assessment of flood defences by the regional water authorities. This work involves some 300 projects ranked by urgency on the basis of non-conformity with the statutory standard, and scheduling is based on the available budget of €360 million a year. The DFPP organisation's primary responsibility is to set up and direct the process of programming and budgeting, to grant subsidies and reporting/accounting. The DFPP also furthers the sharing of knowledge through its own training programme and communities of practice. In addition, the DFPP facilitates the regional water authorities through advisory teams to develop and share knowledge between the DFPP alliance partners. The upgraded flood defences are required to comply with the statutory standards while taking local sustainability and liveability into consideration [2].

Figure 1. Current scope of projects in the Dutch Flood Protection Programme until the year 2027. The expected scope until 2050 is about 1500 km of dikes and about 500 civil-engineering structures.

Innovation is needed to complete the programme on time and within budget, and a knowledge and innovation strategy [3] has been developed for this purpose. The focus is on two aspects: (1) how to develop a learning-by-doing strategy that takes changes in society over the next few decades into account, since the programme will run for almost 40 years, and (2) how to reduce the development time for innovations further so that knowledge and innovation can be used in the flood defence projects, since the development time in the past has been more than 15 years.

(1) Adaptive learning by using a Pyramid of Flood Risk Management [9].

The DFPP developed the Pyramid of Flood Risk Management (Figure 2) to describe the different stages of complexity and the role of flood defence projects in society. The Pyramid of Flood Risk Management identifies four levels, with every level requiring specific knowledge. The next level can only be reached when the preceding level has been adequately mastered. Working towards a new level in the pyramid also requires additional knowledge; the level therefore depends on the context, the actors and the knowledge. The higher the level in the pyramid, the more integrated the approach, which also implies the involvement of an increasing number of actors. Higher up the pyramid, there will be more and more co-creation and actors who reach goals by working together.

Figure 2. The Pyramid of Flood Risk Management.

The lowest level focuses on the behaviour of flood defences, and more specifically, on the different failure mechanisms. Initially, the DFPP focused on developing a better understanding of the different failure mechanisms of flood defences, and on making product innovations for specific failure mechanisms. The recently adopted risk-based approach, as described above, made it possible to move to the second level, 'Managing flood risks'. The evaluations of the successful Room for the River [6] programme contributed skills and knowledge about enhancing the quality of the project area while reducing flood risks. The regional water authorities, especially those active in the Room for the River Programme, adopted the insights acquired in the dike redesign projects under the umbrella of the DFPP. These regional water authorities moved towards the third level in the Pyramid ('Improved quality of environment'). An additional aim of this specific level is to incorporate participatory approaches. Until now, several dike redesign projects have been successful in involving the public [10,11] in order to facilitate the incorporation of ideas and knowledge from local residents and businesses. The top level of the pyramid 'Resilient system' aims to develop adaptive strategies and to reduce future regret relating to investments that must be taken in the coming decades. Linking the implementation of improvements targeting flood risk management to spatial developments and related area processes make synergies and improvements in liveability possible.

Looking through the lens of the pyramid of flood risk management helps the programme to respond to changing boundary conditions associated with society and the climate. This will help to maintain support in Dutch society for flood protection measures. The DFPP aims to develop and further knowledge on all four levels of the pyramid in their innovation portfolio.

(2) The reduction of development time by using a combination of a stage-gate system and a framework of critical success factors.

Before the DFPP began, innovators sometimes felt that they had to conduct similar experiments for different organisations to prove their concepts. This is in contrast with the idea that developing successful new products, services or processes is a gradual process of reducing risk by going through several problem-solving phases [12]. Each phase should therefore be designed to reduce the major uncertainties and risks and build upon the previous one. The information required for this, therefore, determines the activities in each phase.

Based on some analysis, the DFPP agreed with the innovators that clarity in the different phases was needed, not least to know how far an innovation was in its development as an accepted alternative. In order to achieve this clarity, the DFPP adopted the Stage-Gate process [13] to make the innovation process as structured as possible, and it developed a guidance document [14,15] to shed light on which knowledge should be available in each phase. This approach helped the DFPP organisation to determine the stage various innovations had reached, but it still did not explain why these innovations did not become mainstream.

To identify why the development of innovations stopped, the DFPP organisation used the critical success factors framework [15] as derived from [16,17]. This framework (Figure 3) identifies six critical success factors:

1. *People involved*: is sufficient knowledge and expertise available for the project team, and does the involved people have the required competences?
2. *Instruments*: are the knowledge and tools available to design the flood defences in its surroundings?
3. *Organisation*: is there political willingness to take risks, and what boundary conditions have been given to the project team? The officials of the regional water authorities are closely involved in the DFPP, through recurring meetings and their role as ambassadors of innovations, which helps in the formulation of the instructions to, and steering of, the project teams.
4. *Governance*: how is decision-making organised, and how are organisations such as contractors and local stakeholders involved?
5. *Legislation*: how can the project comply with formal and informal regulations?
6. *Discourse*: is there sufficient support in society for strengthening the flood defences?

The three success factors on the lower row of Figure 3 are preconditions located outside the boundaries of influence of dike upgrade projects. The remaining three success factors can be controlled by either the regional water authorities or the project team members. This framework helps the DFPP to formulate activities to enhance, and possibly steer, the development of knowledge and innovations.

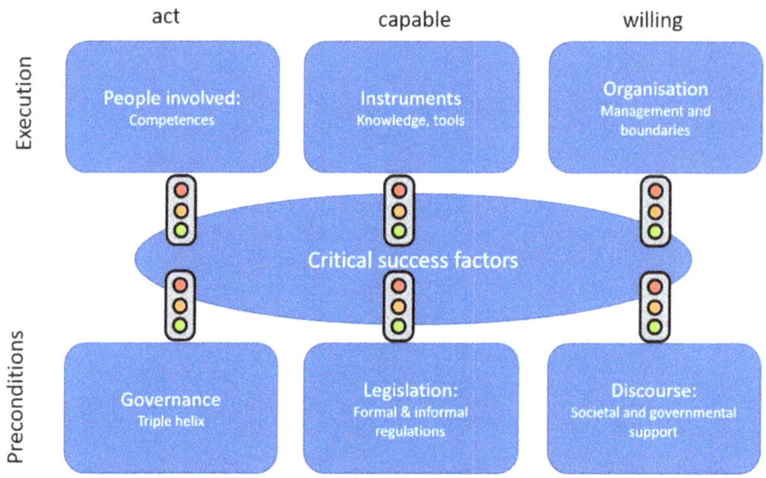

Figure 3. The six critical success factors adopted from [15].

In 2013, the aspects of adaptive learning and reduction of development time, as described above, were used to define the scope of seven technical research and test projects. Most DFPP projects focus on redesigning the dikes to address the two most common factors associated with dike failure: slope instability and seepage/piping. Prevention measures are typically expensive and/or require a lot of space. The DFPP, therefore, conducted an assessment of opportunities [2,18] to see which innovations helped to attain the programme objectives. It asked research institutes, regional water authorities and the private sector to propose potential innovations. The innovations were ranked in terms of both the relative performance of projects and the return investment for the DFPP. In particular, the innovations that scored well in terms of the return investment for the DFPP were considered and incorporated into the scope of these seven projects. Some projects, namely Piping, Slope Instability and Wadden Sea, focused on the first level of the Pyramid. Others, such as Forelands, Central Holland and Vechtstromen, focused more on the second level (managing flood risks). The projects not only covered two different levels of the pyramid, they also looked at different critical success factors. For example, Piping focused on instruments and people; Forelands on legislation and organisation and Slope instability more on procurement strategy, governance and instruments.

Several years after the start of the technical research and test projects, the projects were analysed from a sender–receiver perspective using the FODIKI methodology. The analysis, as shown later in this paper, led to greater awareness of the importance of knowledge uptake and continuous learning.

2.2. Methods

Our aim was not to evaluate the development of the knowledge and innovation strategy but to ascertain whether the generated knowledge is being applied within the dike redesign projects and whether this leads to different end-user action perspectives and, therefore, helps to relay (i.e., communicate) the objectives. To acquire a better understanding of how knowledge transfer and uptake takes place in the design processes of flood defences, the first author [6] developed and validated a conceptual framework called the Framework for Observing, Diagnosing and Intervening in Knowledge Interaction moments (acronym: FODIKI), as depicted in Figure 4. We assume, as inspired by [19], that knowledge transfer and subsequent uptake entail a knowledge supplier and a knowledge user (from sender S to receiver R). We also retained the possibility of failure (semantic distortion due to cognitive barriers). If a transfer interaction succeeds, Knowledge (K) is available to R, meaning that R can choose to use it. By the 'uptake' of K, we mean knowledge utilisation, as defined by [20] on a seven-stage cumulative scale, which ranges from reception via cognition, reference, effort, adoption and implementation to impact.

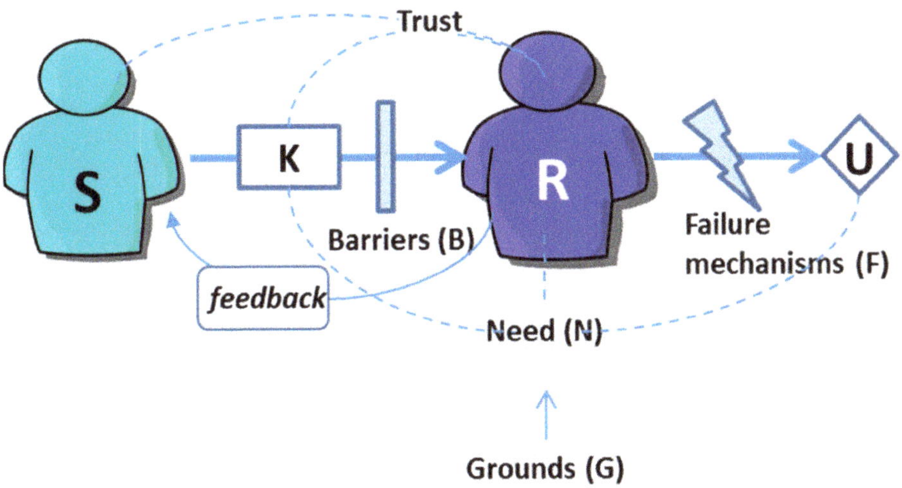

Figure 4. A sender–receiver framework for knowledge transfer and uptake [6].

We distinguish three types of social mechanisms that can explain the success of knowledge transfer and uptake: preconditions, barriers and failure mechanisms. Preconditions for communication and reception include a need for knowledge (for which R may have different grounds) and trust (R trusts that S is competent and acting in R's interest; S trusts that R will make good use of K). Barriers hindering knowledge transfer may be transmission (physical barriers hampering communication or poor communication skills), cognitive (R not properly constructing K) and psychological (K conflicting with existing practices or values of R). When barriers of this kind do not arise or can be overcome, knowledge uptake can still fail due to incorrect use (K is used in different ways than intended by S), diffidence (K is disqualified by a third party, dissuading R from uptake) or a lack of relay (R attempts to transfer K to end users but fails). We have also re-introduced the concept of feedback (fb), as both S and R are aware of their roles and the knowledge they have and lack. Through feedback, R is able to communicate on a meta-level whether he understands the shared knowledge and whether a barrier or failure mechanism occurs, after which, S is able to adapt.

We tested the internal validity of the FODIKI methodology by applying it to a historical case study, demonstrating that it allowed for the categorisation and generalisation of observations about the interaction moments of knowledge transfer in a dike design process and an assessment of the actual transfer and uptake of knowledge [7]. To determine external validity, we field tested the FODIKI methodology in "live" cases: (1) a participatory dike redesign process led by a Dutch regional water authority and (2) the knowledge and innovation strategy and related projects in the Dutch Flood Protection Programme. We followed an action research approach for both types of validity.

Action research is characterised by "the active and deliberate self-involvement of the researcher in the context of her investigation" [21]. Researchers and practitioners jointly act in a particular cycle of activities, including problem diagnosis, action intervention and reflective learning [22,23]. Action research typically follows an iterative approach, as depicted in Figure 5, where each iteration cycles through these four steps: plan, act and observe, reflect and then re-plan.

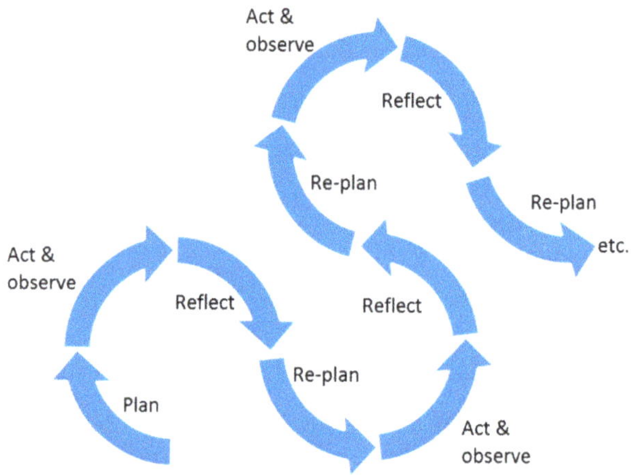

Figure 5. Consecutive action research cycles, based on [24].

Here, we focus on knowledge transfer and uptake in the DFPP knowledge and innovation strategy between the developers of the knowledge in the technical research and test projects and the intended end users, in other words, the dike redesign project teams.

2.3. Data Collection

We used two different datasets for our study:

(1) Knowledge and Innovation Strategy

Seven technical research and test projects, all led by one of the regional water authorities, began in the period between 2012 and 2018. An initial evaluation was made in 2016 of how four of these projects shared their knowledge, result in reflection and re-planning of the knowledge transfer and uptake. The evaluated projects were Piping, Wadden Sea, Central Holland and Slope stability. The state of knowledge creation, transfer and uptake in four projects was analysed by means of a document analysis and in-depth, semi-structured interviews with the professionals involved. For each project, we interviewed four professionals and shared our findings in a concluding workshop to share the insights and jointly draw up recommendations [25]. This was the first action research cycle.

These findings led to a second action research cycle, with a specific focus on the amendments for the knowledge and innovation strategy. The new knowledge and innovation strategy was shaped in a series of 10 meetings with stakeholders from the Ministry, regional water authorities, research institutes and representatives from the contractors and engineering companies. These meetings were well documented and actively shared within the alliance to ensure that feedback and input was given throughout the process. Not only did representatives from the administrative level attend, the officials of the Ministry and the regional water authorities were also informed and challenged to be part of this new strategy in a series of five meetings [26].

The third action research cycle comprised six meetings to develop the current knowledge and innovation strategy. These meetings were documented [6,27,28], and using the FODIKI methodology, the observations were diagnosed and suggestions for interventions were given. Representatives from the regional water authorities, contractors, engineering companies and research institutes were consulted about the different elements of the new knowledge and innovation strategy. Project team members of the seven technical research and test projects reflected on the new elements to pinpoint where further improvement was feasible.

(2) Monitoring data

Beginning in 2018, the DFPP organised annual knowledge and innovation monitoring, in which they reported how much is spent on knowledge and innovation and what the outputs, outcomes and impact are. Monitoring took the form of a survey. In the first years, only the knowledge and innovation projects fill in a questionnaire about their progress, how they share their knowledge and what the outputs and outcomes are. Questions were asked how the created knowledge was shared and if reinforcement projects also used that knowledge. The potential savings of the reinforcement projects were requested, both at the project- and programme-scale. In 2021, the survey was sent to stakeholders from research institutes, regional water authorities and the Ministry but also to contractors and engineering companies. One objective of the survey was to find out whether certain instruments were known. The survey included questions such as:

- Which knowledge and innovations do you know or have you heard of?
- Which innovations are being considered in your project?
- Which innovations are being developed within your project?
- Which new knowledge and innovations are/will be applied in your project?
- What are the main reasons for considering or applying all the knowledge and innovations you mentioned in your project?

In the multiple-choice answers, the six critical success factors could be recognized, without being explicitly mentioned. Furthermore, the respondents were asked whether the application of new knowledge/innovations would lead to savings in time and budget for their project. Finally, the respondents were asked which new knowledge and/or innovations could make the greatest contribution towards achieving the overall programme goals.

The results were used for the action research cycle to define subsequent steps. The results from 2018 [29] and 2019 [30] were used in the evaluation of the strategy in 2020, and so there was no monitoring in 2020. The most recent survey from 2021 [31] was used to pinpoint how to optimise the instruments available at the DFPP to share and absorb knowledge.

3. Results

3.1. Knowledge and Innovation Strategy

In this section, we will describe, using the FODIKI methodology, the three action research cycle loops that we have observed, diagnosed and adapted.

1. Action research cycle: Analysis of four technical research and test projects

Three years after the start of the projects, we conducted our in-depth analysis [6,25], first, by analysing the documents and, then, by conducting in-depth semi-structured interviews with team members and intended end users. Our observations showed that the teams produced different types of knowledge, such as methodological and process knowledge. The knowledge was, in some cases, also developed for, and used in, dike redesign projects. Knowledge uptake in those cases was high. After our first observations and diagnosis, we shared the preliminary findings with the stakeholders involved to contribute to the learning-by-doing approach.

Our diagnosis revealed that the project teams were creating new knowledge but did not see themselves as senders of that knowledge. Some professionals believed that 'the knowledge will sell itself'. However, since the end users had not been identified, it was difficult to let the knowledge flow. The project teams did not involve the end users, which meant that the intended receivers did not automatically trust the knowledge or that the developed knowledge met their needs.

The diagnosis also highlighted the importance of *boundary spanners*, both for the projects and the overall programme. Boundary spanners are [32] "people who proactively scan the organisational environment, employ activities to cross organisational or institutional boundaries, generate and mediate the information flow, coordinate between their 'home' organisation or organisational unit and its environment and connect processes and actors across these boundaries". The boundary spanners play an active role in sense-making and identifying the knowledge needs of the current and future receivers in the DFPP.

Furthermore, we found that in projects where co-creation took place with the dike redesign projects, the knowledge was appropriate to the knowledge needs and was easily absorbed in the project, leading to higher levels of knowledge uptake. In addition, these projects actively shared their insights with interested colleagues, playing their boundary spanning role.

We proposed several interventions to enhance knowledge transfer and uptake, with the most striking being the development of *knowledge strategies* for each project to identify the intended receivers and their potential needs and their knowledge networks. Another intervention was that the project team members also fulfilled an ambassador's role with respect to their own mother organisations in order to share the developed knowledge.

2. Action research cycle: Innovation Next Level

Despite all the efforts made, knowledge uptake from these technical research and test projects remained relatively low. The DFPP organisation undertook action and initiated a new action research cycle, in which the researchers helped to shape the successive steps. In this cycle, 15 meetings were organised, of which five were exclusively for the officials of the Ministry and the regional water authorities. Each meeting was thoroughly prepared using the FODIKI methodology, and each successive meeting used the diagnosis and proposed interventions of the previous meetings [27]. In the first 10 meetings, representatives from regional water authorities, the Ministry, engineering companies, contractors and research institutes were invited to share their experiences and suggestions for improvement. The DFPP organisation was able to identify the critical success factors and needs and to build trust to move the knowledge and innovation strategy towards the 'Innovation Next Level'. The DFPP organisation identified a number of shortcomings resulting in a failure to absorb knowledge. They included the absence of a 'risk safety net' in response to the fact that regional water authorities said that they were hesitant to use innovations when the risk of the innovation not working was not covered by the subsidy arrangements. The DFPP had foreseen this risk and already included this risk in the subsidy, meaning that, if an innovation failed to work, the regional water authority could call on the DFPP to cost the reinforcement of the defence.

In a series of five meetings with the officials of the Ministry and the regional water authorities, the officials came to the understanding that 'the baby should not be thrown out with the bathwater'. They agreed that more efforts had to be made to share the acquired knowledge with current and future projects in order to valorise its potential. This is linked to the 'no relay' failure mechanism in the FODIKI methodology. The intervention proposed was to draw more and more on the different DFPP communities of practices, such as the Technical Manager community and Project managers community, in order to share the developed knowledge and to update the training programme. The officials also foresaw a role for themselves as ambassadors for the developed knowledge.

3. Action research cycle: Towards a new Knowledge and Innovation Strategy

In a third iterative action research cycle, the DFPP organised six consecutive meetings to develop a new knowledge and innovation strategy that focused on transparency, effectiveness and a stronger coupling of the research project and dike redesign projects [27,28]. It built on the insights acquired during the second action research cycle. After each meeting the team reflected and made an analysis based on their observations and diagnosis of how the strategy could be adapted to the needs of the end users. This led to interventions in the strategy (i.e., adaptations), which were discussed with the same end users but also with a new range of end users, slowly converging to a widely accepted approach and resulting in adaptive ongoing development, as depicted in Figure 6. A more detailed description of the new strategy can be found below the figure.

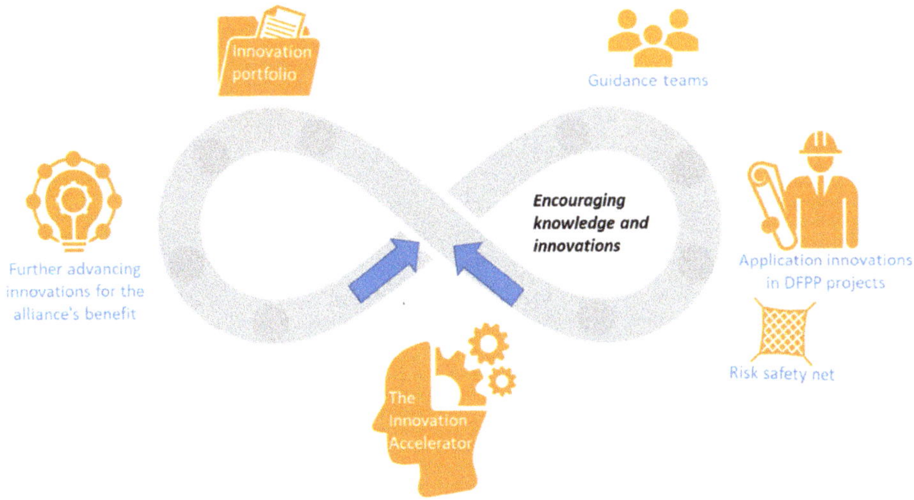

Figure 6. Adaptive ongoing development of knowledge and innovations and encouragement of the use of knowledge and innovation in the Dutch Flood Protection Programme.

In the new knowledge and innovation strategy [9], the DFPP highlights three relevant themes that are linked to the Pyramid of Flood Risk Management. The first theme (*technology)* is linked to the first and second layer of the Pyramid of Flood Risk Management, and it focuses on furthering understanding of how flood defences fail and linking that understanding to the new flood protection standards based on the probability of flooding. These innovations aim to limit the scope of the overall programme and to generate new dike reinforcement techniques [9]. The *integrality* theme (which is linked to the third layer of the Pyramid) tackles broader topics relating to, for example, sustainability, integration and the project approach. The last theme looks further ahead and targets the preparation of the programme for future developments: dealing with increased climate change and new societal challenges. These three themes should help the DFPP to remain adaptive and flexible in a context of change.

The DFPP organisation developed a programmatic approach, orchestrating activities so that the programme objectives would be met. The benefit of the knowledge and innovation developments is often not to be found at the individual project level, but rather at the overall programme level, since successive projects are needed to develop innovations and to reap the benefits in terms of costs, time and quality. To support this programmatic approach, the DFPP initiated (1) a structural alliance with several parties, including research institutes, engineering companies and contractors; (2) the 'risk safety net', referred to above, to ensure that risks are 100% covered by the alliance and (3) an Ambassadors Group to establish support from society at large.

To safeguard an ongoing knowledge and innovation process, several instruments were developed (1) to further the continuous development of knowledge and innovations to meet the overall programme objectives, (2) to encourage the application of new techniques and measures and (3) to share developed knowledge and experiences.

1. Furthering innovations

An R&D budget of €10 million a year is available (2.5% of the total annual budget of DFPP) to encourage the development of innovations. Innovation projects are 100% subsidised. In addition, the risk safety net referred to above ensures that 100% of the risks of innovations are borne by the alliance. Anyone from the regional water authorities, contractors, engineering companies or research institutes can submit ideas to the DFPP organisation to benefit from the R&D budget. For an idea to be included in the innovation portfolio, a regional water authority has to be involved, as only they can apply for funding. The DFPP organisation orchestrates the involvement of one or more regional water authorities on the development of ideas with potential at the programme level. When assessing the potential of ideas, the DFPP organisation focuses on the following criteria:

- Return on investment at the project and programme levels (balance of the costs and benefits between the project and programme);
- Reproduction factor (level of development in relation to applications in the programme);
- Contribution to alliance (builds on previous projects and seeks cooperation with other water authorities and other knowledge developers).

The involvement of the regional water authorities also safeguards ownership and practical application, as innovation is not a goal in itself, but a way of meeting the overall programme objectives. After the approval of an idea for the innovation portfolio, the innovation projects are carried out, often in conjunction with dike redesign projects, as this stimulates the knowledge uptake.

2. Encouraging the application of new knowledge and innovations

The DFPP has developed a *comply or explain* instrument, which consists of an overview of the new knowledge available and the development level of innovations, as well as a framework for considering application in projects. The overview contains a list of knowledge and innovation developments with a potential impact on the programme. When the potential impact has been validated and the applicability of the innovation has been described, projects are asked to contribute to the further development of the innovation. When the application of innovations and knowledge has been demonstrated in practice, projects are asked to take these into account as accepted alternatives. The overview of available knowledge and innovations is managed by the Innovation Accelerator, which consists of professionals from regional water authorities and research institutes. In addition to maintaining an overall perspective, the Innovation Accelerator helps dike redesign projects with the application of accepted innovations, and it translates project results into, and implements, generically applicable tools.

In each phase of the project, the dike redesign project teams and the advisory team examine which knowledge and innovations can be applied or further developed in the reinforcement project. First, the project team selects promising new knowledge and innovations and determines the contribution of the new knowledge and innovations to the project and programme objectives. This is a cyclical process, which means that it is repeated again and again in every project phase. The project team also makes agreements with the advisory team about how risks can be addressed in the reinforcement project but also about how the developed knowledge can be shared widely within the alliance to encourage large-scale application.

After the completion of a dike redesign project, risks may still arise. For example, the costs in the management and maintenance phase may be significantly higher than expected, or the implemented technique may not achieve the intended safety level, and so the primary flood defence will have to be upgraded again with a different technique. The risk safety net referred to above can be used in cases like these.

3. Knowledge transfer and uptake

For the innovations to be effective, the knowledge obtained in the innovation projects must flow to ongoing and upcoming dike redesign projects. Successful knowledge transfer and uptake means that the knowledge becomes common property within the alliance. Different resources are at the disposal of the DFPP to enhance this process. For example, the DFPP organisation increasingly uses boundary spanners to help them and the projects. The lessons learned from the initial period [6,25] helped to further define the role of the boundary spanners, who can act as brokers, translators or synthesisers:

1. The broker role matches the supply and demand of knowledge. This requires a state-of-the-art overview of existing knowledge and the current knowledge gaps.
2. The translator role interprets the knowledge needs of the end users and formulates questions for the knowledge developers. The translator is also capable of translating the knowledge to match the needs of the end users.
3. The synthesiser is capable of synchronising knowledge supply and demand. This requires a broad understanding of knowledge disciplines and sources and of how they can contribute to problem-solving.

Examples of boundary spanners fulfilling all three roles between innovation and strengthening projects are both the members of the Innovation Accelerator and the advisory teams of the DFPP organisation. To further enhance the knowledge transfer and uptake of the gained knowledge in the innovation portfolio, the DFPP organisation started to monitor the innovation projects in the Innovation portfolio in 2018.

3.2. Monitoring Knowledge and Innovation

Before turning to the monitoring of knowledge and innovations, we will start by giving an overview of the knowledge and innovation projects that took place since 2012 in the DFPP. Please note that the focus slowly shifted over the years to other aspects as society also changed over time. The focus of the first DFPP knowledge and innovation projects (Piping, Central Holland) was on gaining a better understanding of the implications of the flood protection standards [2]. A clearer understanding of the behaviour of flood defences was necessary for an effective approach to reinforcement. The slope instability project developed a step-by-step plan for the detailed modelling of the strength of a flood defence. This involved using specific soil investigation and monitoring and advanced calculation techniques, including probabilistic failure probability analyses and failure probability updating, to estimate the actual current strength for the slope-instability failure mechanism as closely as possible [33–37]. Applying the step-by-step plan reduced the need for reinforcement on many projects, and at some locations, reinforcement proved to no longer be necessary. Various studies have also been carried out on piping [38,39] and the deterioration of revetments [40,41] (see also Figure 7b), making safety factors explicit, so that the need for reinforcement can be determined more accurately. Furthering the knowledge uptake will limit the scope of the programme

(a) (b)

Figure 7. Examples of innovations in the Dutch Flood Protection Programme. (**a**) Implementation of the Coarse Sand Barrier in the Waalban dike in Gameren (photo credit E. Tromp); (**b**) testing in the Delta Flume at Deltares for new type of dike revetment in the northern part of the Netherlands (photo credit: Deltares).

In addition, enhancing knowledge uptake paves the way to optimise product innovations. Traditionally, for instance, resistance to piping is reduced by constructing verges or, when there is not enough space, with sheet piling. Various alternative measures that are more economical and sustainable have been developed in the DFPP, e.g., the vertically inserted geotextile. These experiences led to a modified technique, namely the coarse sand barrier [42,43]. This involves introducing a vertical trench with coarse sand at the transition between the clay layer and the sand substrate that replaces the original finer sand. The coarse sand acts as a filter that lets the water through and keeps the sand grains in place, preventing piping. After a successful series of tests, the Rivierenland Regional Water Authority applied a coarse sand barrier for the first time in the summer of 2021 in the Waalban dike in Gameren (Figure 7a). A better understanding of the behaviour of the defences will thereby lead to a further downscaling of the measures to be taken and, therefore, to the optimisation of the budget, the planning and the impact.

From 2016 onwards, the innovation portfolio increasingly included innovations that help to improve the quality of the living environment (slowly moving up the Pyramid of Flood Risk Management). In the Vechtstromen project [44], a system approach was developed that allows measures that retain water upstream to be combined with dike reinforcements downstream. The Forelands project [45] studied how forelands could be used to reduce flood risks. In this project, they not only examined 'physical' measures, but also looked at the opportunities for, and impediments to, legislation and regulations. These two projects led the DFPP organisation to realise that non-technical themes are equally important, resulting in new types of projects: projects with (1) an emphasis on learning and cooperation, such as organising asset management, and (2) reinforcing dikes in areas covered by legislation on nature and biodiversity. In 2020, a roadmap was drawn up to examine a range of sustainability issues in conjunction with each other in the years ahead [46]. The roadmap also opened up the road to the further investigation and use of the potential of nature-based solutions [47–49], as society calls for.

To determine whether the current innovation portfolio will allow the DFPP to meet the overall programme objectives (more economical and faster upgrades), the DFPP organisation engages in monitoring activities. The aim is to evaluate the knowledge and innovation approach and the annual expenditure on, and commitment to, knowledge and innovation in the DFPP programme. In addition, the DFPP organisation wants to determine what is going well, the trends in place and where improvement is possible. Monitoring took place in 2018, 2019 and 2021.

Monitoring has taken the form of a survey since 2018. In the initial years, innovation projects and dike redesign projects with innovations were asked to complete the survey. In 2021, the survey was sent to stakeholders from governmental organisations, research institutes and contractors and engineering companies. This change in the target group led to a higher response rate and a doubling of the number of respondents from dike redesign projects [31].

In 2018, the respondents stated that working across organisations (56%) and putting innovation at the centre (48%) helps to achieve the goals of the DFPP [29]. In 2019, it was concluded that the development of innovations was accelerating [30]. The first successes work like a flywheel: projects and organisations feel more and more responsible for using and further developing the new knowledge from the innovation projects. The mindset is changing: the ever-smarter implementation of dike redesign projects has now become part of the regular activities. The assessment of opportunities is currently a standard component of the dike redesign process by the following regional water authorities: Rivierenland, Aa en Maas, Vallei en Veluwe, Zuiderzeeland and Stichtse Rijnlanden. Moreover, knowledge is being actively disclosed both in- and outside the alliance through reports, articles, contributions to training courses and presentations in the DFPP communities of practice, consisting of representatives from the regional water authorities, engineering companies and research institutes.

On the basis of the monitoring in 2021, it was concluded that innovations are becoming more widely known and being applied more often: approximately 95% of the respondents were familiar with one or more innovations from the Innovation portfolio. On average, current redesign projects are considering nine innovations, developing three and applying four. In 2018 and 2019, innovations were considered and applied less often, and the emphasis was on contributing to the development of knowledge and innovations. In 2021, fewer barriers were identified: if innovations are not applied, it is often because traditional solutions suffice or because time and capacity are lacking.

It was concluded from the 2021 monitoring that innovation continues to have a positive image and is yielding benefits. Over 50% of the respondents in 2021 reported that the knowledge and innovation projects led to a reduction of the amount of upgrade work needed, better integration, more support and savings and an impact on emissions and circularity. Still, innovation takes time: many innovations are known and considered but not yet applied by everyone. This shows that there is potential for the future, but it also demonstrates the importance of the *comply or explain* instrument. However, half of the respondents are unaware that *comply or explain* and the Innovation Accelerator contribute to innovation. Furthering these instruments is required, especially as other instruments, such as 100% subsidies, the use of advisory teams and the risk safety net, are already seen by some 70% as supportive.

The investment in knowledge development and innovations is yielding results. The total investment in the innovation portfolio was €139 million for 2014 to 2021, inclusive. The direct savings reported for 2020 and 2021 amounted to over €200 million. These are in addition to the savings reported in 2019 for the period of 2014 to 2019 of €170 million, and they bring the total of quantitative savings reported to approximately €370 million. However, not all projects report the benefits of innovations for reasons such as not being able to unambiguously attribute the savings to innovation, uncertainty about whether the savings will be realised or the difficulty of quantifying the savings (which are often avoidable costs). Given the actual and planned expenditure of €4.3 billion until 2030, a saving of 20% is expected from the currently known innovations [31]. Continuous emphasis on knowledge transfer and uptake is essential to monetise it.

4. Discussion and Conclusions

In the decades to come, the Dutch regional water authorities and the Ministry face a challenging task to reinforce about 1500 km of dikes and 500 civil-engineering structures. Especially given the recent insights [50] that climate change is accelerating at a higher rate than thought, low-lying populated areas are exposed to an even higher risk than initially believed [51]. The Netherlands, like many other countries, is facing uncertain challenges relating to housing, biodiversity and the energy and agricultural transitions. These challenges, which include climate change, will undoubtedly have an impact on the implementation of the Dutch Flood Protection programme (DFPP). Society's norms and values will also change over the next three decades. The learning-by-doing knowledge and innovation strategy adopted by the DFPP will allow it to be adaptive and flexible. As knowledge is situated and socially constructed, this knowledge must be actively shared and restated after each change in the group of participants. At the same time, a constant awareness of the impact of change on the programme's implementation will be required, as well as additional knowledge. This could also affect how knowledge is shared and absorbed in the future.

Literature shows that various forms of participatory water management [52] have been developed to better handle the wide variety of problem perceptions, values and knowledge. Their effectiveness has been researched from different perspectives, notably social interaction [53,54] and uncertainty [55]. In this paper, we used a knowledge management perspective. Our observations showed that the FODIKI framework helps in detecting and diagnosing social mechanisms in a variety of settings, thus encouraging knowledge uptake within the DFPP, specifically in the knowledge and innovation strategy.

If we interpret our observations on knowledge development, transfer and uptake in terms of team learning and organisational learning [56], we see a lot of substantive team learning and boundary spanning. In recent years, the role of the DFPP as a boundary organisation became apparent, given its links with the regional water authorities and their projects and with policy and practice. Working practices at the DFPP have changed, and knowledge management is being developed further. The role of boundary spanners is being taken more seriously to ensure that the institutional knowledge and information about precedents in dike redesign projects are kept within the DFPP organisation in order to achieve the programme goals. The role of the advisory teams as boundary spanners is still in its infancy: they are still learning on the job, and knowledge sharing between advisory teams is limited. Ongoing attention is needed to further encourage this sharing and help the teams grow into their role. At the same time, the DFPP organisation is encouraging interorganisational learning between the projects and regional water authorities involved. The question arises of how knowledge uptake can be secured for the long term. Staff turnover in a knowledge-intensive process often has a demoralising effect on others. Given the fact that most regional water authorities will work on projects for the next three decades, the question arises of how is the current and future available knowledge being secured for this task, both at the regional water authorities and at the DFPP organisation? This is because like the DFPP organisation, a specific department at a regional water authority carries out the dike redesign projects and also acts as a boundary organisation. This means that regional water authorities need to devote attention to encouraging both organisational and interorganisational learning. The different organisations therefore need to maintain a focus on the role of boundary spanners. Efforts must be made to ensure that the parties involved have the knowledge-uptake capacity that is required. However, earlier research [7] has shown that sharing knowledge between several projects at one regional water authority is already difficult.

In this paper, we have described how the Dutch Flood Protection Programme has adopted a Pyramid of Flood Risk Management, a stage-gate system and a framework for critical success factors. Our study has shown that, given the current time frame, the pyramid of flood risk management and the critical success factors framework are helpful in terms of the observations, diagnosis and interventions for the strategy. However, these concepts have been loosely adopted with no methodological rigour. Further research on the validity and reliability of these concepts is required to use them for the long term.

To conclude, the new knowledge and innovation strategy appears to be working given the increased levels of knowledge uptake. Nevertheless, the DFPP organisation has been monitoring knowledge uptake for only three years, and it has made interventions based on the results. This period is relatively short as a basis for concluding that the current knowledge and innovation strategy is working and that knowledge uptake is sufficient to meet the programme goals, particularly because the pilot paradox [57,58] is still hanging over the DFPP: previous research showed that a successful pilot project (which delivers useful knowledge and results in more knowledgeable participants) is anything but a recipe for successful knowledge uptake and upscaling. To bridge this gap, the DFPP needs to involve future users in early phases of the project, and it should start communicating some of the successes and obstacles to a wider public early in the innovation projects.

Author Contributions: Conceptualization, E.T. and H.K.; methodology, E.T. and H.K.; validation, E.T., H.K. and A.t.N.; resources, E.T., H.K. and A.t.N.; data curation, E.T., H.K. and A.t.N.; writing—original draft preparation, E.T. and H.K.; writing—review and editing, H.K. and A.t.N.; visualization, E.T.; supervision, A.t.N. All authors have read and agreed to the published version of the manuscript.

Funding: This research received no external funding.

Institutional Review Board Statement: Not applicable.

Informed Consent Statement: Not applicable.

Data Availability Statement: Not applicable.

Acknowledgments: The authors would like to thank the reviewers and editors for their critical and targeted comments.

Conflicts of Interest: The authors declare no conflict of interest.

References

1. Most, H.; van der Tánczos, I.; Bruijn, K.M.; de Wagenaar, D. New Risk-based standards for flood protection in the Netherlands. In Proceedings of the 6th International Conference on Flood Management, Sao Paulo, Brazil, 16–18 September 2014.
2. Jorissen, R.; Kraaij, E.; Tromp, E. Dutch flood protection policy and measures based on risk assessment. In Proceedings of the 3rd European Conference on Flood Risk Management (FLOODrisk 2016), Lyon, France, 17–21 October 2016. [CrossRef]
3. DFPP. *1st Knowledge and Innovation Strategy (Kennis en Innovatiestrategie HWBP)*; Report; DFPP: Copenhagen, The Netherlands, December 2012. (In Dutch)
4. Ackoff, R.L. The future of operational research is past. *J. Oper. Res. Soc.* **1979**, *30*, 93–104. [CrossRef]
5. Rittel, H.W.; Webber, M.M. Dilemmas in a general theory of planning. *Policy Sci.* **1973**, *4*, 155–169. [CrossRef]
6. Tromp, E. Enhancing Knowledge Transfer and Uptake in the Design Processes of Flood Defences. Ph.D. Thesis, Delft University of Technology, Delft, The Netherlands, 11 November 2019.
7. Bloemen, P.J.T.M.; Hammer, F.; van der Vlist, M.J.; Grinwis, P.; van Alphen, J. DMDU into Practice: Adaptive Delta Management in The Netherlands. In *Decision Making under Deep Uncertainty: From Theory to Practice*, 1st ed.; Marchau, V.A.W.J., Walker, W.E., Bloemen, P.T.J.M., Popper, S.W., Eds.; Springer: Cham, Switzerland, 2019; pp. 321–355.
8. Taskforce Ten Heuvelhof. *Een Dijk Van een Programma, Naar Een Nieuwe Aanpak Van Het Hoogwaterbeschermingsprogramma*; Report; Ministerie van Verkeer en Waterstaat: Copenhagen, The Netherlands, 16 December 2010. (In Dutch)
9. DFPP. *Knowledge and Innovation Agenda (Kennis en Innovatieagenda HWBP)*; Report; DFPP: Copenhagen, The Netherlands, November 2019. (In Dutch)
10. Van Buuren, M.W.; Edelenbos, J. Kennis en kunde voor participatie. In *Jaarboek Kennissamenleving 'Burger in Uitvoering'*; University Press: Amsterdam, The Netherlands, 2008; Chapter 10.
11. Van Herk, S. Delivering Integrated Flood Risk Management; Governance of Collaboration, Learning and Adaptation. Ph.D. Thesis, Delft University of Technology, Delft, The Netherlands, 14 February 2014.
12. Tidd, J.; Bissant, J.R. *Managing Innovations: Integrating Technological, Market and Organizational Change*; John Wiley & Sons: West Sussex, UK, 2013.
13. Cooper, R. Perspective: The Stage-Gate® Idea-to-Launch Process—Update, What's New, and NexGen Systems. *J. Prod. Innov. Manag.* **2008**, *25*, 213–232. [CrossRef]
14. Knoeff, J.G.; Van Der Meer, M.; Van Nieuwenhuizen, L.; Tromp, E.; Woning, M.; Leeuwdrent, W.; Bizzarri, A. *Guidance on Innovations in Flood Defences Green Version (Handreiking Innovaties Waterkeringen, Groene Versie)*; Hoogwaterbeschermingsprogramma Report; Rijkswaterstaat: Copenhagen, The Netherlands, May 2013. (In Dutch)
15. Tromp, E.; Knoeff, J.G.; Tanis, H.; Heijn, K.; Leung, N. *Guidance on Innovations in Flood Defences (Handreiking Innovaties Waterkeringen)*; Hoogwaterbeschermingsprogramma Report; Rijkswaterstaat: Copenhagen, The Netherlands, 1 November 2016. (In Dutch)
16. Shrivastava, P.; Mitroff, I.I.; Miller, D.; Miglani, A. Understanding industrial crises. *J. Manag. Stud.* **1988**, *25*, 283–303. [CrossRef]
17. Van Staveren, M. *Uncertainty and Ground Conditions, A Risk Management Approach*, 1st ed.; Butterworth-Heinemann Elsevier Ltd.: Oxford, UK, 2006.
18. DFPP. *Assessment of Opportunities (Kansenscan)*; Report; DFPP: Copenhagen, The Netherlands, 2013. (In Dutch)
19. Vlachos, E. A conceptual model of the knowledge transfer process. In Proceedings of the Second International Conference on Water Resources Knowledge, Fort Collins, CO, USA, 29 June 1977.
20. Knott, J.; Wildavsky, A. If dissemination is the solution, what is the problem? *Knowl. Creat. Diffus. Util.* **1980**, *4*, 537–578. [CrossRef]
21. McKay, J.; Marshall, P. The dual imperatives of action research. *Inf. Technol. People* **2001**, *14*, 46–59. [CrossRef]
22. Kemmis, S. Exploring the relevance of critical theory for action research: Emancipatory action research in the footsteps of Jurgen Habermas. In *Handbook of Action Research: Participative Inquiry and Practice*; Reason, P., Bradbury, H., Eds.; Sage: London, UK, 2001; pp. 91–102.
23. Carr, W.; Kemmis, S. *Becoming Critical: Education, Knowledge and Action Research*; RoutledgeFalmer: London, UK, 1986.
24. Kemmis, S.; Taggart, R. Participatory action research. Communicative action and the public sphere. In *Handbook of Qualitative Research*; Denzin, N.K., Lincoln, Y.S., Eds.; Sage: Thousand Oaks, CA, USA, 2005.
25. Duijn, M.; Vreugdenhil, H.; Janssen, S.; Tromp, E.; Ellen, G.J. Organising knowledge generation and dissemination in the Dutch high-water protection programme—A sender-receiver approach. *Knowl. Manag. Res. Pract.* **2021**. [CrossRef]
26. DFPP. *Innovation Next Level (Innovatie Next Level)*; Report; DFPP: Copenhagen, The Netherlands, 2016. (In Dutch)
27. DFPP. *Update Knowledge and Innovation Approach-Executive Level (Process)*; Report; DFPP: Copenhagen, The Netherlands, 2018. (In Dutch)
28. Twynstra Gudde. *Evaluatie POV's, Een Overzicht van Geleerde Lessen*; Report; DFPP: Copenhagen, The Netherlands, 17 December 2019. (In Dutch)

29. DFPP. *Dashboard v0.1 Monitoring of Innovations 2018 (Dashboard Monitoring van Innovaties 2018)*; Report; DFPP: Copenhagen, The Netherlands, January 2018. (In Dutch)
30. DFPP. *Dashboard v0.1 Monitoring of Innovations 2019, (Dashboard Monitoring van Innovaties 2019)*; Report; DFPP: Copenhagen, The Netherlands, 23 January 2019. (In Dutch)
31. Panteia. *Monitor HWBP Kennis en Innovatie—Report Panteia Zoetermeer*; DFPP: Copenhagen, The Netherlands, 4 February 2022. (In Dutch)
32. Van Meerkerk, I.; Edelenbos, J. *Boundary Spanners in Public Management and Governance: An Interdisciplinary Assessment*; Edward Elgar Publishing: Cheltenham, UK, 2018.
33. Konstadinou, M.; Zwanenburg, C. A critical review of membrane and filter paper correction formulas for the triaxial testing of soft soils. *Geotech. Test. J.* **2019**, *43*, 19. [CrossRef]
34. Taccari, M.L.; Zwanenburg, C. Large scale triaxial compression tests on three peat samples from Eemdijk, The Netherlands. In Proceedings of the XVII ECSMGE-2019, Reykjavik, Iceland, 1–6 September 2019; ISBN 978-9935-9436-1-3. [CrossRef]
35. Breedeveld, J.; Zwanenburg, C.; Van, M.; Langkeek, H.J. Impact of the Eemdijk full-scale test programme. In Proceedings of the XVII ECSMGE-2019, Reykjavik, Iceland, 1–6 September 2019. [CrossRef]
36. Van der Krogt, M.G.; Schweckendiek, T.; Kok, M. Do all dike instabilities cause flooding? In Proceedings of the 13th International Conference on Applications of Statistics and Probability in Civil Engineering (ICASP13), Seoul, Korea, 26–30 May 2019. [CrossRef]
37. Van der Krogt, M.G.; Klerk, W.J.; Kanning, W.; Schweckendiek, T.; Kok, M. Value of Information of combinations of proof loading and pore pressure monitoring for flood defences. *J. Struct. Infrastruct. Eng.* **2022**, *18*, 505–520. [CrossRef]
38. Van Beek, V.M.; Van Essen, H.M.; Vandenboer, K.; Bezuijen, A. Developments in modelling of backward erosion piping. *Géotechnique* **2015**, *65*, 740–754. [CrossRef]
39. Robbins, B.A.; Van Beek, V.M.; López-Soto, J.F.; Montalvo-Bartolomei, A.M.; Murphy, J. A novel laboratory test for backward erosion piping. *Int. J. Phys. Model. Geotech.* **2018**, *18*, 266–279. [CrossRef]
40. Van Steeg, P.; Klein Breteler, M.; Provoost, Y. Large-scale physical model tests to determine influence factor of roughness for wave run-up of channel shaped block revetments. In Proceedings of the 6th International Conference on the Application of Physical Modelling in Coastal and Port Engineering and Science (Coastlab 2016), Ottawa, ON, Canada, 10–13 May 2016.
41. Klein Breteler, M.; Provoost, Y.; Van Steeg, P.; Wolters, G.; Kaste, D.; Mourik, G. Stability Comparison of 9 Modern Placed Block Revetment types for Slope Protections. *J. Coast. Eng.* **2018**, *36*, 75. [CrossRef]
42. Rosenbrand, E.; Van Beek, V.M.; Bezuijen Akrami, S.; Terwindt, J.; Koelewijn, A.; Förster, U. Multi-scale experiments for a coarse sand barrier against backward erosion piping. *Géotechnique* **2022**, *72*, 216–226. [CrossRef]
43. Rosenbrand, E.; Van Beek, V.M.; Bezuijen, A. Numerical Modelling of the resistance of the coarse sand barrier against backward erosion piping. *Géotechnique* **2021**, 1–10, ahead to print. [CrossRef]
44. Waterschap Vechtstromen, De Vecht, een Grenzeloze, Halfnatuurlijke Rivier, Grensoverschrijdende Vechtvisie, Report (In Dutch Only). Available online: https://devecht.eu/publish/pages/29934/nl_vechtvisie_internet_1.pdf (accessed on 24 January 2022).
45. Hoogheemraadschap Schieland en de Krimpenerwaard, Guidance document Foreland (Handreiking Voorland), Report (Dutch Only). 2019. Available online: https://povvoorlanden.nl/wp-content/uploads/2019/03/Voorland_lowres_spreads_KL02.pdf (accessed on 24 January 2022).
46. DFPP. Programmatic Approach to Support Sustainability and Spatial Quality Dutch Flood Protection Programme, Report, Dutch Only. Available online: https://www.hwbp.nl/binaries/hoogwaterbeschermingsprogramma/documenten/rapporten/2020/03/24/programmatische-aanpak-duurzaamheid/Programmatische+aanpak+Duurzaamheid+en+Ruimtelijke+kwaliteit.pdf (accessed on 24 January 2022).
47. Faivre, N.; Fritz, M.; Freitas, T.; de Boissezon, B.; Vandewoestijne, S. Nature-Based Solutions in the EU: Innovating with nature to address social, economic and environmental challenges. *Environ. Res.* **2017**, *159*, 509–518. [CrossRef] [PubMed]
48. Bouw, M.; Van Eekelen, E. *Building with Nature—Creating, Implementing and Upscaling Nature-Based Solutions*, 1st ed.; NAi Publishers: Rotterdam, The Netherlands, 2020; ISBN 978-94-6208-582-4.
49. Vreugdenhil, H.; Frantzeskaki, N.; Taljaard, S.; Ker Rault, P.; Slinger, J. Next step in policy transitions: Diffusion of pilot projects. In Proceedings of the 13th Annual Conference of the International Research Society for Public Management (IRSPM XIII), Copenhagen, Denmark, 6–8 April 2009.
50. IPCC Climate Change 2021: The Physical Science Basis, the Working Group I contribution to the Sixth Assessment Report. 2021. Available online: https://www.ipcc.ch/report/ar6/wg1/ (accessed on 24 January 2022).
51. IPCC. The Ocean and Cryosphere in a Changing Climate. A Special Report of the Intergovernmental Panel on Climate Change. Intergovernmental Panel on Climate Change. 2019. Available online: https://www.ipcc.ch/srocc/chapter/summary-for-policymakers/ (accessed on 24 January 2022).
52. Korff, Y.; von Daniell, K.A.; Moellenkamp, S.; Bots, P.W.G.; Bijlsma, R.M. Implementing participatory water management: Recent advances in theory, practice, and evaluation. *Ecol. Soc.* **2012**, *17*, 30.
53. Pahl-Wostl, C.; Tàbara, D.; Bouwen, R.; Craps, M.; Dewulf, A.; Mostert, E.; Ridder, D.; Taillieu, T. Managing change towards adaptive water management through social learning. *Ecol. Soc.* **2007**, *12*, 30. [CrossRef]
54. Edelenbos, J.; van Buuren, A.; van Schie, N. Co-producing knowledge: Joint knowledge production between experts, bureaucrats and stakeholders In Dutch water management projects. *Environ. Sci. Policy* **2011**, *14*, 675–684. [CrossRef]
55. Koppenjan, J.F.M.; Klijn, E.H. *Managing Uncertainties in Networks*; Routledge: London, UK, 2004.

56. Easterby-Smith, M.; Lyles, M.A. (Eds.) *Handbook of Organizational Learning and Knowledge Management*; John Wiley & Sons: West Sussex, UK, 2011.
57. Vreugdenhil, H.; Rault, P.K. Pilot Projects for evidence-based policy-making: Three pilot projects in the Rhine Basin. *Ger. Policy Stud.* **2010**, *6*, 115–151.
58. Van Buuren, A.; Vreugdenhil, H.; Van Popering-Verkerk, J.; Ellen, G.J.; van Leeuwen, C.; Breman, B. The Pilot Paradox: Exploring tensions between internal and external success factors In Dutch climate adaptation projects. In *Innovating Climate Governance: Moving Beyond Experiments*; Cambridge University Press: Cambridge, UK, 2018; p. 145.

water

Article

Insights into Flood Wave Propagation in Natural Streams as Captured with Acoustic Profilers at an Index-Velocity Gaging Station

Marian Muste [1,*], Dongsu Kim [2,*] and Kyungdong Kim [2]

[1] IIHR-Hydroscience & Engineering, The University of Iowa, Iowa City, IA 52242, USA
[2] Civil & Environmental Engineering, Dankook University, 16890 Yongin, Gyeonggi, Korea; kyungdong-kim@dankook.ac.kr
* Correspondence: marian-muste@uiowa.edu (M.M.); dongsu-kim@dankook.ac.kr (D.K.)

Abstract: Recent advances in instruments are transforming our capabilities to better understand, monitor, and model river systems. The present paper illustrates such capabilities by providing new insights into unsteady flows captured with a Horizontal Acoustic Current Profiler (HADCP) integrated at an operational index-velocity gaging station. The illustrations demonstrate that the high-resolution stage and velocity measurements directly acquired during flood wave propagation reveal the intricate interplay among flow variables that are essential for better supporting judicious decision making for river management, flooding, sediment transport, and stream ecology. The paper confirms that the index-velocity method better captures the unsteady flow dynamics in comparison with the stage-discharge monitoring approach. At a time when the intensity and frequency of floods is continuously increasing, a better understanding of the critical features of flood waves during extreme events and the possibility of capturing more accurately their dynamics in real time is of special socio-economic significance.

Keywords: unsteady flows; flood wave propagation; Acoustic Doppler Current Profilers; index-velocity method; stage-discharge method; rating curves

Citation: Muste, M.; Kim, D.; Kim, K. Insights into Flood Wave Propagation in Natural Streams as Captured with Acoustic Profilers at an Index-Velocity Gaging Station. *Water* **2022**, *14*, 1380. https://doi.org/10.3390/w14091380

Academic Editors: Slobodan P. Simonovic, Subhankar Karmakar and Zhang Cheng

Received: 16 February 2022
Accepted: 17 April 2022
Published: 24 April 2022

1. Introduction

Monitoring and predicting streamflows are at the core of the decision making for critical areas of the socio-economic continuum (from water management, energy development, infrastructure design, and recreational uses to forecasting of floods, water quality, and ecosystem viability). While most often collected by specialized agencies, the streamflow data are also used as benchmarks for advancing the understanding of watershed dynamics and underpinning scientific studies on aquatic habitat and climate trends. The measurement protocols used for collecting streamflow data stem from century-long incremental developments [1]. Most of these developments have considered river flows as quasi-stationary processes fluctuating within an unchanging envelope of variability [2]. This view has given rise to empirical or semi-empirical relationships for streamflow estimation complemented by statistical analyses applied to long historical collected during steady and unsteady flows. While this estimation approach might be acceptable for reporting daily discharges for various uses, it is not satisfactory for practical and scientific investigations where the data are desired at sub-daily sampling rates [3]. Such an example is the propagation of flood waves generated by storms that are driven by gradual, time-varying flows with significant hourly variations.

During the gradually varied changes, the flow is affected by both unsteadiness and non-uniformity [4,5]. Depending on the slope of the stream bed and the magnitude and duration of the storm runoff entering the stream, the flood wave can take various forms, i.e., kinematic, diffusion, or full dynamic [6]. The dependencies among the flow variables during the wave propagation are distinct for the rising and falling stages and different from the steady flow relationships. This peculiar dependency is called hysteresis, which is a generic term denoting that the state of a system at any time depends on its past state (i.e., rising or falling). The hysteretic behavior is materialized as "loops" in the relationships between pairs of flow variables [7] and a separation of the flow hydrographs as discussed in [8]. Streams in low- and mild-sloped streams tend to produce considerable hysteresis for large flow changes occurring over short durations [9,10]. Streams on steep slopes display small hysteretic effects [11].

The most reliable method to truthfully capture hysteresis is the direct measurement of the discharge along with other relevant flow variables over the whole duration of the wave propagation. However, direct measurements require personnel deployment for extended time, especially in medium and large-size rivers. Moreover, stream measurements during extreme flows are challenging because of the difficulty accessing the sites and other safety concerns [12,13]. Consequently, streamflow monitoring in natural streams is typically made with indirect methods relying on pre-constructed rating curves used in conjunction with continuously measured variables that can be measured unattended. The most popular methods for real-time monitoring are the stage-discharge and index-velocity methods, which are labeled herein as HQRC and IVRC, respectively. These methods are well documented in multiple sources [12,14–17] therefore, only their salient features will be presented herein. The United States Geological Survey (USGS) publishes real-time discharge data for more than 10,000 sites across the US, with most of them relying upon HQRC method [3]. There are about 500 IVRC-based stations in the US, and their number continues to increase. At about 4000 of the total number of gages, streamflow forecasting is made via hydrologic modeling.

The HQRC streamflow estimation method was developed first, and it is still widely used to date for monitoring steady and unsteady flows. The simple HQRC is governed by equations that are strictly valid for steady flows whereby the energy and river-bed slopes are equal. However, during flood wave propagation, the energy slope is not equal to the bed slope being influenced by additional parameters; therefore, the estimates provided by steady HQRC are not capturing the actual flows [7]. Deviations as large as 9.8%, 15%, and 34% in discharges for the same stage were reported between actual and estimated HQRC flows in large rivers [15,18,19]. Differences up to 40% were found between Acoustic Doppler Current Profiler (ADCP) directly measured discharges and HQRC estimates in a medium-size, lowland river during unsteady flows [20]. A recent USGS evaluation conducted at 5420 HQRC stations found that 67% of them are moderately or strongly affected by hysteresis [3]. Corrections are applied to the simple HQRC for gages located in large rivers exposed to hysteresis [21]. For gages located on medium and small rivers, the correction for hysteresis is not applied, as there is a perception that its impact is small and cannot be discerned from instrument uncertainty [3].

The IVRC method is a better alternative to HQRC when monitoring unsteady flows because it measures directly the index velocity in addition to the stage [8]. This method has been increasingly implemented after the adoption of acoustic-Doppler velocimetry in the late 1990s [22], especially for locations where hysteresis and/or backwater might occur [16]. While the IVRC method can capture hysteresis in real time [23], the hydrometric community continues to test, evaluate, and improve IVRC capabilities (e.g., [24,25]). Due to the high cost of these evaluations and the methods' shorter life span compared with the century-old HQRC, the IVRC method has been less investigated so far.

The motivation for this paper is triggered by the fact that while both HQRC and IVRC have been continuously improved through the incorporation of new measurement technologies (acoustic, radar, image velocimetry), the principles and assumptions associated with the conventional monitoring protocols have not been revised to take full advantage of the superior capabilities of the new instruments. This paper addresses this gap by evaluating how the IVRC method equipped with new measurement technology compares with HQRC (used as reference herein) and revealing new features of the mechanisms in unsteady flows. Exploration of these features is rarely illustrated with data acquired in situ. The data-driven exploration on flood wave propagation presented in this paper is intentionally kept in a simple form to substantiate the hysteresis phenomenology and dependencies rather than cluttering the narration with analytical and statistical analysis.

The paper is organized as follows. First, it reviews the principles and practice of the conventional HQRC and IVRC along with their advantages and limitations. Aspects of flow dependencies during the propagation of flood waves are then illustrated with data acquired at an operational IVRC-based gaging station served by an emerging instrument: the Horizontal Acoustic Doppler Current Profiler (HADCP). Subtle features of the hysteretic behavior are then highlighted to illustrate the potential of the IVRC method to document flow features that might broaden the use of data for practical and scientific purposes. Finally, the paper suggests research for further evaluation of the hysteresis impact on monitoring methods and for developing protocols that can reduce their uncertainty.

2. Principles and Practice of Conventional Monitoring Methods

Conventional monitoring methods are based on semi-empirical relationships that relate in situ discharge measurements with flow variables directly acquired at monitoring site. These relationships (a.k.a., ratings) are valid only for the site where they were constructed. Rating construction assumes that the stream gauge sites are closely following the best practices for gage site selection (e.g., Rantz et al. [14]). Among these guidelines, critical ones include free of obstructions or bed irregularities in the channel reach enclosing the gaging station, quasi-uniform geometry in the streamwise direction, and preservation of the cross-sectional flow distribution for the whole range of flows at the station.

The independent flow variables used in conjunction with the ratings are those that can be continuously measured without operator assistance (i.e., stage and, more recently, velocity distributions along lines or surfaces into the flow). The guidelines for rating construction typically assumes a one-to one relationship among flow variables without specifically distinguishing between the rising and falling phases of the flood wave. The final ratings are established only after an extended dataset of calibration measurements are acquired, processed, and checked for consistency with statistical tools. In many situations, the overall quality of the final rating is highly contingent on the operator's skills and his/her specific knowledge of the measurement site.

The steady HQRCs are obtained through graphical constructions guided by analytical relationships, i.e., flow over weirs at low flows, Manning's equation at higher stages up to the bankful stage, and by other approaches for extreme flows [7,14]. This rating misses the hysteretic signature altogether, as the $Q = f\ (H)$ relationship is the same for the rising and falling limbs of the hydrograph (see dotted line in Figure 1a). The steady HQRC method leads to a maximum value for discharge hydrograph, Q_{max}, when H_{max} is attained. Notably, the discharge corresponding to the maximum cross-section velocity, Q_{Umax}, does not coincide with the position of the actual Q_{max} on the loop. Corrections [21,26,27] can be applied in post-processing to the steady HQRC estimated discharge to uncover the hysteretic loop associated with the passing of flood wave through a gauging site (continuous line in Figure 1a). Given their costs, these corrections are only applied at selected gages, especially for those located in flood-prone areas along large rivers [3].

The IVRC method has become increasingly popular after the adoption of HADCPs for river measurements [16]. The latter authors specify that the method has been developed for sites where "more than one specific discharge can be measured for a specific stage", which are situations specific for flood wave propagation in lowland streams [8]. The main steps involved in the construction and monitoring with the IVRC method are shown in Figure 1b. First, a stage-area (HARC) is constructed to relate the range of stages measured at the station with the corresponding areas for the wetted cross-section (continuous line for HARC in Figure 1b). A second rating is the IVRC, which is obtained with regression equations applied to extensive datasets similarly to the practice of establishing stage-discharge ratings (dotted line for IVRC in Figure 1b).

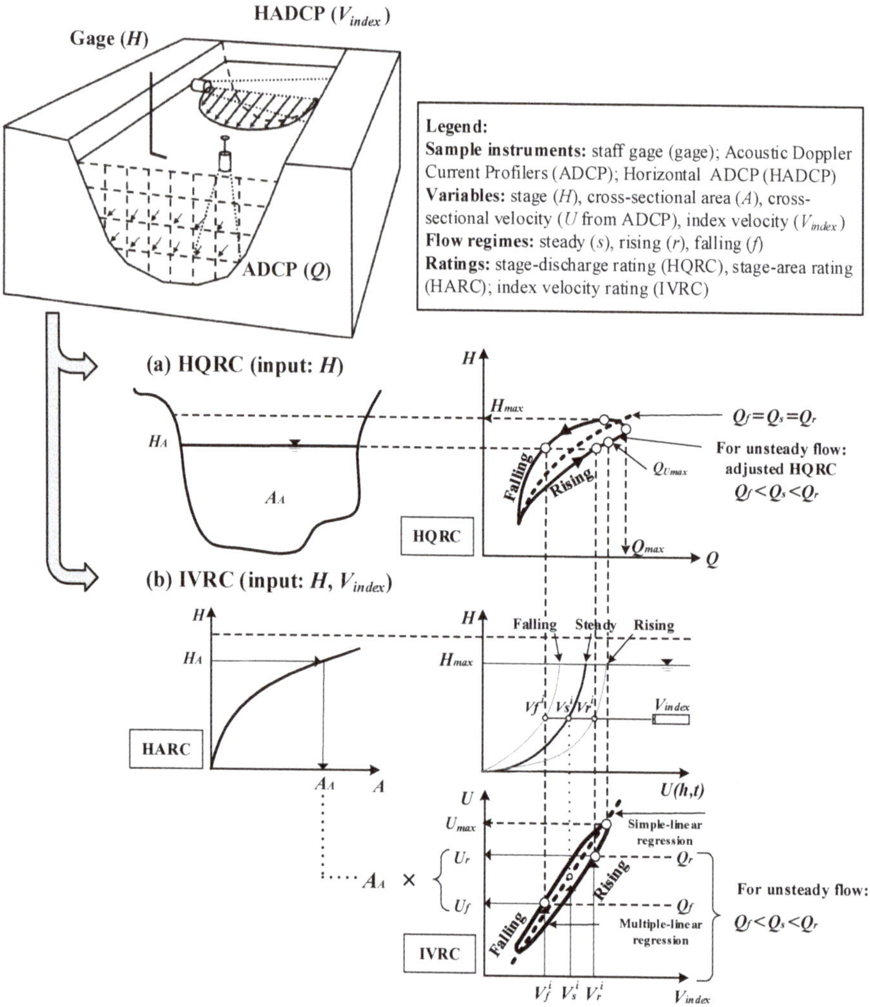

Figure 1. Outcomes of the conventional monitoring methods when measurements were conducted during unsteady flows (adapted from [8]): (**a**) stage-discharge method (HQRC) and (**b**) index-velocity method (IVRC).

The datasets used for constructing IVRC entail direct streamflow measurements (typically acquired with down-looking ADCPs transects) paired with real-time index-velocity measurements (acquired with HADCPs installed within the gaging station cross-section). The index velocity is measured in discrete "bins" along the HADCP instrument axis and subsequently averaged over time and over all bins contained along the acoustical path of the instrument [16]. Similarly to the HQRC protocols, periodic verifications of the validity of the ratings are made following the IVRC rating establishment. Many IVRC stations are equipped with the SonTek/YSI HADCP (i.e., Side-Looker or SL according to manufacturer specifications, San Diego, USA) that measures simultaneously stage and velocities with probes collocated in the same physical enclosure [28].

The index-velocity ratings are considered final when several statistical tests applied to the regression equations are satisfactory, i.e., the coefficient of determination (R^2), standard error, p-values, and residual analysis [16]. Two types of regression equations (i.e., simple-

or multi-linear) are recommended by Levesque and Oberg, as shown on the IVRC rating in Figure 1b. The final rating is a one-to-one relationship that is uniformly used in steady and unsteady flows. A notable skill of the IVRC rating is that it provides different discharges for the same flow depth during wave propagations regardless of the adopted approach for the regression equations, as the vertical streamwise velocity profile is larger on the rising phase than on the falling one (see Figure 1b). Non-unique IVRC ratings are suggested when the flow at a site displays distinct mechanisms during flow variation [29]. These authors developed multi-modal ratings for tidal flows, where they established separate equations for the flood to ebb, ebb to flood, and upper and lower transitions occurring during a full tidal cycle. The same procedure is recommended in guidelines [12]. We argue herein that if this rule would be adopted to flood wave propagation in streams, it will imply that the construction of the IVRC should entail distinct regression equations for the rising and falling phases of the stage hydrograph.

3. Index-Velocity Method Applied to an Actual Gaging Station

3.1. Study Site

The location of the index-velocity gaging site analyzed in this paper is shown in Figure 2a. The drainages are of the station is 35,078 km^2. The station is set on a 2000 m quasi-straight reach of the Illinois River where flood wave propagations is not significantly affected by the presence of man-made structures. The closest river control structures are more than 50 km upstream and downstream from the gage location (i.e., Starved Rock and Peoria lock and dam structures, respectively).

(a)

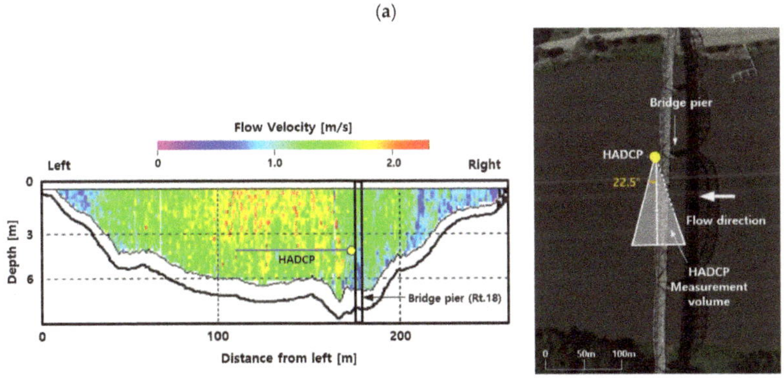

(b)

Figure 2. USGS gaging station #05558300 on Illinois River, at Henry (IL): (**a**) gaging site location; and (**b**) position of the HADCP in the measurement cross-section (the ADCP measurement was taken to an extreme flow event).

The channel is about 300 m wide and 6 m deep for high flows, and it controls the flow through the station at all stages excepting for low flows near the Peoria Lock and Dam (C. Prater personal communication, 2019). The index velocity is measured with a 1500 kHz SonTek-SL HADCP positioned in the cross-section, as illustrated in Figure 2b. The mean bulk channel velocity is obtained with the regression equations provided in Figure 2b. It can be noted that the length of the HADCP acoustic path is about 55 m (including the blanking distance), covering less than half of the width of the river corresponding to the HADCP elevation. This situation is not uncommon [30], and it often occurs in practice, as finding an optimum location for the probe needs to fulfill multiple requirements (e.g., full probe submersion of the instrument acoustic path, traversing a flow area of maximum possible length without encountering large non-uniformities due to the stream boundary presence). The stage, index velocity, and discharge measured at the gages are transmitted every 15 min to the USGS national streamflow system, which is publicly available. All the data reported in this paper are retrieved from the open-accessible gaging station website. Table 1 provided relations built between discharge and index velocity considering stage.

Table 1. IVRC regression equations (source: https://waterdata.usgs.gov/monitoring-location/05558300, accessed on 20 April 2020).

Station	Regression Equation (x_1—Stage; x_2—Index Velocity, ENU)	Rating Validity	Maximum	
			Discharge (m^3s^{-1})	Stage (m)
Henry (USGS #05558300)	$y = (x_2 \times 0.571) + (0.008 \times x_1 \times x_2) + 0.083$	6/2015–2/2018	4560	9.95
	$y = (x_2 \times 0.669) + 0.116$	2/2018 to date		

3.2. Monitoring Datasets

Until recently, the hysteretic behavior in the relationships among the flow variables has rarely been captured in natural streams, because the high-temporal resolution instruments capable of measuring simultaneously more than one flow variable were not available. The introduction of the new generation of acoustic profilers at gaging stations has enabled valuable insights into the dynamics of unsteady open-channel flows. One such site is the USGS gaging station #05558300 analyzed herein. Before selecting this site for illustration, we explored the hysteresis presence over six years of data recorded at the station. The implementation of the hysteresis diagnostic formulas developed by Dottori et al. [27], Mishra and Seth [31], and Fread [32] to the available dataset have compellingly confirmed the possibility of developing diffusion and full dynamic waves even for relatively moderate storms propagating through the station [8].

Our illustration starts with time series of stages and index velocities directly measured at this IVRC-based station for the Water Years (WY) 2014 to 2019 (see Figure 3). The time series in this figure allows identifying flood waves propagating through the station alternating with periods of steady flows (baseflow). We define flood wave duration using the National Oceanic and Atmospheric Administration's (NOAA) terminology, i.e., the contiguous time interval for which the stage is above the one corresponding to baseflow (ncdc.noaa.gov/stormevents). While the most damaging flood impacts are related to the wave peak (i.e., crest), flooding can occur at lower stages, too. To place the magnitudes of the flow waves in the flooding context, we include in Figure 3 the "Action Stage", which is defined by NOAA's National Weather Service (https://w1.weather.gov/glossary, accessed on 20 December 2021) as the first flood warning level. When the stream reaches the Action Stage, communities and agencies in charge of flood mitigation typically initiate preparations for flood response activities. Table 2 lists the numerical values for the maximum of the flow variables during the major flood waves (storm events) of WY 2017. We choose WY 2017 as it is a typical water year with the maximum discharge and stage values (i.e., 2600 m^3s^{-1} and 8.20 m) compared with the top discharge and stage values recorded during the 2014–2019 observation interval (i.e., 4560 m^3s^{-1} and 10 m, respectively). It can be noted that out of the eight storm events in the WY 2017, only three exceeded the "Action Stage" level (i.e., 6.7 m).

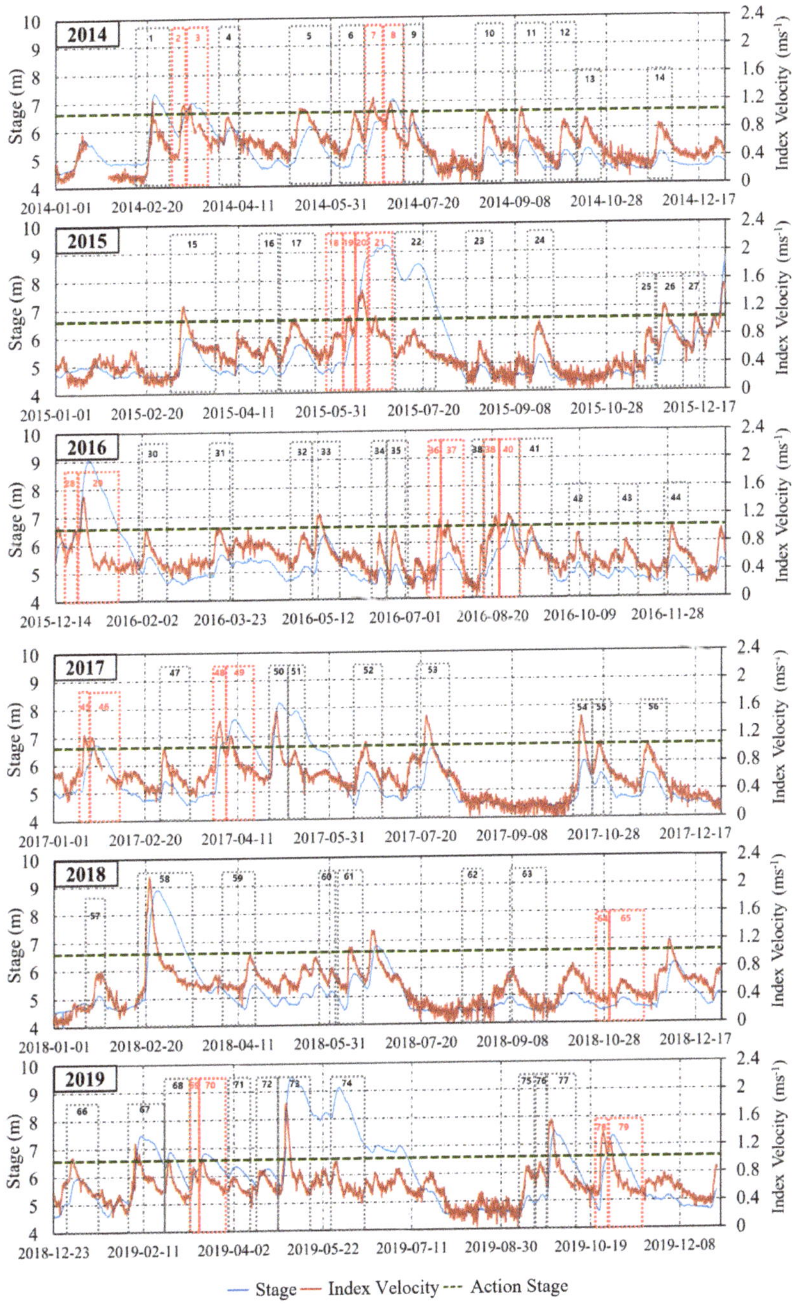

Figure 3. Time series of the stage and index velocity measured at the USGS gaging station #0555830 for supporting IVRC implementation. The numbering in the figure tracks the storm events in chronological order.

Table 2. Characteristics of the major storm events during the Water Year 2017.

Year	Storm Event & Time Interval	Max. Discharge (m^3s^{-1})	Max. Stage (m)	Max. Index Velocity (ms^{-1})
2017	45–46 (16 January; 19:45~7 February; 08:00)	1415	6.75	1.20
	47 (28 February; 09:30~11 March; 16:45)	864	5.43	1.00
	48–49 (29 March; 19:45~29 April; 11:00)	1842	7.61	1.38
	50 (29 April; 11:00~11 March; 16:45)	2599	8.20	1.51
	52 (15 June; 01:00~1 July; 11:15)	968	5.68	1.06
	53 (19 July; 21:30~10 August; 11:15)	1455	6.49	1.45
	54 (13 October; 00:45~23 October; 11:00)	1299	6.00	1.39
	56 (17 November; 01:00~1 December; 23:00)	911	5.55	0.95

The visual inspection of the time series for stage and index velocities in Figure 3 reveals several notable features. First, it can be noticed that the time periods taken by flood wave propagation represent a significant part of the annual records. A second, perhaps less expected feature, illustrated in this figure is the easy-to-detect and persistent trend in the time series: the index-velocity peaks precede the stage peaks. This trend is unequivocally related to the hysteresis effect as will be discussed later in the paper. The non-synchronicity of the hydrographs leads to a situation in which, near the hydrograph peak, the index-velocity time series decreases in magnitude while the stage continues to rise for a short time. This is somewhat conflicting with the conventional description of the flood wave propagation where it is assumed that the rising and falling limbs of variable hydrographs are identical.

In order to make the above-mentioned distinction, we use the terms ascending and descending when referring to the index velocity hydrograph "pulses" and the terms rising and falling for stage hydrograph, since these terms are conventionally used. Pulses are defined herein as the groups of consecutive data points recorded on the index-velocity hydrograph pertaining to an acceleration–deceleration cycle [33]. Pulses are related to changes in the rainfall intensity and/or its spatial distribution over the station's drainage area. As seen in this figure, one storm event can contain a single or multiple pulses. The numbered rectangles in Figure 3 indicate individual storms listed in their chronological order for the observation period. The gradients and magnitudes of the pulses on the rising limb of the stage hydrograph are decisive for determining the severity of the hysteretic loops [34,35]. Severity is used in the present context to indicate the magnitude of the gradients in the changes in the flow variables for a given pulse; the larger the gradients, the more severe the event. The single-pulse storm produces one peak in the stage hydrographs that, in fact, represents the flood crest. For multiple-pulse storm events, the magnitude of the flood crest is impacted by all the pulses occurring on the rising limb of the stage hydrograph.

3.3. *Substantiation of the Hysteretic Features in the Annual Datasets*

The phase difference between the peaks of the index velocity and stage timeseries plotted in Figure 3 is an unmistakable indication of the unsteady flow presence, as for steady flows, the maximum stage coincides with the maximum discharge and bulk flow velocity, as discussed in Figure 1. The trends displayed by the index-velocity hydrographs are the same for the cross-sectional bulk velocity, as the two quantities are related through the one-to-one IVRC relationship for the entire range of flows. Figure 4 displays all the unsteady events within the available dataset plotted in time-independent stage-discharge coordinates to substantiate that each event follows a distinct relationship commensurate with the characteristics of each storm. Collectively, these relationships produce "cloud-like" patterns. Data segments missing in the loops are time intervals when the gaging station temporarily missed or mismeasured data points (typically during high or energic flows). The "Action stage" warning level is also illustrated in this figure to highlight that the flows above this stage are developing the most prominent hysteretic effects.

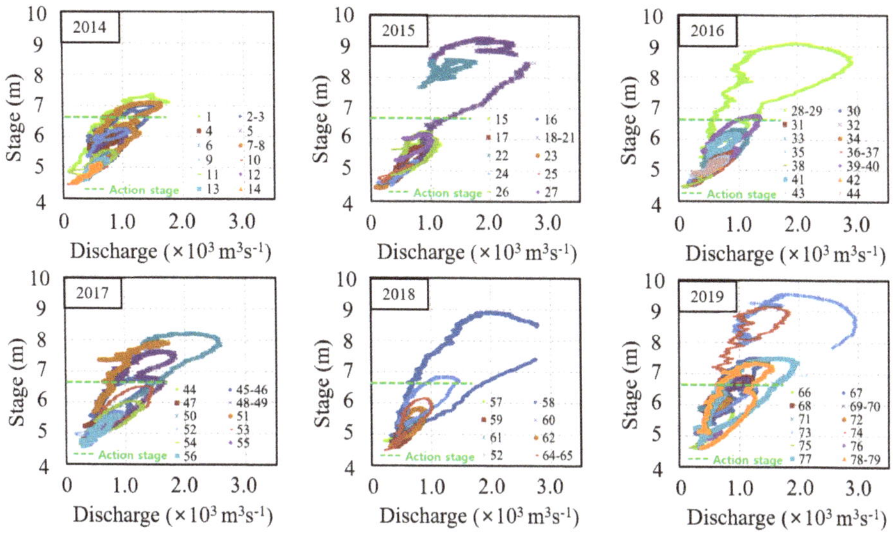

Figure 4. Time-independent stage-discharge relationships for the major storms that occurred during the observation period.

The hysteretic behavior affects not only the stage-discharge relationship but also those among any two of the other flow variables, as indicated in Figure 5 for the major storms of WY 2017. The cloud of loops for each of the dependencies are bundled along inclined lines of different slopes. The thickness of the cloud and their slope are specific for each relationship. The median slope for each cloud corresponds to the steady-flow relationship for the respective variable pairs. More specifically, we argue that the unsteady flow loops follow the Saint-Venant equation while the steady-flow relationships follow the Manning's equation that is a sub-component of the first equation [8]. Visual inspection of the plots in Figure 5 indicates that the most prominent impact of hysteresis for this site is associated with the largest storm of the WY 2017, i.e., storm #50 (see Figure 3). For this event, the largest impact of hysteresis can be noted for the stage vs. index-velocity relationship (Figure 4: 60% at 7.5 m), followed by stage-discharge relationship (Figure 5a: 49% at 7.5 m), and index-velocity vs. discharge relationship (Figure 5b: 36% at 1700 m^3).

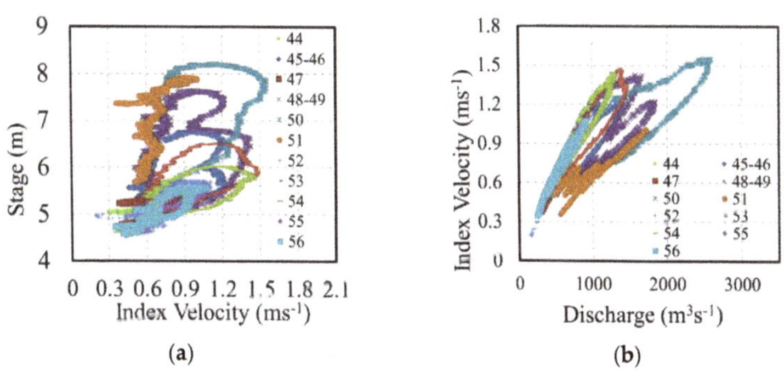

Figure 5. Hysteretic relationships between variable pairs for the Water Year 2017: (**a**) stage vs. index velocity; and (**b**) index velocity vs. discharge.

3.4. Substantiation of the Hysteretic Features in Individual Event Datasets

Ensuing from analytical considerations exposed in Muste et al. [8], the hysteresis associated with unsteady flows propagating through a prismatic channel affects the relationships between flow variables in multiple ways, unravelling the unsteady mechanics from several viewpoints. Some of these views and associated relationships are well documented theoretically (e.g., [36]) and through laboratory experiments (e.g., [37]), but they have been rarely captured in natural streams. With the advent of instruments such as HADCPs, we can observe aspects of the flows that were not possible to document in natural conditions before. Taking advantage of the HADCP data for supporting the index method at USGS #05558300 station, we subsequently illustrate various dependencies occurring between flow variables during single-pulse and multiple-pulse storm events (see Figure 6). The data plotted in this figure, as well as for most of the subsequent plots, are smoothened to aid the visualization of the discussed features. A variable-span smoother based on local linear fit [38] was applied to the 5-point averages dataset collected by the HADCP. It should be noted that the index-velocity time series plotted in Figure 6a are acquired across the stream from fixed positions in the vertical, while bulk channel velocity is derived from this velocity via a rating curve; therefore, there are slight differences between the actual loop shapes for the two related variables. However, the two velocities are closely linked; therefore, the trends visualized during the flood wave propagation are similar.

The plots for the single-pulse event in Figure 6a illustrates that the trends in index velocity and stage hydrographs are closely coupled in time but not synchronous. The time separation (i.e., phase shift) of the two variables' hydrographs leads to different values of the index velocities on the rising and falling limbs for the same depth, as illustrated by points A and B in this figure (and also hinted in the conceptual Figure 1b). The time-independent relationship of the two-directly measured variables is plotted in Figure 6b. The shape of the plot suggests that the flows on the rising and falling limbs of the stage hydrograph display higher velocities during flow acceleration compared with those during decelerated flow. Figure 6c display the often-used representation of the stage-discharge relationship as estimated by the IVRC method for this single-pulse storm event.

Plotted in Figure 6d–f are the same dependencies for a multiple-pulse storm event. While tracing the relationship between the time series for index velocity and stage in Figure 6d is more intricate, a careful inspection of the changes in the time series for the two variables allows us to identify the effect of each individual index-velocity pulse on the stage variation. This impact is visually represented in Figure 6d either as a local stage peak or just as an inflexion point in the stage time series, as indicated with arrows in this figure. Individual pulses contain both ascending and descending phases; hence, we label them as sub-storms contained within the overall storm event (from base flow and back). Each sub-storm produces distinguishable loops within the overall large loop of the multi-pulse storm event, as illustrated in the Figure 6e,f.

Figures 4–6 provide abundant experimental evidence that the hysteretic behavior is materialized through the presence of loops in the time-independent relationships and through a detectable "decoupling" of the flow variable hydrographs in the time-dependent graphical representations. Another feature of interest for characterizing the hysteretic behavior is its severity. One of the built-in factors in determining the hysteresis severity is the slope of the channel bed at the site, with milder slopes producing more severe hysteresis [8]. For the same site, the magnitude of the loops and time lags between hydrographs are determined by the severity of the storm events (i.e., proportional to the gradients for the flow changes), as shown in the measurement summaries presented in Figures 4 and 5. We argue herein that the pulses of unsteady flow with the same starting point, index-velocity magnitude, and gradient on the rising limb of the stage hydrograph produce identical hysteretic loops for a given site, whether the event is single- or multi-pulse.

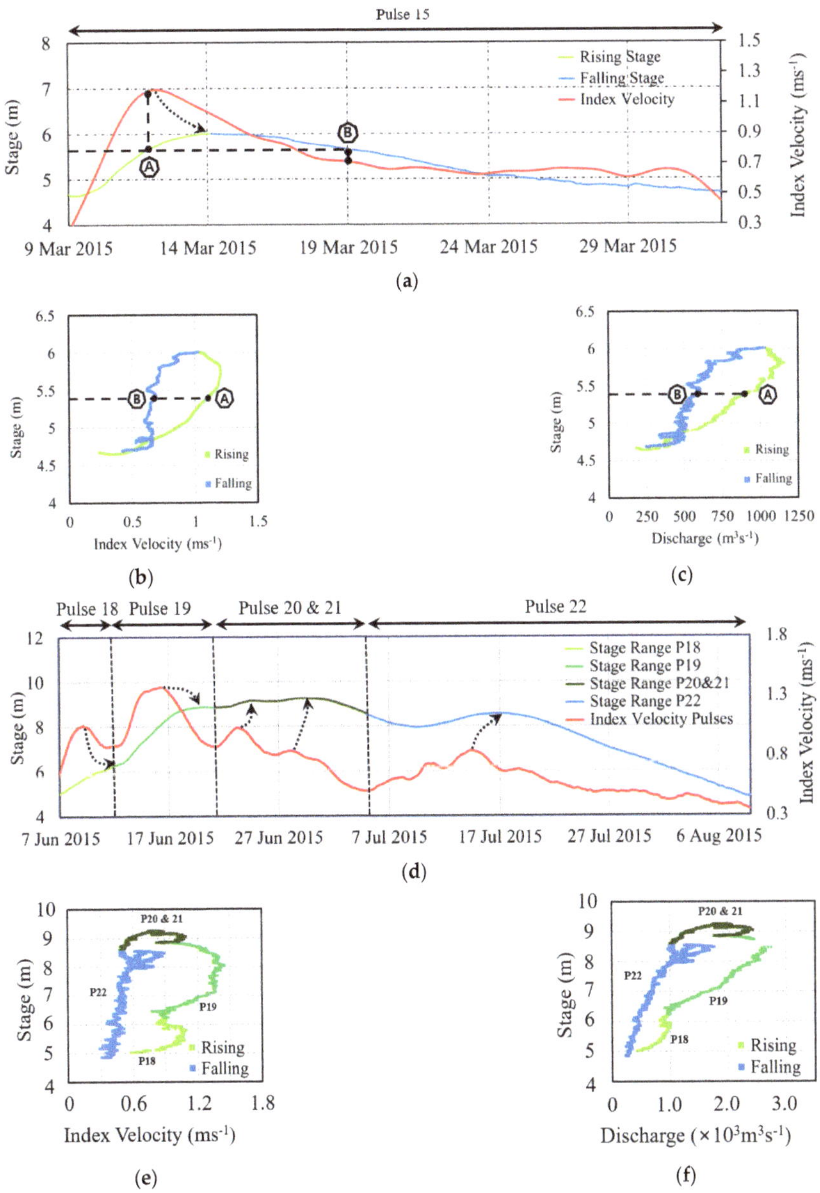

Figure 6. Illustration of hysteretic features in time−dependent and time-independent relationships between flow variables: (**a–c**) single-pulse storm event; (**d–f**) multi−pulse storm event.

Quantification of the storm severity in this discussion is made using the unsteadiness coefficient proposed by Nezu et al. [39]. According to this study, this coefficient is proportional to the lag between the maximum shear stress (that is highly correlated with the maximum velocity peak) and the maximum depth (i.e., stage peak). Assuming a hydrostatic pressure distribution, the unsteadiness coefficient is defined by the ratio between the

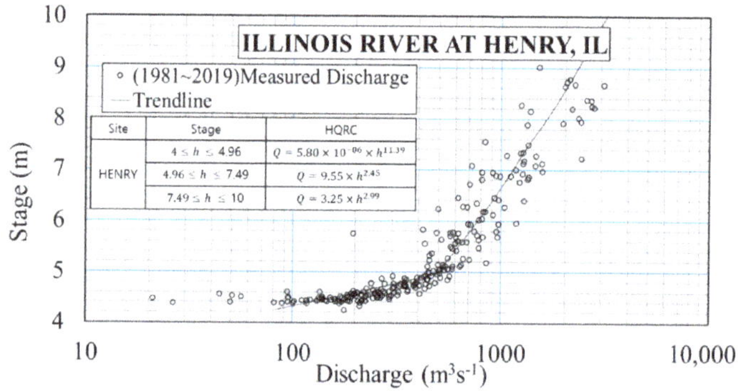

Figure 9. Surrogate HQRC rating constructed using the calibration and validation measurements for supporting the IVRC station operation.

Figure 10a illustrates essential features of the IVRC-HQRC comparison using the stage-discharge relationship commonly used by the hydrometric community to demonstrate hysteresis presence. The storm chosen for comparison is event #15 of the WY 2015 that does not display the largest stage in the analyzed dataset, but it is one of the most intense, as proven by the sharp increase in the index-velocity illustrated in the time-series of the WY 2015 illustrated in Figure 3. The first comment on Figure 10a is the considerable departure of the IVRC estimated values from those of HQRC, i.e., up to 60% on the rising limb of the stage hydrograph, as illustrated by the arrow added in Figure 10b. The second comment is that the HQRC data fall closely to the datapoints on the falling limb of the IVRC estimates. This feature was also noted by the authors in previous studies at other IVRC-based gaging sites (e.g., [20,23]). The most likely explanation for this feature is that while on the rising stage, the flow is marked by a sudden acceleration, on the falling limb, the deceleration of the flow is slower. For this site, the ratios between falling and rising times are anywhere between 4 and 15. We perceive the slower deceleration on the falling limb as being akin to a series of successive step-like steady flows of continuously decreased discharges, therefore being closer to the shape described by a steady HQRC.

Additional inferences can be drawn from the comparison of the time-dependent plots of the discharges estimated by the IVRC and HQRC methods. The stage hydrograph is added in this figure for delineating the rising and falling limbs as defined in the present discussion. The plots in this figure reveal two new inferences. The first one is that the IVRC discharge peak estimate is 8% larger than the one predicted with the HQRC rating. Differences up to 15% between actual flows measured at the site and HQRC estimates were reported by Jarrett [42] and Di Baldassarre and Montanari [19]. Along the same line, Fenton and Keller [43] as well as Henderson [7] explain analytically that the maximum discharge of diffusive flood waves is larger (and occurs earlier) than the flow computed from simple HQRC ratings. An additional factor toward the generalization of this first inference is the fact that the reliability of any type of rating in the higher flow range area is not as robust as for lower flow range, as the density of the calibration/verification measurements is less dense for high flows. The second inference is that the discharges on the rising limb are considerably underestimated by the HQRC method and are not fully compensated over the remaining period of flood wave propagation. While the IVRC and HQRC features are significant and consistent in this study, the numerical values associated with the differences between the two methods cannot be generalized for other sites and types of waves propagating through hysteresis-prone gaging stations.

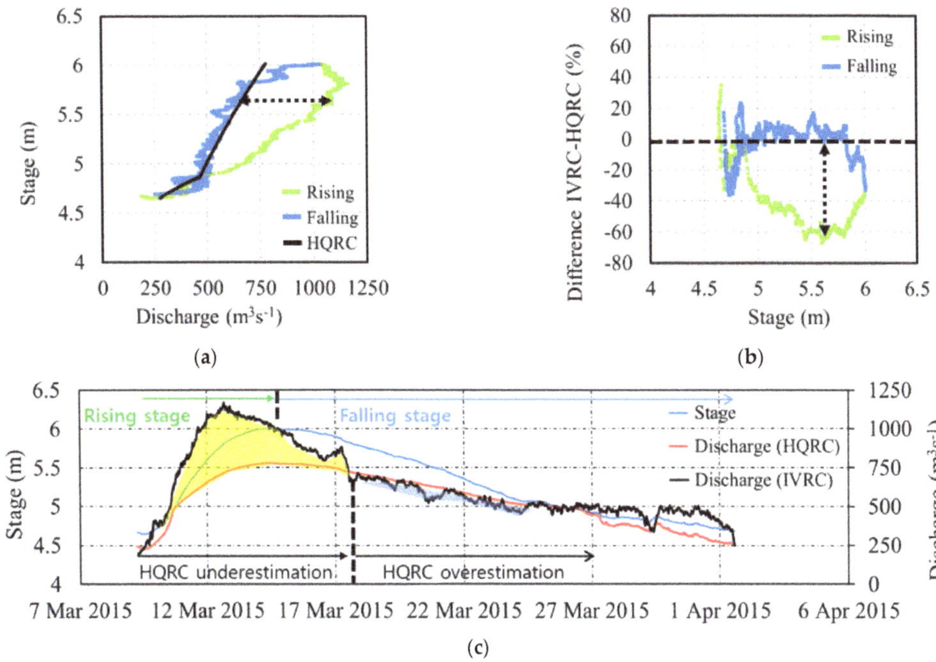

Figure 10. Inferences from the IVRC-HQRC comparison: (**a**) stage vs. discharge relationships; (**b**) numerical values for the differences in Figure 10a; and (**c**) time series for IVRC and HQRC methods.

4. Discussion

The hysteretic behavior at the site of the presented case study is mostly dominated by flow unsteadiness without significant contributions from other potential causes that can lead to hysteresis (e.g., effects of instream vegetation, bedform-induced roughness development, and baseflow–stream interactions). Rantz et al. [14] labels this situation as the one where the energy slope is mostly changed by the variation in discharge to distinguish from the situations when the slope change is mostly driven by the variable backwater. Often, the two effects are both involved. Furthermore, our discussion assumes that the flood wave propagation occurs in channels predominantly controlled by friction, whereby the dominant energy losses are due to channel boundary roughness rather than changes in the channel geometric features (local control) [12]. Moreover, we discuss herein hysteresis impacts for water elevation in the channel that are less than the bankful stage (i.e., as flood stage). When this stage is exceeded, additional flow complexities occur as the mass and momentum exchanges between the main channel and floodplain impede a straightforward interpretation of the hysteretic effects. Finally, it is assumed that there are no issues related to instrumentation deployment and operation, as these factors can also impact the interpretation of the measurements when the sensor positioning and time synchronization are not properly performed. Under the aforementioned conditions, and in the absence of a demonstrable presence of backwater effects, the hysteresis features can be solely attributed to the flow unsteadiness.

While the illustrations of the present study confirm previous conclusions that IVRC better capture unsteady flows, the majority of the discussions in this paper are focused on the exploration of essential features of the hysteretic behavior as reflected by the data collected at an operational IVRC gaging station. The presented results unequivocally demonstrate that the IVRC method: (a) provides distinct relationships between flow variables for the rising and falling limbs of the hydrographs, (b) captures the sequencing of the measured variable hydrographs, which is a feature that is inherent if item (a) above is substantiated at a given site [41], (c) substantiates that the changes in the thickness of the loops and associated hydrograph phasing are commensurate with the magnitude and severity of the hydrographs, and (d) substantiates significant differences in the discharges estimated with the HQRC method.

As mentioned above, the datasets available for the present study do not contain a suitable benchmark to allow a thorough assessment of the IVRC or HQRC performance in steady and unsteady flows. Such benchmark datasets are ideally acquired with high-accuracy alternative instruments, rigorous protocols, and prescribed flow conditions. The latter conditions can be only attained at sites where the river flow can be controlled, and measurements can be repeated a few times to capture the variability of the in-situ conditions. The closest alternative to such an ideal assessment scenario is the acquisition of direct measurements using ADCPs continuously operated in steady and unsteady flows. The above considerations highlight why the IVRC-HQRC performance assessment is difficult to come by and rarely executed. One of these rare studies was, however, conducted by the present authors at a USGS gaging station in Idaho where the flow could be controlled and repeated under the same conditions [44]. In this study, we analyzed data from routine and controlled experiments to illustrate the IVRC and HQRC performance in evaluating streamflow over a range of flow regimes by comparing the monitoring methods' estimates with the measurements acquired with a Winter–Kennedy Meter that are calibrated to relate the pressure losses in the turbine casing to the discharge through the turbine [45]. The WKMs were installed in the hydropower plant's turbine housing located immediately upstream from the gaging station. While we are aware that these conclusions cannot be readily extended to other locations, we deem that it is useful to cite these results in the present context.

The evaluation of the Idaho dataset showed that the flow estimates obtained with the two monitoring methods during steady flows were within 5% from the estimates provided by the WKM. For unsteady flows, it was found that the IVRC method was overall superior to the HQRC performance, but there were differences up to 40% between the IVRC discharge estimates and those provided by WKM during the highest flow-change rates. These large differences were attributed to the fact that the flow changes were considerable in magnitude for both rising and falling phases (from simple to double), and they were made over short durations (less than an hour). Such flow changes are quite typical for hydropower plants where the turbines are set on and off during the day to accommodate the energy consumption in the power distribution grid. There are no specific reasons to expect that the Illinois River dataset analyzed in the present study is widely different in term of overall findings from the Idaho study because the protocols for constructing and using the rating curves are uniformly applied at all USGS gaging stations. Actually, similar findings were inferred by the authors from another IVRC study on controlled flood waves at a gaging station located in Spain [20].

The presented experimental evidence allows us to further explore some issue regarding the IVRC rating construction that are still unsettled. Currently, most of the IVRC ratings are constructed with protocols that do not distinguish between the hydrograph flow phases, similar to the protocols used in stage-discharge approaches. In this approach, the statistical analyses are applied to the whole calibration dataset for establishing one-to-one regression lines that fit best the available data points. The obtained regressions are used for all flow regimes. Similar to HQRC, the graphical construction involves subjective decisions on the parameters and thresholds used in the analysis. In the absence of robust understanding of the flow dynamics or experimental evidence, the use of statistical analyses alone precludes to state when the analysis outcomes are good enough or if the observed disagreements are not merely reflections of the unknown physics processed with statistical tools.

Use of the one-to-one IVRC protocols for the USGS Henry station led to the conclusion that the multiple-linear regression for the IVRC regressions equation was adequate from

June 2015 to February 2018. After this date, a simple linear regression was adopted (see Table 1). The graphical representation of the IVRC ratings using the equations in Table 1 are plotted in Figure 11 for the single-pulse events illustrated in Figure 8. It can be noticed that while for the smaller storms, the rating does not show a clear distinction between the rising and falling stage, for the largest storm (i.e., event #50 in 2017), there is a perceptible and consistent differentiation of the IVRC rating for the rising and falling stages. The thickness of the hysteresis loop in the IVRC for the largest flow events is considerably smaller than that in the stage-discharge relationships shown in Figure 4. Similar findings have been observed through other studies (e.g., [23,46]), therefore signaling the need to further explore the validity of the unicity of the IVRC relationship for all flow regimes.

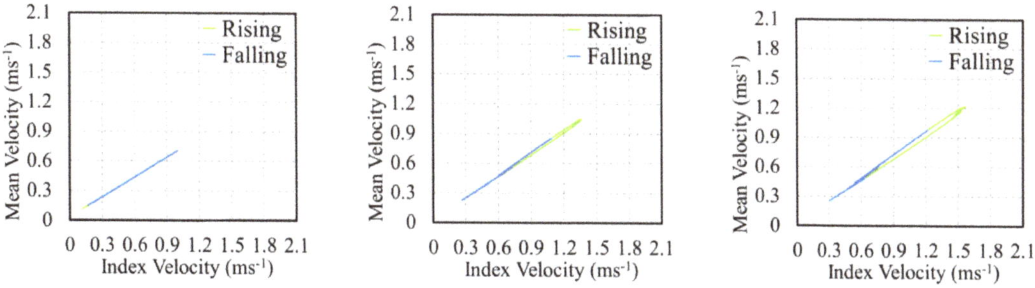

(a) (b) (c)

Figure 11. The graphical representation of the IVRC one-to-one regression equation for the storm events analyzed in Figure 8a, 8b, and 8c, respectively. (**a**) Event #24 (2015); (**b**) Event #60 (2018); (**c**) Event #50 (2017).

In an attempt to test the adequacy of the one-to-one relationship for the IVRC rating during unsteady flows at the Illinois station, an additional check is made using the mean bulk velocities determined from the direct ADCP measurements acquired for validation of the IVRC rating between 2015 and 2018. During this interval, there are 19 ADCP validation measurements, among which only three measurements are taken at stages above the Action Stage indicated in Figure 3. Two of these measurements were acquired during the propagation of Pulses 20 and 21. The comparison between the IVRC rating estimates with the directly measured mean velocity via ADCP is plotted in Figure 12. It can be noted from this figure that the ADCP velocities (acquired on the ascending limb of the stage hydrograph) show larger values than those estimated with the IVRC rating. These differences seem to indicate that the construction of the IVRC relationship using separate relationships for the rising and falling phases of the stage hydrograph is a better alternative than the one-to-one approach provided in Table 1. Definitely, a thorough assessment of this argument requires further analyses on considerable larger datasets of this kind.

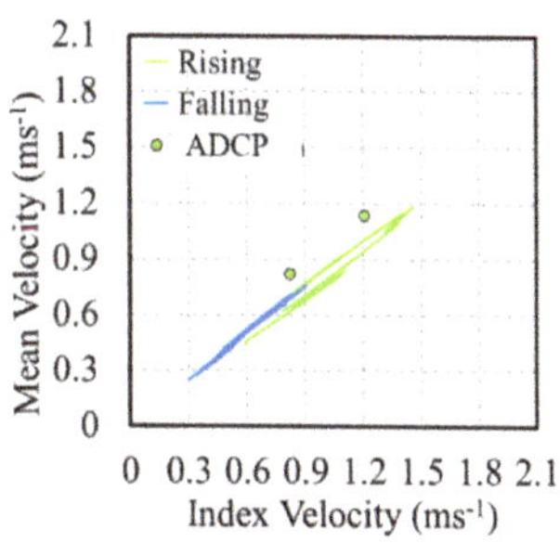

Figure 12. ADCP-derived data compared with IVRC rating for Pulses 20–21 of the WY 2015.

5. Conclusions

The fresh perspectives on the IVRC measured flow variables presented in this paper provide useful insights into the physics of flood wave propagation in natural open channels that are often overlooked in routine monitoring. The experimental evidence presented in this study indicate that:

(a) Hysteresis occurring at monitoring sites located in low-gradient channels exposed to flood waves is significant, being commensurate with the site and flood wave characteristics;
(b) Unsteady flows produce a non-unique relationship between any two of the flow variables and an out-of-phase flow hydrographs during the flood wave propagation;
(c) The index-velocity method can more aptly capture hysteresis compared with the stage-discharge method;
(d) If a discharge monitoring method capture hysteresis, it is implicit that it can also distinguish the peak sequencing and vice versa.

Given that the above findings are based on only a limited dataset, the outcomes of the discussion should be regarded as being indicative rather than confirmative. However, it is our hope that the outcomes of the present study indicate that the effect of hysteresis on monitoring methods requires additional research.

The present study convincingly demonstrates that the acoustic probes such as HADCPs are reliable instruments for providing valuable data and insights in complex measurement situations. The use of directly measured data provided by these instruments are valuable benchmark datasets that can uniquely support multi-purpose investigations based on analytical, data-driven, and numerical simulations. The capabilities of the new generation of instruments can underpin better decision making in water resource management and flood mitigation efforts and open possibilities to use the produced data in novel ways for enhanced scientific investigations in streams and rivers.

Author Contributions: The conceptualization and data processing methodology were developed by M.M. and D.K. The data processing, analysis, and presentation were carried out by K.K. All authors have read and agreed to the published version of the manuscript.

Funding: Most of the illustrations in this paper are outcomes of a research study conducted for SonTek/YSI (I556800-C) by the paper's authors. Center. The first author acknowledges the partial support offered the NSF EAR-HS 2139649 award. The second and third authors were partially funded by Basic Science Research Program through the National Research Foundation of Korea (NRF) funded by the Ministry of Education (2021R1F1A1060295).

Data Availability Statement: Datasets supporting the illustrations presented above will be assembled in an open-access drive for consultation with interested readers.

Acknowledgments: Discussions with Janice Yasui and Xue Fan of SonTek/YSI enriched the horizon of analysis. The analysis presented in this paper have greatly benefitted by constructive discussions with R. Jackson and R. Holmes and additional data offered by J. Duncker, C. Prater, and R. Beaulin with the US Geological Survey Midwest Water Science Center.

Conflicts of Interest: The authors declare no conflict of interest.

References

1. USGS. *A history of the Water Resources Branch, U.S. Geological Survey*; Follansbee, R., Ed.; Volume I, from Predecessor Surveys to 30 June 1919; U.S. Geological Survey: Denver, CO, USA, 1994.
2. Jain, S.; Lall, U. Magnitude and timing of annual maximum floods: Trends and large-scale climatic associations for the Blacksmith Fork River, Utah. *Water Resour. Res.* **2000**, *36*, 3641–3651. [CrossRef]
3. Holmes, R.R. River rating complexity. In *Proceedings River Flow Conference*; Taylor & Francis Group: St. Louis, MO, USA, 2016; ISBN 978-1-138-02913-2.
4. Saint-Venant, A.J.C. Théorie du mouvement non permanent des eaux, avec application aux crues des rivières et a l'introduction de marées dans leurs lits. *Comptes Rendus L'académie Sci.* **1871**, *73*, 147–154, Discussion 237–240.
5. Chow, V.T. *Open Channel Flow*; McGrow Hill: New York, NY, USA, 1959.

6. Hunt, A.E. The Behaviour of Turbulence in Unsteady Open Channel. Ph.D. Thesis, University of Canterbury, Christchurch, New Zealand, 1997.
7. Henderson, F.M.; Open Channel Flow. *Macmillan Series in Civil Engineering*; Macmillan Company: New York, NY, USA, 1966; p. 522.
8. Muste, M.; Lee, K.; Kim, D.; Bacotiu, C.; Oliveros, M.R.; Cheng, Z.; Quintero, F. Revisiting hysteresis of flow variables in monitoring unsteady streamflows. *J. Hydraul. Res.* **2020**, *58*, 867–887. [CrossRef]
9. Ponce, V.M. *Development of Physically based Coefficients for the Diffusion Method of Flood Routing*; Contract No. 53-3A75-3-3; U.S. Soil Conservation Service: Lanham, MD, USA, 1983.
10. Ferrick, M.G. Analysis of River Wave Types. *Water Resour. Res.* **1985**, *21*, 209–220. [CrossRef]
11. Aricò, C.; Nasello, C.; Tucciarelli, T. Using unsteady-state water level data to estimate channel roughness and discharge hydrograph. *Adv. Water Resour.* **2009**, *32*, 1223–1240. [CrossRef]
12. WMO. Manual on stream gauging, Volume I, Field Work; World Meteorological Organization Report WMO No. 1044. 2010. Available online: www.wmo.int/pages/prog/hwrp/publications/stream_gauging/1044_Vol_I_en.pdf (accessed on 20 December 2021).
13. Muste, M.; Hoitink, T. Measuring Flood Discharge. In *Oxford Research Encyclopedia of Natural Hazard Science*; Subject: Case Studies, Risk Assessment, Vulnerability, Floods; Oxford University Press: Oxford, UK, 2017. [CrossRef]
14. Rantz, S.E. *Measurement and Computation of Streamflow*; US Geological Survey Water Supply Paper 2175, Vol.1 and 2; US Department of the Interior, Geological Survey: Reston, VA, USA, 1982.
15. Herschy, R. *Streamflow Measurement*, 3rd ed.; Taylor & Francis: Oxford, UK, 2009.
16. Levesque, V.A.; Oberg, K.A. *Computing Discharge Using the Index Velocity Method*; U.S. Geological Survey Techniques and Methods 3–A23; U.S. Geological Survey: Reston, VA, USA, 2012; 148p. Available online: http://pubs.usgs.gov/tm/3a23/ (accessed on 20 April 2020).
17. Rennie, C.; Hoitink, A.J.F.; Muste, M. *Chapter 6 in Experimental Hydraulics*; Aberle, J., Rennie, C.D., Admiraal, D.M., Muste, M., Eds.; Taylor & Francis: New York, NY, USA, 2017; Volume II.
18. Faye, R.E.; Cherry, R.N. *Channel and Dynamic Flow Characteristics of the Chattahoochee River, Buford Dam to Georgia Highway 141*; Geological Survey Water-Supply Paper 2063; U.S. Government Printing Office: Washington, DC, USA, 1980.
19. Di Baldassarre, G.; Montanari, A. Uncertainty in river discharge observations: A quantitative analysis. *Hydrol. Earth Syst. Sci.* **2009**, *13*, 913–921. [CrossRef]
20. Muste, M.; Lee, K. Evaluation of Hysteretic Behavior in Streamflow Rating Curves. In Proceedings of the 2013 IAHR Congress, Chengdu, China, 8–13 September 2013; Tsinghua University Press: Beijing, China, 2013.
21. Schmidt, A.R. Analysis of Stage-discharge Relations for Open Channel Flows and their Associated Uncertainties. Ph.D. Thesis, University of Illinois at Urbana-Champaign, Champaign, IL, USA, 2002.
22. Morlock, S.E.; Nguyen, H.T.; Ross, J. *Feasibility of Acoustics Doppler Velocity Meters for the Production of Discharge Records from U.S. Geological Survey Stream-Flow-Gaging Stations*; U.S.G.S. Water-resources Investigations Report; USGS: Indiannapolis, IN, USA, 2002.
23. Cheng, Z., Lee, K.; Kim, D.; Muste, M.; Vidmar, P.; Hulme, J. Experimental Evidence on the Performance of Ratting Curves for Continuous Discharge Estimation in Complex Flow Situations. *J. Hydrol.* **2019**, *568*, 959–971. [CrossRef]
24. Le Coz, J.; Pierrefeu, G.; Paquier, A. Evaluation of river discharges monitored by a fixed side-looking Doppler profiler. *Water Resour. Res.* **2008**, *44*, 1–13. [CrossRef]
25. Jackson, P.R.; Johnson, K.K.; Duncker, J.J. *Comparison of Index Velocity Measurements Made with a Horizontal Acoustic Doppler Current Profiler and a Three-Path Acoustic Velocity Meter for Computation of Discharge in the Chicago Sanitary and Ship Canal Near Lemont, Illinois*; U.S. Geological Survey Scientific Investigations Report. 2011–5205; US Department of the Interior, US Geological Survey: Reston, VA, USA, 2012; 42p.
26. Kennedy, E. *Discharge Ratings at Gaging Stations*; US Geological Survey Techniques of Water-Resources Investigations, Book 3, Chap. A10; U.S. Geological Survey: Reston, VA, USA, 1984; 59p.
27. Dottori, F.; Martina, M.L.V.; Todini, E. A dynamic rating curve approach to indirect discharge measurement. *Hydrol. Earth Syst. Sci.* **2009**, *13*, 847–863. [CrossRef]
28. SonTek/YSI. *Acoustic Doppler Profiler Principles of Operation*; SonTek/YSI: San Diego, CA, USA, 2000; 28p.
29. Ruhl, C.A.; Simpson, M.R. *Computation of Discharge Using the Index-Velocity Method in Tidally Affected Areas*; Scientific Investigations Report 2005-5004; U.S. Geological Survey: Reston, VA, USA, 2005.
30. Hoitink, A.J.F. Monitoring and analysis of lowland river discharge. In Proceedings of the River Flow Conference, IAHR, Lyon, France, 5–8 September 2018. [CrossRef]
31. Mishra, S.K.; Seth, S.M. Use of hysteresis for defining the nature of flood wave propagation in natural channels. *Hydrol. Sci. J.* **1996**, *41*, 153–170. [CrossRef]
32. Fread, D.L. *Channel Routing*; Anderson, M.G., Burt, T.P., Eds.; Hydrological Forecasting; Wiley: New York, NY, USA, 1985.
33. Muste, M.; Kim, D.; Kim, K.; Ehab, M. Monitoring streamflow pulses. In Proceedings of the 39th IAHR World Congress, Granada, Spain, 19–24 June 2022.
34. De Sutter, R.; Verhoeven, R.; Krein, A. Simulation of sediment transport during flood events: Laboratory work and field experiments. *Hydrol. Sci. J.* **2001**, *46*, 599–610. [CrossRef]

35. Mrokowska, M.M.; Rowiński, P.M. Impact of Unsteady Flow Events on Bedload Transport: A Review of Laboratory Experiments. *Water* **2019**, *11*, 907. [CrossRef]
36. Kozak, M. *A Szabadfelsonu Nempermanens Vizmozgasok Szamitasa*; Academia Kiado: Budapest, Hungary, 1977.
37. Graf, W.; Qu, Z. Flood hydrographs in open channels. In Proceedings of the Institution of Civil Engineers-Water Management; Thomas Telford Ltd.: London, UK, 2004; Volume 157, pp. 45–52. [CrossRef]
38. Friedman, J.H. *A Variable Span Smother*; Technical Report PUB-3477; Laboratory for Computational Statistics, Stanford University: Stanford, CA, USA, 1984.
39. Nezu, I.; Kadota, A.; Nakagawa, H. Turbulent Structure in Unsteady Depth-Varying Open-Channel Flows. *J. Hydraul. Eng.* **1997**, *123*, 752–763. [CrossRef]
40. Muste, M.; Kim, D.; Kim, K. A flood-crest forecast prototype for river floods using only in-stream measurements. *Commun. Earth Environ.* **2022**, *3*, 1–10. [CrossRef]
41. Muste, M.; Kim, D. Augmenting the Operational Capabilities of SonTek/YSI Streamflow Measurement Probes. Sontek/YSI-IIHR Collaborative Research Report. 2020. Available online: https://info.xylem.com/rs/240-UTB-146/images/augmenting-capabilities-sontek-probe.pdf (accessed on 20 December 2021).
42. Jarrett, R.D. Errors in slope-area computations of peak discharges in mountain streams. *J. Hydrol.* **1987**, *96*, 53–67. [CrossRef]
43. Fenton, J.D.; Keller, R.J. *The Calculation of Streamflow from Measurement of Stage*; Technical Report; Cooperative Research Centre for Catchment Hydrology and Centre for Environmental Applied Hydrology, Department of Civil and Environmental Engineering, The University of Melbourne: Melbourne, Australia, 2001.
44. Muste, M.; Cheng, Z.; Vidmar, P.; Hulme, J. Considerations on Discharge Estimation Using Index-Velocity Rating Curves. In Proceedings of the 36th IAHR World Congress, The Hague, The Netherlands, 28 June–3 July 2015.
45. ASME. *Hydraulic Turbines and Pump-Turbines*; Power Test Code (PTC)—18-2011; American Society of Mechanical Engineers (ASME): New York, NY, USA, 2011.
46. Nihei, Y.; Kimizu, A. A new monitoring system for river discharge with horizontal acoustic Doppler current profiler measurements and river flow simulation. *Water Resour. Res.* **2008**, *44*, 1–15. [CrossRef]

water

Article

Extreme Flood Disasters: Comprehensive Impact and Assessment

Qian Yu [1,2,*], Yanyan Wang [1,2] and Na Li [1,2]

[1] State Key Laboratory of Simulation and Regulation of Water Cycle in River Basin, China Institute of Water Resources and Hydropower Research, Beijing 100038, China; wangyy@iwhr.com (Y.W.); lina@iwhr.com (N.L.)
[2] Research Center on Flood and Drought Disaster Reduction of the Ministry of Water Resources, Beijing 100038, China
* Correspondence: qcherie@126.com or yuqian@iwhr.com

Abstract: Evaluating extreme flood disasters is a prerequisite for decision making in flood management. Existing extreme flood disaster assessments fail to either consider or evaluate comprehensive impacts from social, economic, and environmental aspects. This study first analyzes the causes of extreme flood disasters and subsequently the potential flood consequences in depth. On the basis of this comprehensive analysis, an extreme flood disaster indicator system is developed by taking into account social, economic, and environmental consequences. To assess the comprehensive impacts, we propose a refined social and economic impact evaluation method and a semi-quantitative environmental impact evaluation method, which are applied to Jingjiang Flood Diversion District (JFDD) located in the Yangtze River Basin, and analyze two extreme flood scenarios. The results show that almost all of the JFDD area is flooded with inundation areas of 901.36 km^2 and 879.49 km^2, respectively. The corresponding affected populations are 0.51 million and 0.5 million. The direct economic losses are 18.83 billion and 14.33 billion, respectively. Moreover, 5 potential pollutant sources and 11 protected areas are inundated under two scenarios. Extreme floods have relatively serious impacts on local ecology and the environment. The proposed methodology can provide effective support for decision makers.

Keywords: extreme floods; disaster chain; impact assessment; flood damage; environmental impacts

Citation: Yu, Q.; Wang, Y.; Li, N. Extreme Flood Disasters: Comprehensive Impact and Assessment. *Water* **2022**, *14*, 1211. https://doi.org/10.3390/w14081211

Academic Editors: Slobodan P. Simonovic, Subhankar Karmakar and Zhang Cheng

Received: 27 February 2022
Accepted: 5 April 2022
Published: 9 April 2022

1. Introduction

Flood is one of the most serious natural disasters worldwide [1–3], accounting for 44% of all disaster events from 2000 to 2019 [2]. Floods usually cause huge economic losses and severe human casualties. With the implementation of flood control and disaster reduction systems, the human death caused by floods is, to some extent, reduced. However, flood is still one of the main global risks according to the Global Risks Report [4] under the double pressure of climate change and socio-economic developments. With climate change, the frequency of heavy precipitation events increases [5,6]. Extreme floods are usually caused by extreme rainstorms [7], such as the one across central Europe in 2021 and the 20 July 2021 heavy storm in Zhengzhou, China. The total number of flood disaster events in 2000–2019 was more than twice as high as that in 1980–1999 [2]. In addition, with the socio-economic developments, extreme floods may also lead to more disastrous consequences, such as human injury, livelihood being seriously damaged or destroyed, and even the disruption of the global supply chains. From 2000 to 2019, floods affected 1.6 billion people worldwide [2]. Moreover, floods contributed to USD 651 billion of economic losses according to EM-DAT data [2]. Besides, floods may also have potentially adverse impacts on the environment and ecology when polluted floods flow into surface drinking water sources or other protected areas [8,9]. Therefore, it is significant to conduct refined and comprehensive extreme flood disaster assessment to better support flood prevention decision making and further lessen the negative impacts. However, the existing extreme flood disaster assessments fail to either consider or evaluate comprehensive impacts from social, economic, and environmental aspects.

Extreme and disastrous floods can cause various and comprehensive adverse consequences [10–12], including economic losses, social impacts, and environmental impacts. Previous studies of flood disaster assessments mainly focused on monetary loss evaluation and the affected population assessment. Environmental impacts are rarely assessed not only because of the lack of public awareness but also the fact that they are intangible and hard to quantify. Indeed, they are very important to society [9,13]. Therefore, it is important to integrate the social, economic, and environmental impacts together for the flood management decision making.

Compared to the rare study on floods' impacts on the environment, there are many studies on flood loss assessments [14,15]. Flood loss assessment is multidimensional, complex, and interdisciplinary. It is affected by multiple factors, including regional rainfall characteristics, geographical features, flood control systems, and socio-economic conditions [15,16]. Therefore, the assessment methods are diverse [17]. Previous efforts to estimate flood economic losses have focused on depth-damage functions [18] to integrate flood depths, land use types, and economic factors. With the fast development of GIS, spatial analysis technology has also been introduced into the loss assessment method. Then, the accuracy of the assessment employing this method is determined by the exposure data. The exposure data can be represented in terms of each affected object, such as buildings, critical infrastructures, etc., at the microscale and/or land use types, such as farmlands, urban areas, etc., at the mesoscale. However, the detailed data of each affected object are not easy to acquire. The land use data can be easily obtained through remote sensing (RS) interpretation [14], and it is effectively used to extract the spatial distribution of buildings, but it is difficult to establish the relationship between flood loss data and land use types. These problems lead to high uncertainties and disparities in flood loss assessments [19,20]. In this case, we should propose a refined method to evaluate the detailed flood loss of each affected object.

To address the above-mentioned problems, the present study analyzes the potential negative impacts caused by extreme floods and draws disaster chains, which provides a comprehensive insight into extreme flood disasters. Specifically, an extreme flood disaster assessment index system is established by incorporating social, economic, and environmental indicators. We then develop a refined method to evaluate social and economic impacts and also put forward a semi-quantitative method to assess environmental impacts. Last, the newly established methods are employed in the Jingjiang flood diversion district (JFDD) to demonstrate the effectiveness of our assessment system.

2. Causes and Impacts of Extreme Floods

Extreme floods are caused by the interactions of atmospheric processes, basin hydrological processes [21,22], and human activities (refer to Figure 1). Heavy and/or persistent precipitations are usually the very first cause of extreme floods, which is not the case for small floods. The land use changes [23], due to fast urbanizations or geographic variations, will increase the runoff coefficient, enhance runoff generations, and then increase flood peaks and volumes in the rivers. In this case, fluvial flooding occurs when rivers flow overtop their banks or dams/dikes break. Dam overtopping is one of the main reasons for dam break [24]. Pluvial flooding occurs when rainfall volumes exceed the infiltration capacity and the drainage capacity and result in the inundation of urban areas. Besides, the collapse of the barrier lakes in mountainous areas, formed by landslides, is likely to cause extreme floods downstream.

Extreme flood disaster is a threat caused by extreme flood to human life, property, sensitive ecological and environmental areas, etc. Extreme flood is characterized by a wider inundation range, higher hazards, and therefore, probably leads to a wider range of affected areas. Correspondingly, the category of expected disaster-bearing bodies is much wider. For example, flood control structures, such as dikes and dams as part of flood disaster prevention systems, are not only flood prevention forces but also disaster-bearing bodies under extreme floods [25]. In addition, extreme floods also likely cause negative impacts on sensitive ecological areas, such as protected surface and groundwater as well as nature reserves. On this basis, we draw the extreme flood disaster chain in Figure 2. The direct impacts include economic, social, and environmental consequences. Among them, the economic impacts include the damage to houses, agriculture, industry and commercial assets, and damage to critical infrastructures, such as transportation, water supply, power supply, and communication systems. Social impacts mainly refer to people affected by the evacuation, homelessness or injury, and potential human casualties [26]. Environmental impacts include scouring the sensitive ecological and environmental areas or polluting sensitive ecological and environmental areas after extreme floods flow through potential pollutants. Moreover, in modern society, since the construction of urban flood control and drainage facilities lags behind the construction speed of urbanization, and lifeline projects, such as communication, power supply and water supply, have more weaknesses, extreme floods are likely to result in much larger flood impacts than flooded areas. Then, extreme floods probably further lead to indirect impacts beyond the geographical area or after the flood events. For example, the disastrous river floods that occurred in Thailand in 2011 seriously disrupted the global supply chains [27].

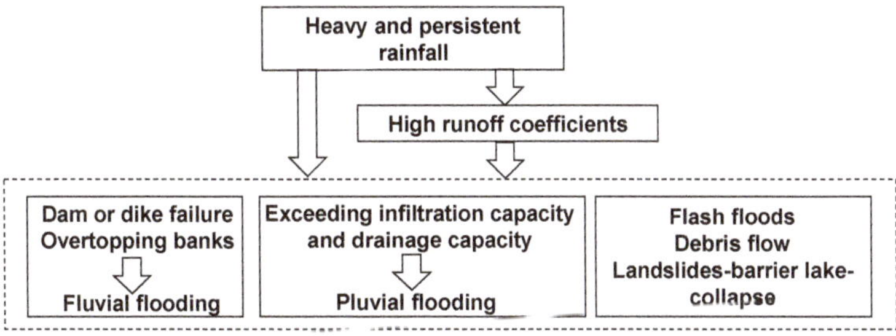

Figure 1. The causative mechanisms of extreme floods.

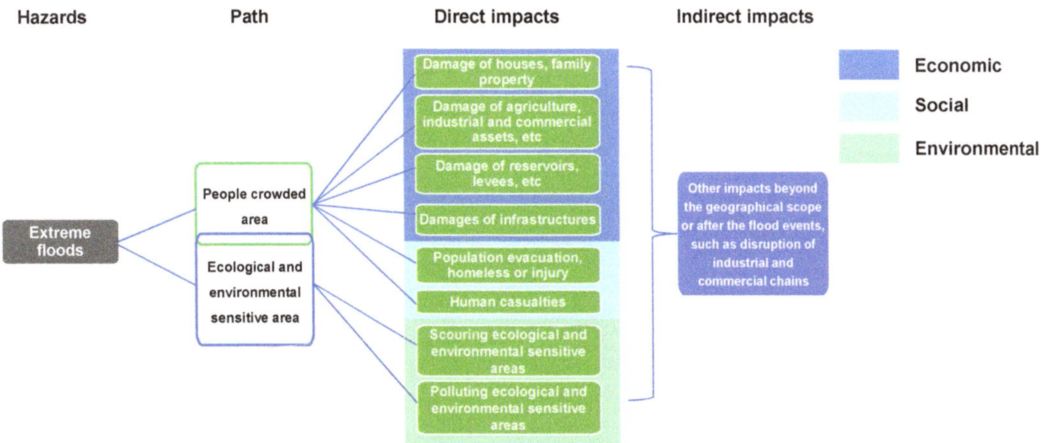

Figure 2. Extreme flood disaster chains (direct and indirect impact refer to the consequences occurring within the inundated region and far away from the flooded region during and/or after a flooding event, respectively).

3. Materials and Methods

3.1. Extreme Flood Disaster Indicator System

We establish the index system of extreme flood disasters (refer to Figure 3). The index system contains 10 assessment indicators from three aspects, i.e., social impacts, economic impacts, and environmental impacts.

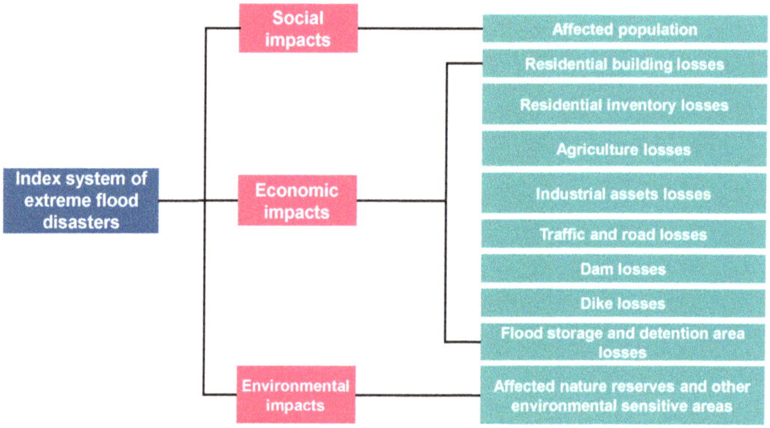

Figure 3. Extreme flood disaster indicator system.

(1) Social impacts are described by the expected number of people affected within the flooded area because ensuring the safety of people's lives is the primary task of flood prevention and emergency response. Understanding the number of affected populations is a prerequisite for flood management decision making and then providing support for personnel evacuation, emergency rescue, provision of medical and food relief, etc. In addition, various types of losses are closely related to human production and life, and the distributions of the affected population can further imply the impacts and losses of socio-economic activities and properties that are closely related to them in the floods. We do not consider the index of the number of people injured or dead because it is hard to quantify it accurately.

(2) Economic impacts are reflected by direct economic losses, including residential building losses, residential inventory damage, agriculture losses, losses of industrial enterprises, traffic and roads, as well as the losses of hydraulic engineering in the flooded area.

(3) Environmental impacts include the affected nature reserves and other affected environmental sensitive areas caused by extreme floods.

3.2. Flood Impact Assessment

3.2.1. Economic Impact Assessments

The economic impact evaluation is to measure the consequences of extreme flood disaster based on economic value data, including the losses of the residential buildings, inventories, agriculture, industry, infrastructure, etc. The assessment process is divided into 5 major steps as shown in Figure 4.

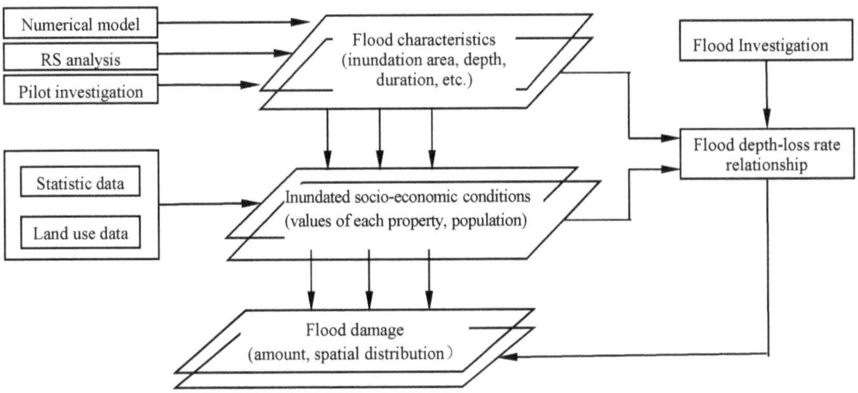

Figure 4. The flow chart of flood damage assessment.

(1) Analyze the flood inundation areas, flood depth, and durations using flood simulation model or RS analysis or field investigations.

(2) Collect economic data, statistical data, and land use spatial distribution data, and then use the area weighting method or regression analysis method to generate an economic database with spatial attributes that can reflect the distributions and differences of economic indicators.

(3) Overlay the flood inundation layer with the economic layer according to their spatial relationships using GIS analysis tools and then manage to obtain the values and distributions of economic properties under different flood depths.

(4) Collect historical flood loss data of typical regions, units, or industry sectors, etc., estimate the flood loss rate of each property under different flood depths, and establish the flood depth-loss rate relationship of each property.

(5) Based on the property type and flood depth-loss rate relationship in a flooded area, calculate flood economic losses according to Equation (1).

$$D = \sum_i \sum_j Vij\eta(i,j) \quad (1)$$

where V_{ij} is the value of ith property under jth water depth in a flooded area, $\eta(i,j)$ is the flood loss rate of ith property under jth water depth.

V_{ij} is mainly calculated using the current market value method. Housing property is calculated using the replacement cost method, i.e., based on the cost of a newly built house in the local area, and then obtaining the values after depreciation. Residential household property is discounted according to the current market value method based on the number of durable goods owned per 100 households in the administrative region involved. The values of other industrial and commercial assets are directly derived from the statistical yearbooks of the administrative regions. The cost of roads is considered in accordance with the repair costs by referring to the relevant national budgetary quotas for highway and railroad projects.

The flood loss rate represents the vulnerability of different properties impacted by extreme floods. The flood loss rate of different properties varies with flood depth. It is a relative index to depict direct economic losses and is described as the ratio of the loss values to pre-flood values of each property.

$$\eta = \frac{V_b - V_a + F}{S_b} \quad (2)$$

where η refers to flood loss rate, V_b is pre-flood properties' value, V_a is post-flood properties' value, F is rescue cost of properties.

3.2.2. Social Impact Assessments

The main assessment process is similar to that of economic impact assessment (refer to Figure 4). The total number of the affected population is estimated by overlaying the flood inundation area layer with the population distribution layer. To more accurately and comprehensively characterize the social impacts, the number of affected populations under different water depths and inundation duration can also be evaluated by overlaying the flood inundation depth/duration layer with the population distribution layer. In addition, if distribution data of the vulnerable population (the elderly, children, etc.) are available, the number of affected vulnerable populations under different depths/durations can also be estimated correspondingly.

The main focus of social impact assessment is to create a spatial distribution layer of population data. In China, the population data are usually presented in terms of administrative units, which express differences between statistical units. However, there are no differences within the statistical units. In order to conduct accurate statistical analyses of the affected population, spatial analysis of demographic data is employed.

(1) With the residential land layer

If the residential land layer is available, the spatial analysis of the demographic data is presented using the residential land method. We assume that the distribution of population is discrete and uniform within the residential land area of the administrative unit. The affected population is calculated by Equation (3).

$$P_e = \sum_i \sum_j A_{i,j} \cdot d_{i,j} \tag{3}$$

where P_e represents the affected population, $A_{i,j}$ is the flooded area of residential land in jth block of ith administrative unit, $d_{i,j}$ is the population density of the residential area in jth block of ith administrative unit.

If two administrative units are affected by extreme floods (gray part in Figure 5), we obtain the areas of flooded residential lands in each administrative unit (shaded parts in Figure 5) by superimposing the residential map layer, the flooded area layer, and the administrative division layer. It is worth noting that there is a big difference between urban population density and rural population density in a certain administrative unit. In practice, the smaller the level of an administrative unit (e.g., town/street), the smaller the differences in corresponding population densities on different rural residential plots or urban residential plots. Hence, we use the population density data of the smaller units and assume that the rural population or the urban population density of small administrative units can be considered equal.

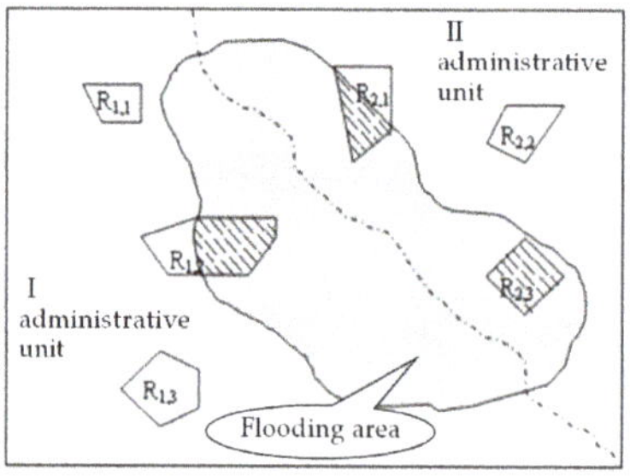

Figure 5. Diagram of flooded residential areas (the residential lands $R_{1,2}$ and $R_{2,1}$ are partially flooded, and the residential land $R_{2,3}$ is completely flooded).

(2) Without the residential land layer

If the residential land layer is not available, we use an approximate estimation to calculate the number of affected populations. We assume the population is evenly distributed in the whole administrative area, and the proportion of the affected population is the same as that of the flooded administrative area. The number of affected populations can then be calculated according to the total population of the administrative region using Equation (4).

$$P_e = \frac{PA_f}{A} \tag{4}$$

where P is the total population of an administrative region, A_f is the flooded area of an administrative region, A is the total area of an administrative region.

This approximate estimation method gives reasonable results only if the population is actually distributed evenly. In areas where that is not the case, the calculation results should be corrected. For example, flash floods usually occur in river valleys. However these are mountainous areas where the population is concentrated in low and flat areas on both riversides. In this case, this method usually underestimates the actual number of affected populations and should then be corrected using appropriate methods.

3.2.3. Environmental Impact Assessment

We evaluate the potential ecological and environmental impacts caused by extreme floods considering three aspects, i.e., flood inundation, the hazards of potential pollution sources, and the sensitivity of the protected areas. The assessment process concludes with 5 steps.

(1) Analyze the flood inundation areas.

(2) Collect and identify the potential pollutant sources and the protected areas within flooded areas, and then create corresponding distribution maps.

(3) Superimpose the flood inundation layer and the potential pollutant source layer to estimate the total dangerous scores of pollutant sources within inundation areas using Equation (5). Each pollutant source considers the hazard level of potential released contaminants, the numbers, and the scales of pollutant sources with the same hazard level. We analyze the typical contaminants of each pollutant source and then divide all of the potential sources into five levels according to the hazards or the toxicity of the typical pollutants (refer to Table 1).

$$S = \sum_{i=1}^{5} L_i \sum_{j=1}^{n} M_{ij} \tag{5}$$

where S is the total dangerous score of the potential pollutants within the flood extent; L_i is the hazard score of the pollutant source (see Table 1), $I = 1, 2, 3, 4, 5$; N_i is the corresponding number of pollutant sources with the same hazard score; M_{ij} is the scale of the jth pollutant source with the score L_i.

Table 1. Scores of potential pollutant sources with typical pollutants at different hazard levels.

Score L	Potential Pollutant Sources	Possible Typical Contaminants
5	Chemical plants	COD, NH_4^+, SO_2, toxic organic pollutants, solid waste, etc.
4	Metal plants, metal recovery works	COD, SO_2, metal, oxygen-demanding pollutants, solid waste, etc.
3	Factories	Solids, oxygen-demanding pollutants, solid waste, etc.
2	Landfills	Solids, solid waste, etc.
1	Farmlands	TN, TP, etc.

(4) Superimpose the flood inundation layer and the protected area distribution layer to estimate the total protection scores of the protected areas within inundation areas using Equation (6). We identify three kinds of protected areas, i.e., nature reserves/tourist attractions, protected surface water (e.g., surface drinking water source area), and protected groundwater (e.g., ground drinking water source area). Among them, nature reserves are divided into 3 categories according to the National Nature Reserve List, i.e., ecosystem type reserves, biological species reserves (e.g., wild animal protected areas), and natural

heritage reserves. According to their sensitivity levels, we assign a score to each protected area (refer to Table 2).

$$T = A + B + kC \tag{6}$$

where T is the total protection scores of the protected areas within the inundation area; A, B, and C are corresponding scores of nature reserves, protected surface water, and protected groundwater (refer to Table 2); k is 0.5 in this study.

Table 2. Scores of protected areas at different sensitivity levels.

Nature Reserves/Tourist Attractions		Surface Water		Ground Water	
Sensitivity Level	**Score *A***	**Sensitivity Level**	**Score *B***	**Sensitivity Level**	**Score *C***
World-class/national-level	10	Class I	10	Class I	10
Provincial-level	5	Class II	5	Class II	5
City-level	1	Class III	1	Class III	1

(5) Determine the classification criteria of dangerous level and protection level (see Table 3) and then combine the two aforementioned levels using the matrix method (see Figure 6) to determine the severity level of the environmental impacts.

Table 3. Classification criteria of dangerous level and protection level.

Pollutant Sources		Protection Areas	
Dangerous Level	**Score *S***	**Protection Level**	**Score *T***
5	20 and above	5	10 and above
4	15~19	4	8~9
3	10~14	3	5~7
2	5~9	2	3~4
1	0~4	1	0~2

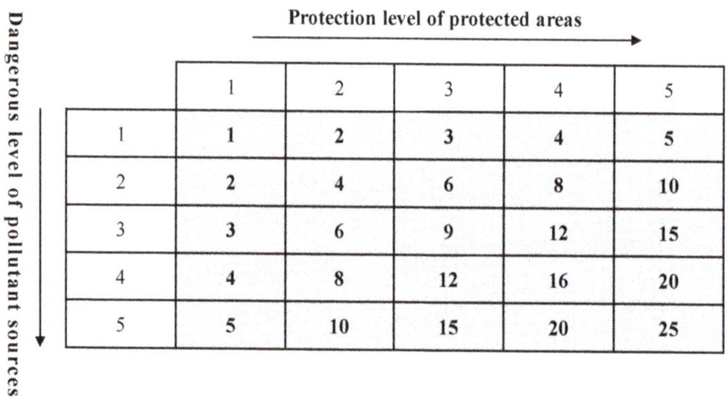

Figure 6. Matrix method used to determine the severity level of environmental consequences.

4. Case Study

4.1. Study Area

JFDD is a flood storage and detention area (FSDA), the low-lying area for the temporary storage of floods outside the river levees, along the Yangtze River. It is located in Gongan County, Hubei Province (Figure 7), covering 920.6 km^2. The topography within JFDD is higher in the north and lower in the south. The average width from north to south is about 70 km and 13 km wide from east to west. The narrowest width is 2.7 km at the neck area. JFDD, constructed in 1952, is one of the most important flood control works in the Yangtze

River Basin. Its function is to temporarily store floods from the Yangtze River. The design flood storage level at the golden mouth is 42.00 m, the design flood storage volume is 5.4 billion m^3, and the flood diversion flow is 7700 m^3/s [28]. JFDD was used only once in 1954 to prevent extreme floods of the Yangtze River. With the socio-economic development, JFDD has become the political, economic, and cultural center of Gongan County. There are 8 towns and 211 villages in JFDD [29], and the whole population is about 0.61 million up to 2019. There are more than 70 large-scale enterprises in the county, and most of them are located in the JFDD.

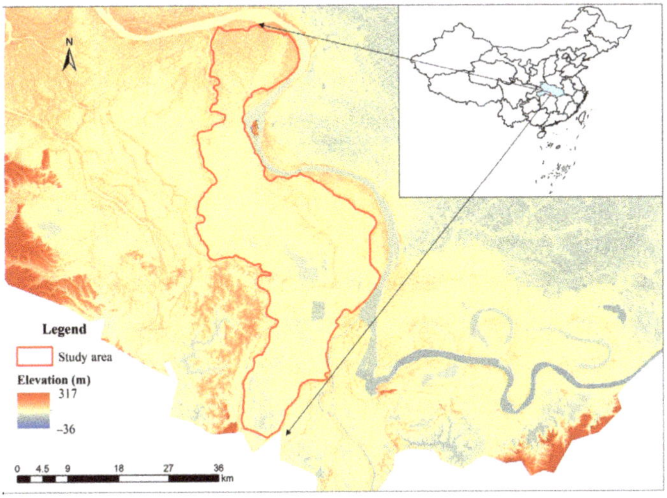

Figure 7. Location of the study area (JFDD), which is located in Hubei province (the light blue area on the map of China) and along the Yangtze River. The reference coordinate system is WGS84.

4.2. Data

The town-level socio-economic data in the study area were collected from open data and field investigation (Table 4). The economic classification data considered the main affected types, namely residential buildings and inventories, agriculture, industrial and commercial assets, and roads. The spatial distribution of socio-economic statistical data was based on 1:50,000 geographic data. The environmental data, including pollutant sources and protection levels, were collected from investigation and open data.

Table 4. Socio-economic and environmental data.

Data	Sources
Geographic data	Investigation
GDP	Statistical Yearbook of Hubei Province
Population	Statistical Yearbook of Hubei Province Statistical Yearbook of Gongan County
Industrial and commercial assets, agriculture areas and productions, and other socio-economic data	China County (City) Socio-economic Statistics Yearbook Statistical Yearbook of Gongan County Field Investigation
Pollutant sources (names, types)	Field Investigation
Protected areas (names, types, and sensitivity level)	National Nature Reserve List Field Investigation

5. Results

In this study, we set two extreme flood scenarios, i.e., 1000-year return period flood under 1954 flood type (hereinafter referred to as 54_1000) and 1000-year return period flood under 1998 flood type (hereinafter referred to as 98_1000). Under these two scenarios, we evaluated the number of affected populations, direct economic losses (residential building and inventory losses, agriculture losses, industry and commercial assets losses, and road losses), and potentially affected nature reserves and other sensitive areas. The flood hazards, including flood depths, durations, and velocities, were simulated using a 2D hydro-hydraulic model. Since flood hazard evaluation is not the focus of this paper, it was not elaborated on.

5.1. Social and Economic Impacts

We analyzed the historical data regarding flood economic losses in 1954 and referenced nearby or similar areas to determine the flood depth-loss rate relationship. In addition, we adjusted the flood loss rate according to the socio-economic conditions and disaster characteristics of the study area. On this basis, we established the flood loss rate relationship between flood depths and different affected objects in JFDD (refer to Table 5). The flood economic losses and affected populations are shown in Table 6.

Table 5. Flood depth-loss rate (%).

Flood Depth	Residential Building	Residential Inventory	Agriculture	Industry Assets	Commercial Assets	Railway	National and Provincial Roads	Roads below Provincial Level
0.05~0.5m	2	3	8	5	7	3	3	3
0.5~1.0m	10	15	41	10	27	12	9	15
1.0~1.5 m	18	22	49	15	34	17	15	20
1.5~2.0m	24	29	60	21	42	22	20	25
2.0~2.5m	30	34	72	30	49	27	23	30
2.5~3.0m	36	42	78	39	57	32	26	35
>3.0m	45	60	88	55	64	40	33	42

Table 6. The assessment results of social and economic impacts.

Scenario	Inundation Area (km^2)	Affected Population (million)	Economic Losses (billion RMB)				
			Residential Building and Inventory	Agriculture	Industry and Commercial Assets	Roads	Total
54_1000	901.36	0.51	13.27	3.88	1.13	0.54	18.83
98_1000	879.49	0.5	9.45	3.52	0.94	0.42	14.33

When facing extreme floods, almost the whole area of JFDD was flooded (see Table 6). Hence, the difference between the inundation areas is not obvious. However, the flood depth differences are obvious. The inundation area under the 54_1000 scenario is 901.36km^2, and the inundated area with a flood depth of more than 3m accounts for 95.6%. By contrast, the inundation area under the 98_1000 scenario is 879.49 km^2 and 62.4% flood depths area deeper than 3 m.

Almost all the population in the JFDD has been affected by floods, except those who live in some upland areas and safety zones. The floods also caused serious economic losses of RMB 14.332 billion and RMB 18.828 billion in the 98_1000 flood and the 54_1000 flood, respectively. The flood losses caused by the 54_1000 flood were more severe than the 98_1000 flood because the 54_1000 flood caused more severe inundation in the more densely populated and asset-rich areas.

5.2. Environmental Impacts

There are five potential chemical enterprises located in JFDD. All of them were flooded under the two scenarios. The hazard score of each pollutant source is 5 according to Table 1, and the corresponding hazard levels under two flood scenarios are 5, respectively. Besides, there are two national-level protected areas, i.e., Chonghu national wetland park and Jingjiang flood diversion project, and nine city-level protected areas within the flooded areas. Hence, the expected sensitivity levels of protected areas are 5 under the two scenarios according to Table 3. In this case, the severity levels of environmental impacts under the two scenarios are 5. Hence, extreme floods have relatively significant impacts on local ecology and the environment in JFDD (refer to Table 7).

Table 7. The assessment results of environmental impacts.

Scenario	The Hazard Level of Pollutant Sources			The Sensitivity Level of Protected Areas			Severity Level
	The Number of Pollutant Sources within the Flooded area	Hazard Scores	Hazard Level	The Number of Protected Areas	Protection Scores	Protection Level	
54_1000	5	25	5	2 national-leve 19 city-level	29	5	25
98_1000	5	25	5	2 national-level 9 city-level	29	5	25

6. Discussion

Several reasons explain the serious social, economic, and environmental impacts under two extreme flood scenarios. First, JFDD's function is to temporarily store floods and then lower the flood levels in the Yangtze River. If extreme floods exceed the flood prevention standard of the Yangtze River levees, the JFDD will be used and divert a large number of flood volumes. Floods will soon fill the whole area, except for the safety areas and other highlands [29]. Second, socio-economic development has been fast in the past few decades, since JFDD has never been used since 1954. The population density was 668 people/km^2 and the GDP per capita was 3.05 RMB per person in 2019, higher than in the surrounding areas, according to our investigation. The average annual growth rate of GDP in JFDD was 14.6% from 2004 to 2018. In this case, despite the management methods to restrict polluting enterprises, there are still several enterprises with potential contaminants, such as chemical plants and wastewater treatment plants. Extreme floods may lead to contaminant leakage or a release of organic loads [9] when flowing through those exposed potential pollutant sources. The polluted floods can result in surface and groundwater pollution, as well as soil contamination [8]. Hence, the population living or working, the assets, and the protected surface/groundwater in JFDD, exposed to extreme floods, are vulnerable. They are seriously affected when experiencing extreme floods.

In this study, we only estimated the total number of affected populations within the inundation areas and used this indicator to represent the social impact due to the special characteristics of JFDD. Once JFDD, as an FSDA, is decided to be used, most of the population living or working in it will be evacuated to the safety zones, highlands, or outside the JFDD, according to the plan. In addition, the duration of flood diversion will last for more than 120 h. In this case, the total number of affected populations can represent the social impacts in this study area.

Flood loss rates are influenced by many factors, such as topography, flood characteristics (flood depth and duration, etc.), property categories, flood season, and emergency measures. Generally, the flood depth-loss rate relationship is established according to the conditions of study areas and property categories. A certain number of flood-affected regions (or similar areas affected by the flood disasters in recent years) should be selected to investigate the flood loss rate of different properties under different inundation ranges. Since JFDD was only used in 1954, it lacks historical data of extreme flood disasters in the past few decades. Hence, the flood depth-loss rate relationship in this paper was established by referring to the data from similar areas. In addition, we adjusted and verified the flood depth-loss rate relationship by consulting stakeholders, such as construction engineers, enterprises owners, and related managers. Besides, we further compared the flood economic losses with those in similar areas. For example, in terms of the classification of flood economic losses, residential building and inventory losses account for the largest share of the total economic losses, followed by agriculture losses, which is consistent with other studies in similar areas [30]. The comparisons show that the flood consequences of this study are basically scientific and reasonable, which can provide a reference for the flood regulation decision making of JFDD.

One of the aims of this study is to propose a methodology that can assess the comprehensive consequences, social impacts, economic losses, and environmental impacts, caused by extreme floods. Limited by data, there are several limitations both in the social and economic impact evaluation and the environmental impact assessment.

This study established the relationship between flood depth and flood loss rate. Other characteristics of flood hazards, such as flood velocity, duration, and arrival time, may also influence the final flood economic losses, which should be studied in depth and be incorporated into the loss evaluation model as future work. In addition, with the variations in characteristics of flood economic losses, underground space, vehicles and other seriously affected properties in recent flood disasters should also be involved in flood economic losses evaluations. In terms of social impacts, the human casualty assessments based on the assessment method of the affected population should be studied in the future as well.

The environmental impact assessment is based on the assessment of potential contaminant release from pollutant sources and the sensitivity of protected areas with a semi-quantitative method. A more accurate simulation requires a micro-scale approach [9] to describe the transport of pollutants. Such evaluation needs detailed data on each pollutant source. However, these data are not open and are hard to acquire in practice. The micro-scale evaluation methods of environmental impacts will be studied when detailed data are available. In addition, the distances between the potential pollutant sources and the protected objects have different levels of impact on the protected areas, which is not considered in this study. For example, the pollutant sources within the flood extent located in the vicinity of the protected areas or close to aquifers used for domestic water supply have a greater impact than those located far away. The pollutant sources within the flooded area located upstream of the protected areas have a greater impact than those located downstream.

7. Conclusions

The causes of extreme floods and subsequently caused potential comprehensive influences of extreme flood disasters were analyzed in depth in this study. The causative mechanisms and extreme disaster chains were drawn. Heavy and/or persistent precipitations are usually the very first cause of extreme floods. Extreme flood disasters form when extreme floods cause human life loss, property loss, and sensitive ecological and environmental area damage. The potential consequences can be divided into three categories, i.e., social impacts, economic impacts, and environmental impacts. Social impacts are the potentially affected (injured, homeless, etc.) or dead people caused by extreme floods. Economic impacts reflect the direct/indirect economic losses, including residential building losses, agriculture losses, industrial losses, as well as the losses of hydraulic engineering caused by floods. Environmental impacts reveal that the protected areas are polluted due to possible contaminants spread from different types of pollutant sources, or extreme floods directly scouring the sensitive ecological and environmental areas.

On this basis, an extreme flood disaster index system and the corresponding evaluation methods, i.e., a refined social and economic impact evaluation method and a semi-quantitative environmental impact evaluation method, were proposed. The proposed methods were then applied to the JFDD to evaluate the economic impacts, social impacts, and environmental impacts under two extreme flood scenarios. The result indicates that almost all of the JFDD area is inundated by extreme floods with inundation areas of 901.36 km^2 and 879.49 km^2, respectively. The potentially affected populations are 0.51 million and 0.5 million. In addition, the possible direct economic losses are RMB 18.83 billion and RMB 14.33 billion, respectively. Among them, the residential building and inventory losses are the largest, followed by agriculture losses. It is likely that the nature reserves within JFDD are seriously influenced due to possible contaminant spread from five potential pollutant sources. The detailed analysis results provide effective information for decision making in flood management.

We focused mainly on the direct flood impact assessments. However, under rapid socio-economic developments, indirect impacts become more and more significant. Therefore, we intend to study indirect flood impacts in future work. Furthermore, we could explore artificial intelligent technology to assess flood impacts in a data-driven manner, such as learning about the relationship between flood hazards and flood losses. In addition, future work should also include underground space, vehicles, and other seriously affected properties during recent extreme flood disasters in flood damage assessments.

Author Contributions: Conceptualization, Q.Y. and Y.W.; methodology, Q.Y. and Y.W.; formal analysis, Y.W. and Q.Y.; data curation, Q.Y.; investigation, Y.W. and Q.Y.; writing—original draft preparation, Q.Y. and Y.W.; writing—review, N.L. All authors have read and agreed to the published version of the manuscript.

Funding: This research was funded by the National Natural Science Foundation of China [No. 51909273], the National Key Research and Development Program of China [No. 2021YFC3001404], the Yangtze River Water Science Joint Fund Project [No. U2240203], and Talent Innovation Team for the Strategic Research on Flood and Drought Disaster Prevention of the Ministry of Water Resources [No. WH0145B042021]. The support provided by the IWHR Talented International Expert Program is also acknowledged.

Institutional Review Board Statement: Not applicable.

Informed Consent Statement: Not applicable.

Data Availability Statement: The data presented in this study are available on request from the corresponding author.

Acknowledgments: The authors appreciate the editors and anonymous reviewers for their great efforts on the manuscript.

Conflicts of Interest: The authors declare no conflict of interest.

References

1. Tanoue, M.; Taguchi, R.; Alifu, H.; Hirabayashi, Y. Residual flood damage under intensive adaptation. *Nat. Clim. Change* **2021**, *11*, 823–826. [CrossRef]
2. UN Office for Disaster Risk Reduction. *The Human Cost of Disasters: An Overview of the Last 20 Years (2000–2019)*; UN Office for Disaster Risk Reduction: Geneva, Switzerland, 2020.
3. Agrawal, N.; Elliott, M.; Simonovic, S.P. Risk and Resilience: A Case of Perception versus Reality in Flood Management. *Water* **2020**, *12*, 1254. [CrossRef]
4. WEF. *The Global Risks Report*, 14th ed.; World Economic Forum: Davos, Switzerland, 2019.
5. Summary for Policymakers. In *Climate Change 2021: The Physical Science Basis*; IPCC: Geneva, Switzerland, 2021.
6. Schiermeier, Q. Increased flood risk linked to global warming. *Nature* **2011**, *470*, 316. [CrossRef]
7. Winsemius, H.C.; Aerts, J.C.; Van Beek, L.P.; Bierkens, M.F.; Bouwman, A.; Jongman, B.; Kwadijk, J.C.; Ligtvoet, W.; Lucas, P.L.; Van Vuuren, D.P.; et al. Global drivers of future river flood risk. *Nat. Clim. Change* **2016**, *6*, 381–385. [CrossRef]
8. Lynch, S.F.; Batty, L.C.; Byrne, P. Critical control of flooding and draining sequences on the environmental risk of Zn-contaminated riverbank sediments. *J. Soils Sediments* **2017**, *17*, 2691–2707. [CrossRef]
9. Arrighi, C.; Masi, M.; Iannelli, R. Flood risk assessment of environmental pollution hotspots. *Environ. Model Softw.* **2018**, *100*, 1–10. [CrossRef]
10. Bhowmik, N.G. Impacts of the 1993 Flood on the Mississippi River in Illinois. In *Hydraulic Engineering*; ASCE: Reston, VA, USA, 1994; pp. 613–617.
11. Scawthorn, C.; Flores, P.; Blais, N.; Seligson, H.; Tate, E.; Chang, S.; Mifflin, E.; Thomas, W.; Murphy, J.; Jones, C.; et al. HAZUS-MH flood loss estimation methodology. II. Damage and loss assessment. *Nat. Hazards Rev.* **2006**, *7*, 72–81. [CrossRef]
12. Kreibich, H.; Van Den Bergh, J.C.; Bouwer, L.M.; Bubeck, P.; Ciavola, P.; Green, C.; Hallegatte, S.; Logar, I.; Meyer, V.; Schwarze, R.; et al. Costing natural hazards. *Nat. Clim. Change* **2014**, *4*, 303–306. [CrossRef]
13. Allaire, M. Socio-economic impacts of flooding: A review of the empirical literature. *Water Secur.* **2018**, *3*, 18–26. [CrossRef]
14. Merz, B.; Kreibich, H.; Schwarze, R.; Thieken, A. Review article "Assessment of economic flood damage". *Nat. Hazards Earth Syst. Sci.* **2010**, *10*, 1697–1724. [CrossRef]

15. Wang, Y.Y.; Wang, J.; Hu, C.W.; Liu, L.; Yu, Q. Benefit simulation of flood control project in Taihu Lake basin under extreme floods. *Adv. Water Sci.* **2020**, *31*, 885–896. (In Chinese)
16. Jin, Y.; Zhang, J.; Liu, N.; Li, C.; Wang, G. Geomatic-Based Flood Loss Assessment and Its Application in an Eastern City of China. *Water* **2022**, *14*, 126. [CrossRef]
17. Carisi, F.; Schröter, K.; Domeneghetti, A.; Kreibich, H.; Castellarin, A. Development and assessment of uni-and multivariable flood loss models for Emilia-Romagna (Italy). *Nat. Hazards Earth Syst. Sci.* **2018**, *18*, 2057–2079. [CrossRef]
18. Figueiredo, R.; Schröter, K.; Weiss-Motz, A.; Martina, M.L.; Kreibich, H. Multi-model ensembles for assessment of flood losses and associated uncertainty. *Nat. Hazards Earth Syst. Sci.* **2018**, *18*, 1297–1314. [CrossRef]
19. Gerl, T.; Kreibich, H.; Franco, G.; Marechal, D.; Schröter, K. A review of flood loss models as basis for harmonization and benchmarking. *PLoS ONE* **2016**, *11*, e0159791. [CrossRef] [PubMed]
20. Pinelli, J.P.; Da Cruz, J.; Gurley, K.; Paleo-Torres, A.S.; Baradaranshoraka, M.; Cocke, S.; Shin, D. Uncertainty reduction through data management in the development, validation, calibration, and operation of a hurricane vulnerability model. *Int. J. Disast. Risk Reduct.* **2020**, *11*, 790–806. [CrossRef]
21. Merz, B.; Blschl, G.; Vorogushy, N.S.; Dottori, F.; Macdonald, E. Causes, impacts and patterns of disastrous river floods. *Nat. Rev. Earth Environ.* **2021**, *2*, 592–609. [CrossRef]
22. Rosenzweig, B.R.; McPhillips, L.; Chang, H.; Cheng, C.; Welty, C.; Matsler, M.; Iwaniec, D.; Davidson, C.I. Pluvial flood risk and opportunities for resilience. *Wiley Interdiscip. Rev. Water* **2018**, *5*, e1302. [CrossRef]
23. O'Connell, P.E.; Ewen, J.; O'Donnell, G.; Quinn, P. Is there a link between agricultural land-use management and flooding? *Hydrol. Earth Syst. Sci.* **2007**, *11*, 96–107. [CrossRef]
24. Sharafati, A.; Yaseen, Z.M.; Shahid, S. A novel simulation–optimization strategy for stochastic-based designing of flood control dam: A case study of Jamishan Dam. *J. Flood Risk Manag.* **2021**, *14*, e12678. [CrossRef]
25. Yu, Q.; Li, N.; Wang, Y.Y. Review of flood management based on the concept of resilience. *China Flood Drought Manag.* **2021**, *31*, 19–25. (In Chinese)
26. Mohanty, M.P.; Simonovic, S.P. Understanding dynamics of population flood exposure in Canada with multiple high-resolution population datasets. *Sci. Total Environ.* **2021**, *759*, 143559. [CrossRef] [PubMed]
27. Haraguchi, M.; Lall, U. Flood risks and impacts: A case study of Thailand's floods in 2011 and research questions for supply chain decision making. *Int. J. Disast. Risk Reduct.* **2015**, *14*, 256–272. [CrossRef]
28. Huang, Y.; Li, C.W.; Li, A.Q.; Wang, Q.; Zhu, S.R. Research on decision support techniques of over-standard flood emergency evacuation. *J. Hydraul. Eng.* **2020**, *51*, 805–815. (In Chinese)
29. Guo, P.; Xia, J.Q.; Chen, Q.; Li, N. A mechanics-based model of flood risk assessment and its application in a flood diversion zone. *Adv. Water Sci.* **2017**, *28*, 858–867. (In Chinese)
30. Wang, J.; Wang, Y.Y.; Li, N.; Zhang, N.Q. Study on flood losses and disaster reduction benefits assessment of key flood control projects inTaihu Basin in 2016. *Water Resour. Hydropower Eng.* **2019**, *50*, 176–183. (In Chinese)

water

MDPI

Article

Co-Design for Enhancing Flood Resilience in Davao City, Philippines

Mamoru Miyamoto [1,*], Daiki Kakinuma [1], Tomoki Ushiyama [1], Abdul Wahid Mohamed Rasmy [1], Masaki Yasukawa [2], Della Grace Bacaltos [3], Anthony C. Sales [4], Toshio Koike [1] and Masaru Kitsuregawa [5]

1 International Centre for Water Hazard and Risk Management, Public Works Research Institute, Tsukuba 3058516, Japan; kakinuma-d977bt@pwri.go.jp (D.K.); ushiyama55@pwri.go.jp (T.U.); abdul@pwri.go.jp (A.W.M.R.); t-koike@pwri.go.jp (T.K.)
2 Earth Observation Data Integration and Fusion Research Initiative, The University of Tokyo, Tokyo 1538505, Japan; yasukawa@iis.u-tokyo.ac.jp
3 Davao del Sur State College, Digos City 8008, Philippines; bacaltosdella@gmail.com
4 Department of Science and Technology Region XI, Davao City 8000, Philippines; dr.acsales@region11.dost.gov.ph
5 National Institute of Informatics, Tokyo 1018430, Japan; kitsure@tkl.iis.u-tokyo.ac.jp
* Correspondence: mmiyamoto@pwri.go.jp; Tel.: +81-29-879-6779

Abstract: Enhancing flood resilience, including the development of social capacity and early warning systems, in addition to structural measures, is one of the key solutions to mitigating flood damage, which will be more intensified in the future due to climate change. This study was conducted to develop a comprehensive methodology for enhancing flood resilience by improving society-wide disaster literacy under the governance formed through the active participation of all levels of stakeholders in Davao City, Philippines. Specifically, the development of the Online Synthesis System for Sustainability and Resilience, which integrates different disciplines, and the fostering of Facilitators, whose role is to interlink the science community and society, were implemented in a co-designing manner by the collective governance body. The development of basin- and barangay-scale hydrological models realized real-time flood forecasting and climate change impact assessment to identify intensified flood risk under the future climate. Co-designed e-learning workshops were held to foster about thirty Facilitators and help them produce twenty-one risk communication plans and workshop designs for fourteen barangays considering geographic, demographic, economic, and social features that they can utilize for public dissemination related to climate change adaptation to the target audiences in society. This paper presents a practical method to enhance flood resilience, demonstrating that the synthesis of science-based knowledge and human resource development can fill the gaps between the science community and society.

Keywords: OSS-SR; facilitator; flood resilience; disaster literacy; community-based; e-learning

Citation: Miyamoto, M.; Kakinuma, D.; Ushiyama, T.; Rasmy, A.W.M.; Yasukawa, M.; Bacaltos, D.G.; Sales, A.C.; Koike, T.; Kitsuregawa, M. Co-Design for Enhancing Flood Resilience in Davao City, Philippines. *Water* **2022**, *14*, 978. https://doi.org/10.3390/w14060978

Academic Editors: Slobodan P. Simonovic, Subhankar Karmakar and Zhang Cheng

Received: 28 February 2022
Accepted: 16 March 2022
Published: 20 March 2022

1. Introduction

Given the floods that have occurred around the world in recent years, flood risk management is undoubtedly an urgent and major social issue. The World Meteorological Organization has reported that the number of weather-related disasters that have hit the world has increased five-fold over the past 50 years [1]. The trend is due to the increased water vapor in the atmosphere, exacerbating extreme rainfall and deadly flooding, as a result of climate change. More than 90% of the mortalities due to weather-related disasters have occurred in developing countries, though the death number has decreased significantly over the last 50 years. This fact implies that strengthening resilience, such as the development of social capacity and early warning systems, is one of the key solutions to averting water-related disasters, which will be more intensified in the future due to climate change.

In international settings on disasters, the term "resilience" is quoted with the UNISDR's definition in the Hyogo Framework for Action 2005–2015 [2]: "The capacity of a system, community or society potentially exposed to hazards to adapt, by resisting or changing in order to reach and maintain an acceptable level of functioning and structure." However, Fisher (2015) pointed out that more than 70 definitions of resilience existed across the scientific literature [3] and that they roughly fall into two categories: "the ability of a system to bounce back after stress" and "the capacity of social-ecological systems to adapt or transform in response to unfamiliar, unexpected and extreme shocks". Kerri (2020) systematically reviewed the literature on flood risk management and clarified the position and importance of flood risk management in the interdisciplinary understanding of resilience [4]. Efforts have also been made to verify and compare the effect of resilience by devising different approaches, for example, developing frameworks, indices, and models to quantitatively assess flood resilience [5–11]. For the quantitative assessment of flood resilience, while hydrological models or indicators are employed, a unified approach has not yet been established because adjustments to the study area are often necessary. In addition to the importance of resilience assessment, an emphasis should also be placed on how to enhance flood resilience from the practical perspective of flood risk mitigation. Some case studies of qualitative approaches have attempted to enhance the flood resilience of specific communities through particular methods of risk communication or awareness-raising activities using web-based systems [12–15]. However, the existence of various stakeholders and communities in society and the segmentalization of scientific disciplines nowadays make it difficult to realize the actual enhancement of the flood resilience of society as a whole. Morrison et al. (2017) presented five themes—stakeholder engagement; policy effectiveness; research in practice; tools; and frameworks—by investigating literature on governance for flood resilience [16]. They also concluded that research on governance for flood resilience lacks integration of themes, and methods of mitigating this lack of integration are poorly studied. Hence, the development of holistic methods for substantially enhancing flood resilience at all levels of stakeholders and communities is an indispensable research theme for society and has not been addressed in previous studies.

This study aims to develop a comprehensive methodology to enhance flood resilience in society as a whole by improving society-wide disaster literacy based on the governance in which all stakeholders engage. This study, therefore, demonstrates the importance of co-design by the science community and all levels of society in Davao City, Philippines, for beneficial and sustainable achievements. Figure 1 shows the overall structure that this paper uses to formulate a method to enhance society-wide flood resilience based on co-design among relevant stakeholders. We begin with the localization of a general concept of flood resilience enhancement in the study area. In line with the localized concept design, we developed an integrated system, Online Synthesis System for Sustainability and Resilience (OSS-SR), for consilience and fostered Facilitators, which are detailed in the Methodology presented in Section 3. We also produced an institutional design of interactive risk communication by Facilitators to fill gaps between the science community and society for the enhancement of flood resilience.

2. Study Area

Davao City has 182 barangays, the smallest political unit in the Philippines [17]. Its population growth rate from 1990 to 2010 was 2.4%, with the total population rising to 1,449,296 in 2010. The population density was 5.9 persons per hectare in 2010. Among the urban barangays, the population density reached 43 persons per hectare, whereas the population density among the rural barangays was estimated to be 1.5 persons per hectare. Agricultural land accounts for the largest share in the land use classification, with 30% of the total land area of Davao City. The second highest land type, forest areas, occupies 16.4%, and urban use 5.4%.

The Davao River is one of the 18 major rivers in the Philippines and is prioritized for master planning because of its socio-economic importance. Severe floods often occur in January, June, and July. In 2011, more than 10,000 families were affected by floods in Davao City. Out of its 130 km^2 built-up area, 10.9 km^2 is considered highly flood-prone, and 42.1 km^2 is classified as residential areas. Figure 2 shows the study area of 3644 km^2, covering eight river basins surrounding Davao City, including the Davao River basin. As part of the Hydrology for the Environment, Life and Policy (HELP)-UNESCO Program, the HELP Davao Network, which has been one of the most active local initiatives, was established for better informing complex decisions and hard choices concerning the wise management and use of water.

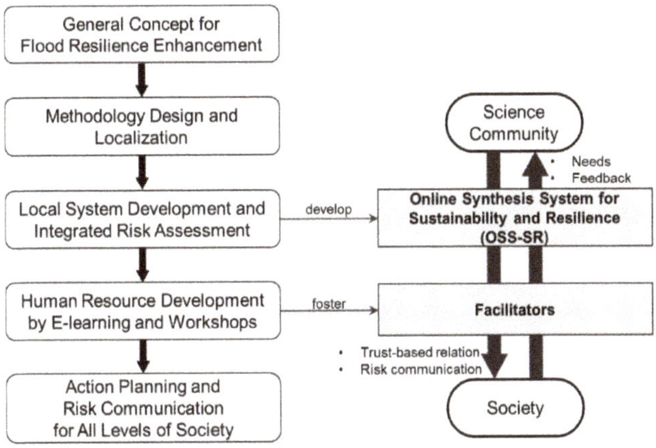

Figure 1. Structure of this study for formulating a method to enhance society-wide flood resilience based on co-design among relevant stakeholders.

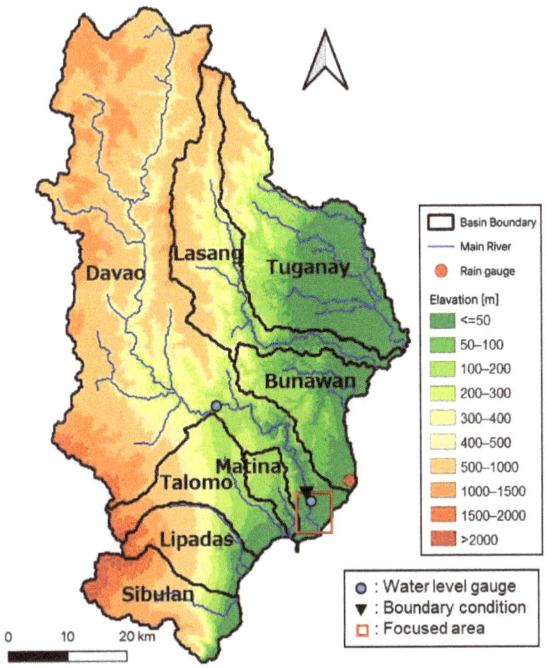

Figure 2. Eight river basins around Davao City, including the Davao River basin. The locations of rainfall gauges, water level gauges, boundary conditions, and the focus area are also shown.

3. Methodology

3.1. General Concept for Disaster Resilience Enhancement

Major challenges that science faces in order to improve disaster literacy and enhance disaster resilience at all levels of society include how to integrate different disciplines of science and how to fill the gaps with local communities. In its recommendation published in 2020 [18], the Committee on International Cooperation for Promoting Science-Based Disaster Risk Reduction of the Science Council of Japan addressed necessary elements to enhance disaster resilience: consilience and human resources. Methodologies to strengthen consilience and human resources are described concerning developing the OSS-SR, which is designed to integrate knowledge, information, experience, awareness, and models under different disciplines, and to foster catalytic Facilitators, whose role is to interlink the scientific community and the local society. This study employs the methodologies in the recommendation and implemented them as an end-to-end science covering cutting-edge science and community practices to enhance resilience to water-related disasters in the Philippines.

During the eighth meeting of High-level Experts and Leaders Panel on Water and Disasters (HELP) on 1 November 2016 [19], the International Flood Initiative (IFI) discussed flood resilience under climate change by organizing a side event, and the HELP-IFI Jakarta Statement based on discussions in the side event was adopted. The statement highlights the interdisciplinary and transdisciplinary partnership to establish a platform as part of a national platform for facilitating dialogue among all stakeholders from national, local, and community levels [20]. With this statement and the recommendation in the outcome document of High-Level Panel on Water released in 2018 [21], Platforms on Water Resilience and Disasters (PLATFORM), which are institutional frameworks involving all levels of stakeholders to enhance the resilience of the whole society for water-related disasters, have been established since 2017 in the Philippines, Sri Lanka, Myanmar, Indonesia, and other countries. PLATFORM also focuses on scientific themes of data integration, flood forecasting, contingency planning, climate change impact, economic assessment, and agricultural productivity, as well as capacity development, in order to contribute to the creation of social benefits in such areas as decision-making, policy-making, local practice, and investment principles. The governance of PLATFORM helps social implementations to develop OSS-SR and foster Facilitators, as well as support them in starting cooperation with Davao City in the Philippines.

3.2. Development of OSS-SR for Davao City

The general concept for enhancing disaster resilience has been localized in Davao City by accumulating discussions with stakeholders and considering climatic, geopolitical, and social aspects. Figure 3 shows the concept design and OSS-SR for Davao City, highlighting two focuses of real-time flood forecasting and climate change impact assessment. OSS-SR also has an e-learning function to provide users with ten introductory lectures, examinations, and four classes of hands-on training, widely covering issues in climate change, flood management, and disaster risk reduction. The e-learning function enables fostering Facilitators, who translate scientific knowledge and information for actors in society, including decision-makers, policymakers, government officers, DRR practitioners, civil society organizations, the private sector, the media, and local communities.

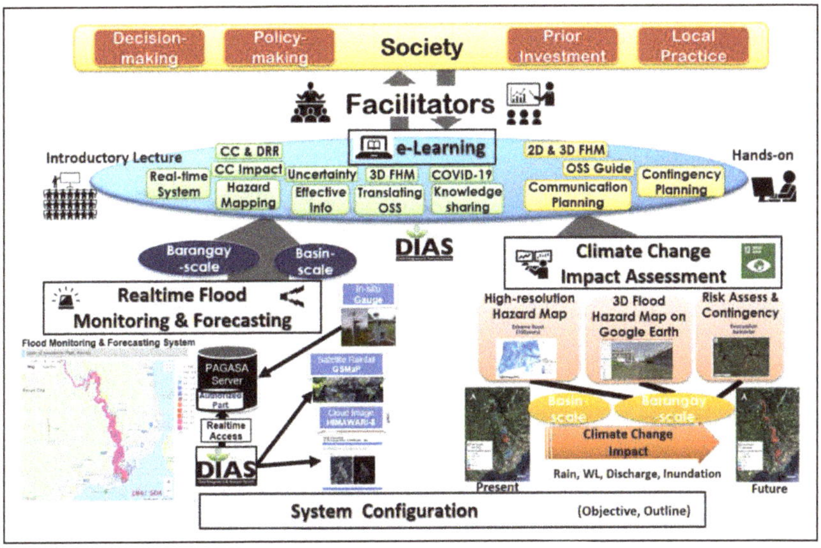

Figure 3. Concept design of the Online Synthesis System for Sustainability and Resilience (OSS-SR) and the Facilitator fostering for Davao City in the Philippines.

3.3. Hydrological Model Development

We developed two scales of hydrological inundation models: a basin-scale model with 1 km resolution for creating an understanding of the overall picture of the basin and a barangay-scale model with 40 m resolution to investigate the detailed conditions of critical infrastructure in the focus area. The focus area and locations of gauging stations and boundary conditions for the barangay-scale model are also shown in Figure 2.

The basin-scale model employed the WEB-RRI model, a water and energy budget-based Rainfall-Runoff-Inundation model applicable to wet- and dry-climate river basins [22]. The model was calibrated by the river discharge of the annual simulation of 2002 at the Lacson station, the upstream water-level station. The model was also validated by the river discharge of the annual period and the biggest flood event of 2008, as shown in Figures 4 and 5, respectively. Due to the low availability of long-term hydrometeorological gauge data, we employed satellite-based rainfall data, Climate Hazards Group InfraRed Precipitation with Station data (CHIRPS) [23], for calibration and validation. Although there are still challenges in calibration and validation because of the uncertainty of the rainfall data themselves, Figure 4 shows a reasonable agreement between the observed and simulated discharge. The event-based comparison provided in Figure 5 also shows rough agreement, though the Nash–Sutcliffe efficiency [24] is a modest value of 0.32. While the cause of this relatively low index value has not been clearly identified, it may be due to the insufficient reliability of the satellite-based rainfall input. The barangay-scale model employs the RRI model, which focuses on floods and simultaneously simulates rainfall-runoff and flood inundation processes on a 2-D basis at a river basin scale [25].

These models are connected by inputting the river discharge simulated by the basin-scale model into the barangay-scale model as a boundary condition. The synthesis of these two models with different spatial scales in the Data Integration Analysis System (DIAS) realized real-time local flood forecasting and a climate-change impact assessment. DIAS is a system for the creation of an information storage infrastructure for applications of public benefits and the deepening of scientific knowledge in the areas of the climate and water cycle for application in fisheries, agriculture, and biodiversity management, particularly through the linking of information across disciplines [26].

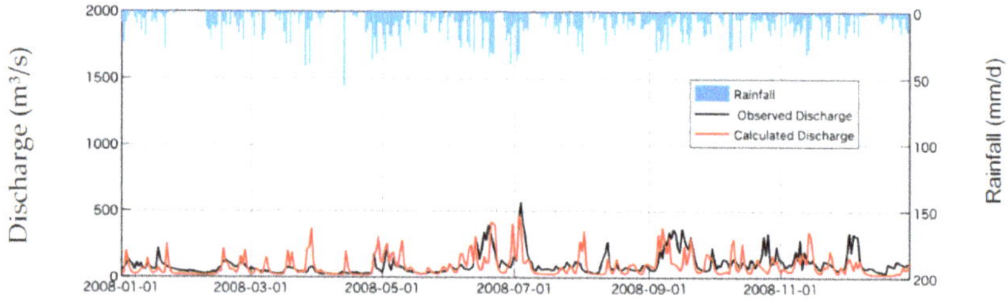

Figure 4. Validation results of river discharge at the Lacson station with CHIRPS, satellite-based rainfall.

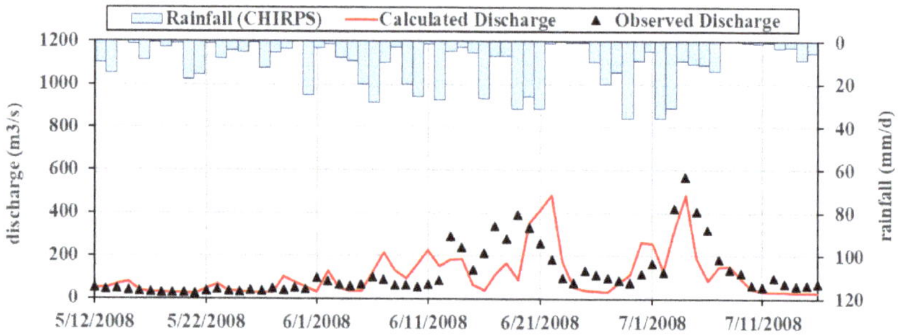

Figure 5. Event-based validation of river discharge at the Lacson station with CHIRPS, satellite-based rainfall during the annual maximum flood.

3.4. Co-Design of E-Learning Workshops

E-learning workshops were proposed as an effective means of fostering Facilitators and undertaken after co-designing the structure of the workshops among relevant stakeholders of academia and all levels of society. PLATFORM fulfilled its role as the governance body and was able to gather cooperation from various stakeholders. Through the co-designing of the e-learning workshops, one of the most important yet controversial points was how to gather candidates for Facilitators from different disciplines and sectors in society. Based on the past efforts and future direction related to disaster resilience, we came up with four selection criteria, as shown in Table 1: direct disciplines, a good mix of sciences, representation from different levels of governance, and local initiative. In line with the results of the co-designing effort, two e-learning workshops were held in April 2021 and January 2022.

Table 1. Criteria for gathering Facilitator candidates from different disciplines and sectors in the society.

	Category	Expertise
Criteria 1	Direct disciplines	disaster risk reduction, flood management, meteorology, climate change, integrated water resources management
Criteria 2	Good mix of sciences	natural science, engineering, social science, ICT
Criteria 3	Representation from different levels of governance	city/municipality office, national government, community leader, private sector, media, academia
Criteria 4	Local initiative	civil society organizations, non-governmental organizations

4. Results

4.1. Development of OSS-SR

OSS-SR for Davao City, which was developed in DIAS, stores and displays observed data, such as satellite-based rainfall, GSMaP [27], and cloud images of Himawari-8. Advances in the integration of remote sensing technologies strengthen the effectiveness of flood monitoring functions, contributing to timely and proper decision-making [28]. Observed data not only help to understand spatial rainfall distribution but can also be used as input data for simulation models. OSS-SR also illustrates the extent of inundation and depth from the results of real-time forecasting and climate-change impact assessment conducted using the basin- and barangay-scale models with rainfall data stored in DIAS. Furthermore, OSS-SR delivers ten materials of introductory lectures and four materials of hands-on training. In addition to the information from numerical models and analysis, on-site information and experiences can also be archived and shared among users with photos and texts on local information maps. Figure 6 shows an example of a local information map. The accumulation of local information, fed by residents from various local spots in the area, contributes to the advancement of science.

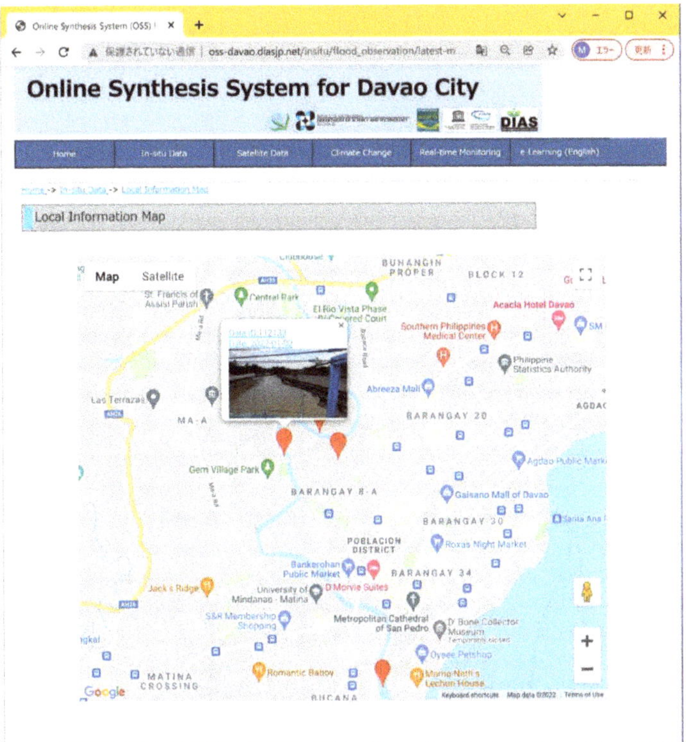

Figure 6. Example window of local information map in the OSS-SR for Davao City.

4.2. Climate Change Impact Assessment

Tropical cyclones are likely to become more intense in the future, as evidenced by recorded tropical cyclones in several cities and municipalities in Mindanao such as Davao City, which used to be known as a typhoon-free metropolis. Cabrera and Lee (2018) reported the necessity of immediate actions of decision-makers to develop a community-based disaster risk plan under future climate in Davao Oriental [29].

To assess the climate change impact in the Davao region, we employed a GCM output of MRI-AGCM 3.2H developed by the Meteorological Research Institute of Japan, because this paper focuses on communities' understanding of hazard estimation due to past and future floods for saving lives, not the facility planning based on a statistical approach with ensemble GCM outputs. MRI-AGCM 3.2H was selected because of the advantage of its high resolution and four kinds of multi-SST. The RCP8.5 Scenario of MRI-AGCM 3.2H was dynamically downscaled to 5 km by the WRF model with no convection scheme. Since the probability result of past climate after downscaling showed gaps with observed data, CHIRPS was used to perform bias correction based on monthly factors of cumulative probabilities for daily rainfall, excluding the top 0.5%, as was applied to the Pampanga River basin in the Philippines [30]. Regarding the downscaled GCM outputs, the worst extreme events in the past climate from 1979 to 2003 and the future climate from 2075 to 2099 were identified in terms of 24 h maximum rainfall and used for the impact assessment of the hazard estimation.

The rainfall data sets of the worst extreme events in the past and future climate were firstly input to the basin-scale WEB-RRI model. Figure 7 shows the river discharge of the worst flood events for the past and future climate, simulated by the WEB-RRI basin-scale model. Although the response of river discharge depends on the spatiotemporal distribution of extreme rainfall, the peak discharge of the worst event in the future is about 2.5 times that of past events. These time series of river discharge were applied to the barangay-scale model as the boundary condition at the upstream end. Figure 8 compares the basin-scale inundation maps of the worst extreme events in the past and future climate. Due to the difference in rainfall intensity and peak discharge, the inundation extent and depth are expected to be more significant in the future. The area more than 0.3 m deep under the future climate was more than four times as large as that under the past climate. Figure 9 compares the barangay-scale inundation maps of the worst extreme events in the past and future climate. The barangay-scale inundation maps also illustrate the significance of the future climate in terms of flood extent and depth with the spatial distribution of flood risk indicated at a resolution that enables users to learn to what risk level each critical infrastructure, evacuation facility, and evacuation route is exposed. Table 2 summarizes the comparisons between the past and future climate in terms of 24 h maximum rainfall, peak discharge, inundation extent, and inundation volume. The inundation area and volume in the table were calculated for the areas more than 0.3 m deep. All items indicate an increase in future flood disaster risk.

Table 2. Comparisons between the past and future climate from the aspects of 24 h maximum rainfall, peak discharge, inundation extent, and inundation volume.

		Past	Future	Future/Past
24 h maximum rainfall (mm)		114.0	198.1	1.7
Peak discharge (m^3/s)		1029.8	2564.7	2.5
Whole basin	Inundation area (km^2)	35.0	144.1	4.1
	Inundation volume (10^6 m^3)	176.7	456.2	2.6
Focus area	Inundation area (km^2)	2.8	7.1	2.5
	Inundation volume (10^6 m^3)	2.7	10.7	4.0

OSS-SR-based e-learning also provided information about a climate change scenario of MRI-AGCM output, uncertainties, and bias correction. Importantly, the results presented in Table 2 only concern the worst event cases in one GCM output. Flood inundation does not always occur as simulated because the magnitude and spatiotemporal distribution of floods vary from event to event. Hence, OSS-SR also has the function of real-time flood forecasting to contribute to reducing the risk of damage for a specific upcoming flood.

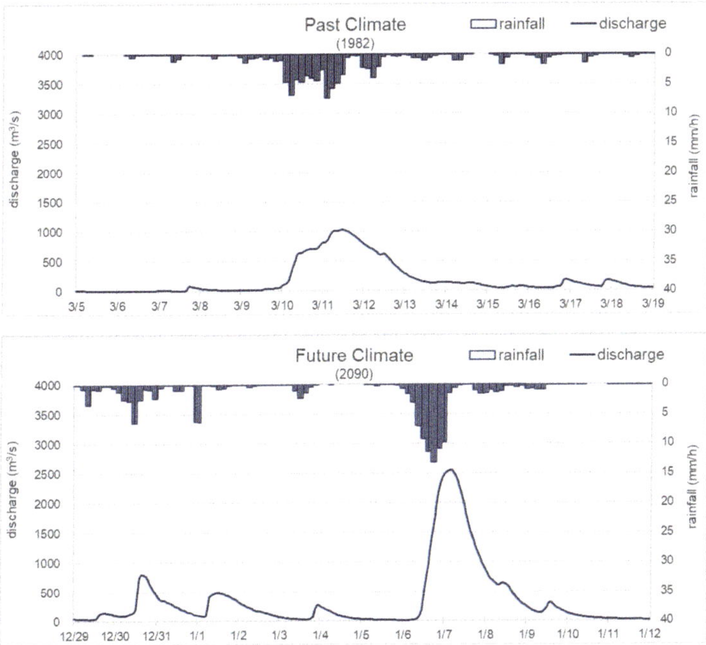

Figure 7. Results of river discharge at the upstream end of the focus area during the worst flood events for the past (**upper**) and future (**lower**) climate, simulated by the basin-scale model. The peak discharge in the future is about 2.5 times that of the past. These time series of river discharge were applied to the barangay-scale model as the boundary condition.

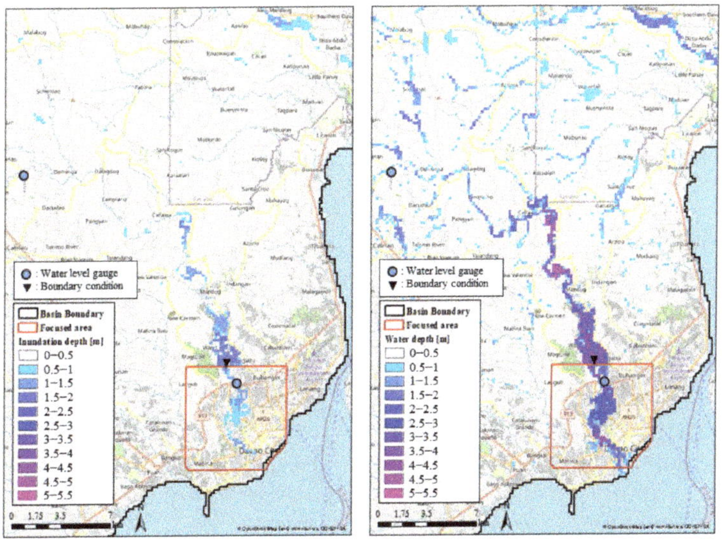

Figure 8. Basin-scale inundation maps of the worst extreme events in the past (**left**) and future (**right**) climate. The inundation extent in the future is more than 4 times that of the past.

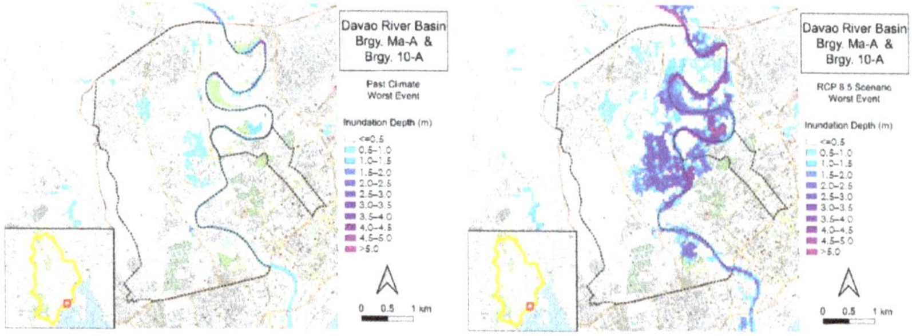

Figure 9. Barangay-scale inundation maps of the worst extreme events in past (**left**) and future (**right**) climate. The inundation extent in the future is more than 2.5 times that of the past.

4.3. E-Learning Workshops for Fostering "Facilitators"

Two e-learning workshops, introductory lectures in April 2021 and hands-on training in January 2022, were conducted to promote fostering Facilitators in Davao City. In line with the selection criteria produced in a co-designing manner, the workshops gathered about thirty candidates of Facilitators specializing in six different disciplines from national and local governments, academia, CSO and NGO, the private sector, and the media, as shown in Table 3. While the participants were instructed to download lecture materials from OSS-SR and learn them by themselves, the first workshop consisted of a month-long series of interactive sessions, such as an opening session, lecture introduction sessions, Q&A sessions, an examination, assignment submission, and a closing session. The hands-on training also included interactive sessions of an opening session, lecture introduction sessions, Q&A sessions, and a closing session. During the workshops, the participants were given assignments to produce deliverables for public dissemination, such as hazard maps considering climate change impact, local contingency plans, and action plans. The produced deliverables, which are twenty-one risk communication plans and workshop designs for fourteen barangays considering geographic, demographic, economic, and social features, can be particularly useful for the next actions, in which Facilitators will play the leading role in delivering the information to all levels of actors in society through workshops, seminars, training, focus group discussions, fact sheets, posters, and many other opportunities and tools.

Table 3. Disciplines of Facilitators who have participated in the e-learning workshops.

Discipline	Number of Participants	
	1st Workshop for Introductory Lectures	2nd Workshop for Hands-on Training
National government	11	10
Local government	2	4
Academe	11	13
Civil Society Organizations, Non-Governmental Organizations	1	2
Private sector	2	1
Media	2	1
TOTAL	29	31

4.4. Dissemination by Facilitators

After the training through the e-learning workshops, Facilitators are expected to undertake the role of disseminating their learnings to improve disaster literacy in society as a whole and collect feedback from the public. However, since society consists of different actors, it is important to decide what content to deliver and what approach to take for effective communication according to the target audiences. In Davao City, PLATFORM identified six target audience types by referring to the disciplines of Facilitators and co-designed the contents and communication methods considered effective for each target audience, as shown in Table 4. Although the contents and communication methods greatly depend on various aspects such as climatology, sociality, and geopolitics, this research was able to find a valid practice.

Table 4. Contents and methods of science communication with local society. Knowledge learned from OSS-SR and e-learning workshops is disseminated and translated in different ways according to the target audiences.

Target Audience	Contents to Disseminate/Translate	Effective Communication Methods
Local community	- Causes of flood - Climate change impacts - Contingency planning for DRR	- Focus group discussions - Radio/TV programs - Posters
DRRM team	- Flood monitoring - Flood hazard mapping - Disaster risk management cycle	- Training - Handouts
Government Agency	- Flood monitoring - Flood hazard mapping - Vertical and horizontal integration of DRRM plans and development plans	- Focus group discussions - Fact sheets
Policymaker	- Causes of flood - Climate change impacts - Contingency planning for DRR - Vertical and horizontal integration of DRRM plans and development plans	- Policy briefs - Policy recommendations - Fact sheets
Private sector	- Causes of flood - Climate change impacts	- Fact sheets - Posters
Media	- Causes of flood - Climate change impacts	- Media releases
NGO and CSO	- Disaster risk management cycle - Contingency planning for DRR	- Focus group discussions - Handouts - Posters

5. Discussion

Although this paper successfully illustrates a practical implementation method to enhance flood resilience in a co-designing manner among relevant stakeholders, the applied method has not been quantitatively assessed for the improvement of flood resilience. Qualitative approaches account for 61% of studies on flood resilience [4]; nevertheless, the quantitative assessment of enhanced resilience will be demanded in case of comparison with other approaches. A study on quantitative evaluation of enhanced resilience by the hydrological model, agent-based model, or indicator should be further addressed.

The impact of climate change in Davao City was assessed using a comparison of the worst flood events in past and future climates, produced from a single GCM. For a more elaborate and reliable impact assessment with various flood magnitudes, a statistical method based on ensemble outputs from multiple GCMs is expected to be applied as a further assessment.

6. Conclusions

This study demonstrated a methodology for enhancing flood resilience by improving society-wide disaster literacy through the development of OSS-SR and the fostering Facilitators under the framework of PLATFORM in Davao City, Philippines. The OSS-SR development and the Facilitator training have been implemented through the co-design of governance in which all stakeholders engage. OSS-SR integrates knowledge, information, experiences, awareness, and models under different disciplines for consilience. Facilitators play a vital role, as a catalytic existence, in interlinking the scientific community and all levels of local society.

The localized OSS-SR for Davao City highlighted real-time flood forecasting and climate-change impact assessment, as well as the e-learning function. The development of two flood models with different spatial scales (basin and barangay scales) realized real-time flood forecasting, which can contribute to flooding risk mitigation by early warning, and climate change impact assessment, which can clarify the magnitude and frequency of future floods. In Davao City, increases in 24 h maximum rainfall, peak discharge, inundation extent, and inundation volume are confirmed at both basin and barangay scales in the case of the expected future climate. In particular, the simulation found that the basin-scale inundation area and the barangay-scale inundation volume under the expected future climate will be about four times as large as under the past climate.

Two e-learning workshops, which were co-designed among all levels of stakeholders, gathered about thirty candidates of Facilitators from six different disciplines according to the four selection criteria. During the e-learning workshops, the participants learned to skillfully operate OSS-SR and produced deliverables for public dissemination, such as hazard maps considering the impact of climate change, local contingency plans, and action plans. The deliverables, which comprise twenty-one risk communication plans and workshop designs for fourteen barangays considering geographic, demographic, economic, and social features, can be utilized at various opportunities at which Facilitators will take the initiative in deliver them to all levels of the target audiences in society. PLATFORM also co-designed the contents to disseminate and the communication methods that are effective for each target audience.

This study presented a practical method to enhance flood resilience, demonstrating the synthesis of science-based knowledge, such as the integration of different scales of hydrological models and climate-change impact assessment, and human resource development, such as fostering Facilitators capable of translating knowledge for communities, can fill the gaps between the science community and local society.

Author Contributions: Conceptualization, M.M., A.C.S. and T.K.; methodology, M.M., T.U., A.W.M.R., D.G.B., A.C.S. and T.K.; software, D.K., T.U., A.W.M.R., M.Y. and M.K.; formal analysis, M.M., D.K. and T.K.; investigation, M.M., D.G.B. and A.C.S.; resources, M.Y., A.C.S., T.K. and M.K.; data curation, M.M., D.K., A.W.M.R. and M.Y.; writing—original draft preparation, M.M.; writing—review and editing, D.K., T.U., A.W.M.R., D.G.B., A.C.S. and T.K.; visualization, M.M., D.K., M.Y. and M.K.; supervision, D.G.B., A.C.S. and T.K.; project administration, D.G.B., A.C.S. and T.K. All authors have read and agreed to the published version of the manuscript.

Funding: This research was financially supported by the Integrated Research Program for Advancing Climate Models (TOUGOU) Grant Number JPMXD0717935457 from the Ministry of Education, Culture, Sports, Science, and Technology (MEXT), Japan. Additionally, DIAS has been continuously received financial support from MEXT.

Acknowledgments: This research activity was supported by the Platform on Water Resilience and Disasters in the Philippines and the International Flood Initiative (IFI).

Conflicts of Interest: The authors declare no conflict of interest.

References

1. World Meteorological Organization. WMO Atlas of Mortality and Economic Losses from Weather, Climate and Water Extremes (1970–2019) 2021, WMO-No.1267. Available online: https://library.wmo.int/doc_num.php?explnum_id=10902 (accessed on 28 February 2022).
2. International Strategy for Disaster Reduction. Hyogo Framework for Action 2005–2015. World Conference on Disaster Reduction. 2005. Available online: https://www.unisdr.org/2005/wcdr/intergover/official-doc/L-docs/Hyogo-framework-for-action-english.pdf (accessed on 28 February 2022).
3. Fisher, L. Disaster Responses: More than 70 Ways to Show Resilience. *Nature* **2015**, *518*, 35. Available online: https://www.nature.com/articles/518035a (accessed on 28 February 2022). [CrossRef] [PubMed]
4. Kerri, M.C.; David, M.; Lindsay, B.; Esther, C. Flood resilience: A systematic review. *J. Environ. Plan. Manag.* **2020**, *63*, 1151–1176. [CrossRef]
5. Moustafa, N.A.; Ahmed, Y.; Wael, E.; Paulin, C. Community Flood Resilience Categorization Framework. *Int. J. Disaster Risk Reduct.* **2021**, *61*, 102349. [CrossRef]
6. Jorge, L.; Kai-Feng, C.; Raul, R.W.; Ralph, L. A scalable flood-resilience-index for measuring climate change adaptation: Munich City. *Water Res.* **2020**, *173*, 115502. [CrossRef]
7. Keating, A.; Campbell, K.; Szoenyi, M.; McQuistan, C.; Nash, D.; Burer, M. Development and testing of a community flood resilience measurement tool. *Nat. Hazards Earth Syst. Sci.* **2017**, *17*, 77–101. [CrossRef]
8. Woodruff, S.; Meerow, S.; Gilbertson, P.; Hannibal, B.; Matos, M.; Roy, M.; Malecha, M.; Yu, S.; Berke, P. Is flood resilience planning improving? A longitudinal analysis of networks of plans in Boston and Fort Lauderdale. *Clim. Risk Manag.* **2021**, *34*, 100354. [CrossRef]
9. De Bruijn, K.M. Resilience indicators for flood risk management systems of lowland rivers. *Int. J. River Basin Manag.* **2004**, *2*, 199–210. [CrossRef]
10. Kai-Feng, C.; Jorge, L. A Conceptual Time-Varying Flood Resilience Index for Urban Areas: Munich City. *Water* **2019**, *11*, 830. [CrossRef]
11. Schinke, R.; Kaidel, A.; Golz, S.; Naumann, T.; López-Gutiérrez, J.S.; Garvin, S. Analysing the Effects of Flood-Resilience Technologies in Urban Areas Using a Synthetic Model Approach. *ISPRS Int. J. Geo-Inf.* **2016**, *5*, 202. [CrossRef]
12. O'Sullivan, J.J.; Bradford, R.A.; Bonaiuto, M.; De Dominicis, S.; Rotko, P.; Aaltonen, J.; Waylen, K.; Langan, S.J. Enhancing flood resilience through improved risk communications. *Nat. Hazards Earth Syst. Sci.* **2012**, *12*, 2271–2282. [CrossRef]
13. Albano, R.; Sole, A.; Adamowski, J. READY: A web-based geographical information system for enhanced flood resilience through raising awareness in citizens. *Nat. Hazards Earth Syst. Sci.* **2015**, *15*, 1645–1658. [CrossRef]
14. Schelfaut, K.; Pannemans, B.; van der Craats, I.; Krywkow, J.; Mysiak, J.; Cools, J. Bringing flood resilience into practice: The FREEMAN project. *Environ. Sci. Policy* **2011**, *14*, 825–833. [CrossRef]
15. Prerna, S.; Adjo, A.; Brian, W.; Sanskruti, J. Lessons from case studies of flood resilience: Institutions and built systems. *Transp. Res. Interdiscip. Perspect.* **2021**, *9*, 100297. [CrossRef]
16. Morrison, A.; Westbrook, C.; Noble, B. A Review of the Flood Risk Management Governance and Resilience Literature. *J. Flood Risk Manag.* **2017**, *11*, 291–304. [CrossRef]
17. *Philippine Climate Change Adaptation Project Cities and Climate Change Initiative*; Vulnerability and Adaptation Assessment Report: Davao City, Philippines, 2018.
18. Website of Science Council of Japan, Committee on International Cooperation for Promoting Science-Based Disaster Risk Reduction. Building a Sustainable Global Society by Strengthening Disaster Resilience-Developing an "Online Synthesis System (OSS)" and Fostering "Facilitators" to Realize Consilience. Available online: https://www.scj.go.jp/ja/info/kohyo/pdf/kohyo-24-t298-1en.pdf (accessed on 28 February 2022).
19. Website of High-Level Experts and Leaders Panel on Water and Disasters (HELP). Available online: https://www.wateranddisaster.org/ (accessed on 28 February 2022).
20. Website of International Flood Initiative. HELP-IFI Jakarta Statement. Available online: http://www.ifi-home.info/documents/pdf/Jakarta-Statement.pdf (accessed on 28 February 2022).
21. High-Level Panel on Water Outcome Document, Making Every Drop Count—An Agenda for Water Action. 2018. Available online: https://reliefweb.int/sites/reliefweb.int/files/resources/17825HLPW_Outcome.pdf (accessed on 28 February 2022).
22. Rasmy, M.; Sayama, T.; Koike, T. Development of water and energy Budget-based Rainfall-Runoff-Inundation model (WEB-RRI) and its verification in the Kalu and Mundeni River Basins, Sri Lanka. *J. Hydrol.* **2019**, *579*, 124163. [CrossRef]
23. Funk, C.; Peterson, P.; Landsfeld, M.; Pedreros, D.; Verdin, J.; Shukla, S.; Husak, G.; Rowland, J.; Harrison, L.; Hoell, A.; et al. The climate hazards infrared precipitation with stations—a new environmental record for monitoring extremes. *Sci. Data* **2015**, *2*, 150066. [CrossRef]
24. Nash, J.E.; Sutcliffe, J.V. River Flow Forecasting through Conceptual Model. Part 1—A Discussion of Principles. *J. Hydrol.* **1970**, *10*, 282–290. [CrossRef]
25. Sayama, T.; Tatebe, Y.; Iwami, Y.; Tanaka, S. Hydrologic sensitivity of flood runoff and inundation: 2011 Thailand floods in the Chao Phraya River basin. *Nat. Hazards Earth Syst. Sci.* **2015**, *15*, 1617–1630. [CrossRef]
26. Website of DIAS. Available online: https://diasjp.net/en/ (accessed on 28 February 2022).

27. Kubota, T.; Aonashi, K.; Ushio, T.; Shige, S.; Takayabu, Y.N.; Kachi, M.; Arai, Y.; Tashima, T.; Masaki, T.; Kawamoto, N.; et al. Global Satellite Mapping of Precipitation (GSMaP) Products in the GPM Era. In *Satellite Precipitation Measurement. Advances in Global Change Research*; Springer: Cham, Switzerland, 2020; Volume 67. [CrossRef]
28. Hadi, F.; Ali, E.; Mohammad, N. Flood monitoring by integration of Remote Sensing technique and Multi-Criteria Decision Making method. *Comput. Geosci.* **2022**, *160*, 105045. [CrossRef]
29. Cabrera, J.S.; Lee, H.S. Impacts of Climate Change on Flood-Prone Areas in Davao Oriental, Philippines. *Water* **2018**, *10*, 893. [CrossRef]
30. Ushiyama, T.; Hasegawa, A.; Miyamoto, M.; Iwami, Y. Dynamic downscaling and bias correction of rainfall in the Pampanga River Basin, Philippines, for investigating flood risk changes due to global warming. *Hydrol. Res. Lett.* **2016**, *10*, 106–112. [CrossRef]

water

MDPI

Article

Application of the TOPKAPI Model in Flood Forecasting of the Upstream of the Zhenjiang River in China

Lili Liang [1,*], Yufeng Hu [2], Zhiwu Liu [1], Yuntao Ye [2,*], Kuang Li [2], Kexin Liu [2], Haiqing Xu [2] and Xiquan Liu [1]

1 Institute of Science and Technology, China Three Gorges Corporation, Beijing 100038, China; liu_zhiwu@ctg.com.cn (Z.L.); liu_xiquan@ctg.com.cn (X.L.)
2 China Institute of Water Resources and Hydropower Research, Beijing 100038, China; hyf6969@163.com (Y.H.); likuang0000@163.com (K.L.); liukexin0412@163.com (K.L.); feralstone@163.com (H.X.)
* Correspondence: liangli0921@163.com (L.L.); yeyuntao@iwhr.com (Y.Y.)

Abstract: The lumped hydrological model and empirical model have the problems of low accuracy and short forecasting period in real-time flood forecasting of small- and medium-sized rivers in a mountainous watershed. The sharing of underlying surface data such as high-resolution DEM, land use data, soil data, and the popularization and application of the Internet of Things, big data, cloud computing, and intelligent calculation methods makes distributed hydrological model an effective method for real-time runoff simulation and prediction. The topographic, kinematic, approximation, and integration (TOPKAPI) model is a distributed hydrological model whose physical mechanism developed gradually in the late 20th century. It has great advantages in real-time flood forecasting in small- and medium-sized watersheds. Based on the data required by the TOPKAPI model, in this study, 26 selected flood events were simulated from 2000 to 2013 at the outlet section of the upper reach of the Zhenjiang River in Guangdong Province, and the effect of application of the model in flood forecasting of small- and medium-sized rivers was evaluated. The results show that the pass rate (considering the peak discharge as the evaluation item) of 18 flood events in the calibration period was 66.67%, and that of 8 flood events in the validation period was 75%, while the mean Nash efficiency coefficient of the selected 26 flood events was 0.789. According to the simulation results, real-time flood forecasting should be closely combined with the dispatching of the small- and medium-sized reservoirs in the basin. The application of the TOPKAPI model can make a scientific and rapid analysis of the flood control situation in the whole basin and provide accurate information and maximum convenience for flood forecasting consultation and decision making. Additionally, it can improve the efficiency of disaster prevention and mitigation work in small- and medium-sized river basins, and has a major significance in enhancing the modernization level of flood forecasting.

Keywords: distributed hydrological model; flood forecasting; TOPKAPI; Zhenjiang River; small- and medium-sized river

Citation: Liang, L.; Hu, Y.; Liu, Z.; Ye, Y.; Li, K.; Liu, K.; Xu, H.; Liu, X. Application of the TOPKAPI Model in Flood Forecasting of the Upstream of the Zhenjiang River in China. *Water* **2022**, *14*, 618. https://doi.org/10.3390/w14040618

Academic Editors: Slobodan P. Simonovic, Subhankar Karmakar and Zhang Cheng

Received: 17 January 2022
Accepted: 10 February 2022
Published: 17 February 2022

1. Introduction

Rainstorm and flash floods in small and medium-sized mountainous watersheds (with an area less than 3000 km^2) are greatly controlled by terrain, so the flash floods have short times of runoff generation and routing but large temporal and spatial variation [1]. It is difficult to accurately monitor and forecast the mountainous flash floods due to the characteristics and technological problems, such as the large topographic fluctuation, complex vegetation types and underlying surface conditions of the mountainous watershed, general lack of hydrological measured data, significant nonlinear characteristics of runoff and concentration, steep rise and fall of the floods, and so on, leading to great difficulty for accurate analysis, simulation and early warning [2–4]. In recent years, the World Meteorological Organization (WMO), the Global Water Partnership (GWP), the International Association of Hydrological Sciences (IAHS) [5], the International Association for water environmental engineering and Research (IAHR), and the National Oceanic and Atmospheric Administration (NOAA) of the United States are devoting more attention to the forecasting of mountainous flash floods [2].

The integrated hydrological and empirical models have the disadvantages of low prediction accuracy and short prediction period in flood forecasting of small- and medium-sized mountainous watersheds, which makes it difficult to meet the service requirements of current accurate hydrological prediction [6,7]. In recent years, with the development of remote sensing (RS) [8], geographic information system (GIS) [9], and computer technology [10], distributed hydrological model has become an effective method to simulate and predict the short-term runoff of mountainous flash floods. The RS technology provides a considerable amount of underlying surface information, such as land use, soil types, and estimated precipitation based on radar and satellite images. The GIS provides a means of processing geographic information, while computer technology solves a large number of computing problems in a distributed model. The combination of a distributed hydrological model and digital elevation model (DEM) provides conditions for making full use of spatial distribution data to analyze the hydrological processes [11–13].

The topographic, kinematic, approximation, and integration (TOPKAPI) model was developed on the basis of two hydrological models—Arno and TOPMODEL—by Professor Ezio Todini, in 1995 [14,15]. It is a fully distributed hydrological model with a physical basis, relatively simple structure, clear parameter meaning, and large spatial scale applicability [16], so it can fully consider the spatial variability of rainfall, terrain, vegetation, soil, and other elements. In real-time flood forecasting applications, it has been widely used for many rivers in Italy, Spain, China, and other countries. In Italy, it had been incorporated into real-time flood forecasting systems of some major watersheds such as Po, Arno, Tiber, Adige, and Reno [17]. In Spain, it had been used in the real-time flood forecasting system of the Segura basin and Jucar basin, which was under the responsibility of the government department SAIH [18]. In America, in the second phase of the distributed model intercomparison project (phase 2, DMIP 2), it was applied to the Americana basin (humid area) and Carson basin (high cold and snow melting area) in the Sierra Nevada, respectively [19]. In China, it was applied in the ungauged arid region of Nalinggele River and its ending salt lake with complex hydrological conditions, and the results showed that the TOPKAPI model was suitable for watersheds with large spatial scales, and more importantly, suitable for flood forecasting of ungauged basins [20]. It was selected as a tool to simulate flood events of a small river basin—Chengcun Watershed—in China and was compared with the Xin'anjiang model in terms of model structure, flood characteristic values, and simulation results; the results revealed that the TOPKAPI model could be used in flood forecasting, land use, environmental impact assessment and ungauged hydrological simulation calculation [21]. Jian et al. [22] applied TOPMODEL, TOPKAPI, and CASC2D model to simulate floods of Banqiao and Maduwang watersheds in China, and the results showed that the TOPKAPI model utilizing a saturated runoff mechanism was more suitable for the small- and medium-sized Maduwang watershed. In order to improve the level of flood forecasting for small- and medium-sized mountainous watersheds in China, and to further promote the application and improvement of the distributed hydrological model, taking the Zhenjiang River basin above the Xiaogulu hydrological station in Guangdong Province as the study area, in this paper, the application of the TOPKAPI model in flash floods forecasting was discussed, and it was found to be of great significance to the flood control of the whole basin: This model could make a scientific and rapid analysis of the flood control situation of the whole basin, provide accurate information and maximum convenience for the flood control consultation and decision making, and could improve the efficiency of flood disaster prevention and reduction.

2. Materials and Methods

2.1. Study Area

Zhenjiang River is upstream of the Beijiang River of the Pearl River system. The study area, with an area of about 1922.5 km^2, is the upper basin of the Zhenjiang River watershed above Xiaogulu hydrological station. It is a mountainous watershed, and the elevation of the study area ranges from 98 m to 1271 m; it is high in the elevation in the middle area from northeast to the southwest (mountains and hills) and low on both sides. It has low vegetation coverage in the Nanxiong basin, and the forest coverage rate is 64%, belonging to the area of soil erosion. There are 12 tributaries in this basin, with a total river length of 356.7 km. Xiaogulu hydrological station is located at the boundary line of Nanxiong City and Shixing County, Guangdong Province. From 2000 to 2013, the maximum peak flow of the Xiaogulu hydrological station was 1250 m^3/s. There are 5 medium reservoirs and 13 small reservoirs in the basin; the basic information of the 18 reservoirs can be seen in Tables A1 and A2 of Appendix A, but their operation data are not collected. Figure 1 shows the DEM, river system, hydrometeorological stations, and reservoirs of the study area.

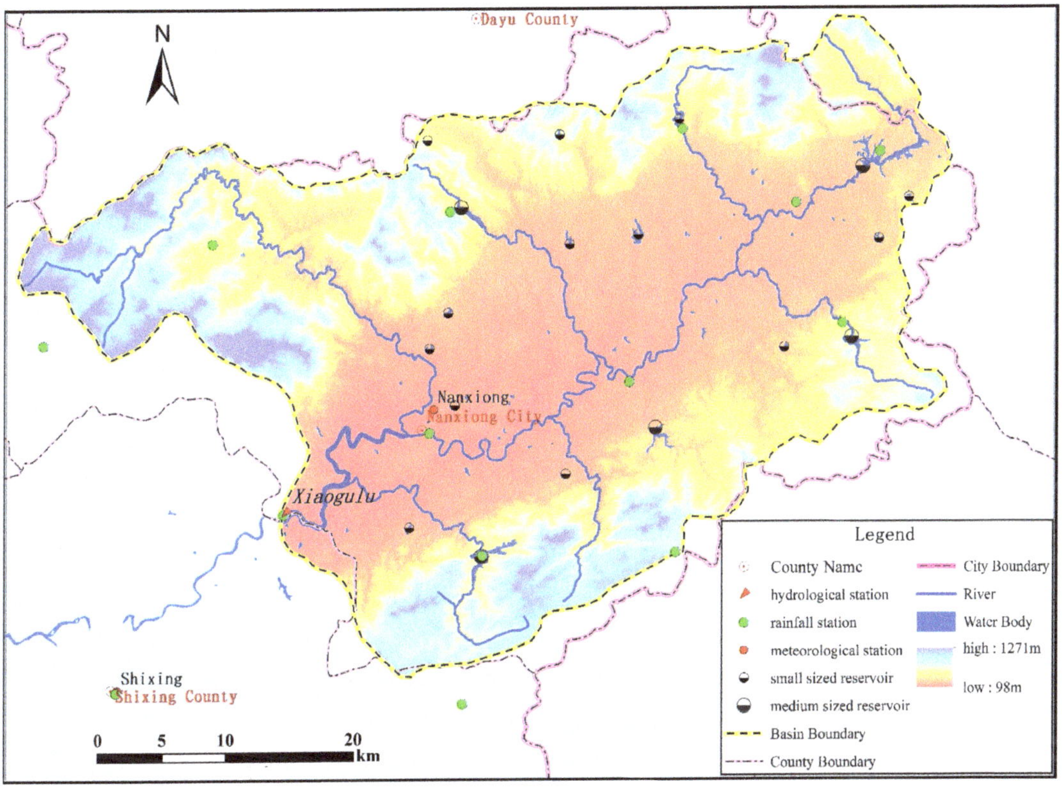

Figure 1. DEM, river system, hydrometeorological stations, and reservoirs of the study area.

The watershed belongs to the subtropical monsoon climate area. The multi-year average temperature in the study area is 19.9 °C, the highest monthly average temperature is 28.8 °C, and the lowest monthly average temperature is 9.5 °C. The study area has abundant rainfall but uniform rainfall distribution: The average annual rainfall is about 1500 mm, and the average rainfall from May to October accounts for 66.9% of the annual average rainfall. The measured average annual evaporation of the Nanxiong meteorological station in the 30 years from 1981 to 2010 was 1066.2 mm (the evaporator was E601). The annual average runoff depth was about 782 mm, and the annual average sediment transport modulus was 245 t/km^2.

The floods in the basin are formed by rainstorms, which mainly occur during the flood season from May to September. It has the characteristics of rainstorms and floods in small- and medium-sized mountainous watersheds—rapid concentration of flow, sharp rise and fall of the flood, high flood peak but small flood volume, wide coverage, and long duration. Most floods have single peaks or double peaks, and only a few floods have multiple peaks.

2.2. A Real-Time Flood Forecasting Method Based on the TOPKAPI Model

TOPKAPI is a fully distributed, physically based hydrological model with a simple structure and parameter scheme, which makes it one of the several operational distributed hydrological models in the world at present. It is also one of the core modules of the European Flood Forecasting System (EFFORTS). The latest model package consists of four parts—PreTPK, ITOPKAPI, TPKVIEW, and TPKMAPS. All stages, including the processing of DEM, the input of initial parameters, model operation, and graphical representation of the intermediate and final results, are completed in the same environment, which greatly improves the efficiency of data analysis and decision making of model application.

The model is generally divided into components of plant interception, snow melting, evapotranspiration, infiltration, percolation, surface flow, subsurface flow, underground flow, channel flow, reservoir/lake, and discharge. The model structure and simulation steps are shown in Figure 2.

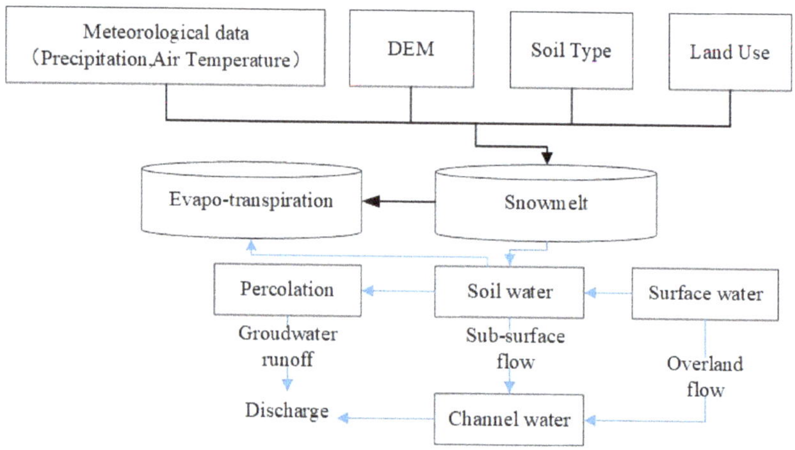

Figure 2. Simulation steps of the TOPKAPI model.

We used the software of Version 2.3 (released on 7 January 2016) of the TOPKAPI in this paper. The main characteristics of the model are as follows:

(1) It has reduced execution times suitable for distributed model calibrations and real-time operational applications and flood forecasting; (2) the model can be run at different time scales, from very fine temporal (few minutes) to daily simulations, and can resolve basin hydrologic response at spatial scales (100–1000 m) in both small and large catchments; (3) it can be easily calibrated due to physically meaningful parameters whose values can be retrieved from digital elevation maps, soil maps, land use, and vegetation maps, in terms of measurable physical quantities such as slope, soil permeability, surface roughness, etc.; (4) it tracks the spatial variability of runoff conditions in the catchment getting flow predictions at any point of the channel network (1D outputs) and explicitly considers the spatial variability in precipitation fields, fully exploiting distributed rainfall estimates such as the ones produced by RADAR networks [23]; (5) it represents the behavior of the main components of the hydrological cycle, thus producing stream flow forecast, as well as

distributed information on soil moisture, evapotranspiration, snow accumulation, etc. (2D output maps); (6) the TOPKAPI model can run independently offline, but also be used as a flood forecasting model in real-time flood operation system for continuous simulations or prediction calculation. It can be coupled with HEC–RAS hydraulic model (directly) or with Delft–SOBEK hydraulic model (via FEWS); (7) with the help of 3S technology, the model can be applied to the basin without data and can run basing events and continuously simulate different scenarios (water balance analysis, climate change, water resource condition change, etc.).

2.3. Materials

The application of the TOPKAPI model requires GIS data such as DEM, land use map, soil map, water system, maps of reservoirs/power stations and their physical attribute information, hydrological and meteorological data, and water resource utilization information of the basin. The specific process for the determination of its parameters can be seen in related studies or literature [24]. In addition, in order to better calibrate the model, it also needs to understand and master the basin characteristics, hydrometeorology, historical flood characteristics, etc. Some basins may have weather radar rain data to input to make up for the lack of rainfall station data, such as time series of rainfall distribution map, which includes geographic location, spatial coverage, time, and spatial resolution of observation data.

2.3.1. DEM

DEM of the upper reaches of the Zhenjiang River in Guangdong Province was extracted from the GDEM data (http://www.gscloud.cn/listdata/showinfo_new.Shtml?From=&id=304, accessed on 20 April 2020). After downloading the data and simple processing, the PreTPK interface can help the user to add the main rivers of the basin, set the minimum threshold value of the simulation area, and modify the DEM using the tools provided. Then, the DEM was converted into the data that can be directly used by the model, and the resolution was reduced to 500 m × 500 m (it can be set and modified according to the threshold of simulation area).

2.3.2. Soil Data

The soil map with the resolution of 1:1,000,000 was obtained from the website of the National Special Environment and Function of Observation and Research Stations Shared Service Platform (http://www.ncdc.ac.cn/portal/metadata/navigator, accessed on 20 April 2020). There were 8 soil subclasses in the basin, and the names of each soil subclass are shown in Figure 3a and Table 1.

Table 1. Name and area of each soil subclass in the study area.

Soil Name	Area/km^2	Area Percentage/%	Soil Name	Area/km^2	Area Percentage/%
Limestone soil	4.062	0.21	Red soil	979.80	50.96
Brown calcareous soil	5.141	0.27	Yellow and red soil	163.21	8.49
Purple soil	242.929	12.64	Yellow soil	43.36	2.25
Paddy soil	466.55	24.27	Water	17.435	0.91

Red soil had the largest area in this basin, accounting for 50.96% of the basin area, followed by paddy soil, accounting for 24.27%, and purple soil, accounting for 12.64% of the basin area. These three soil subclasses were the main soil types in the basin. We queried the attribute information of each soil subclass needed by the model in the literature [25,26] and made a CSV type file named soil.csv to store these data. The file must have columns of data as follows: the first column was ID (vector of ID codes for soil types); the second column was DEPTH (vector of soil depth), followed by KSH (vector of hydraulic permeability (horizontal)), KSV (vector of hydraulic permeability (vertical)), THETA_R (vector of residual water content values), THETA_S (vector of saturation water content values), EXPH (vector of soil reservoir exponent), EXPV(vector of percolation law exponent), and NAME(vector of soil type names), respectively.

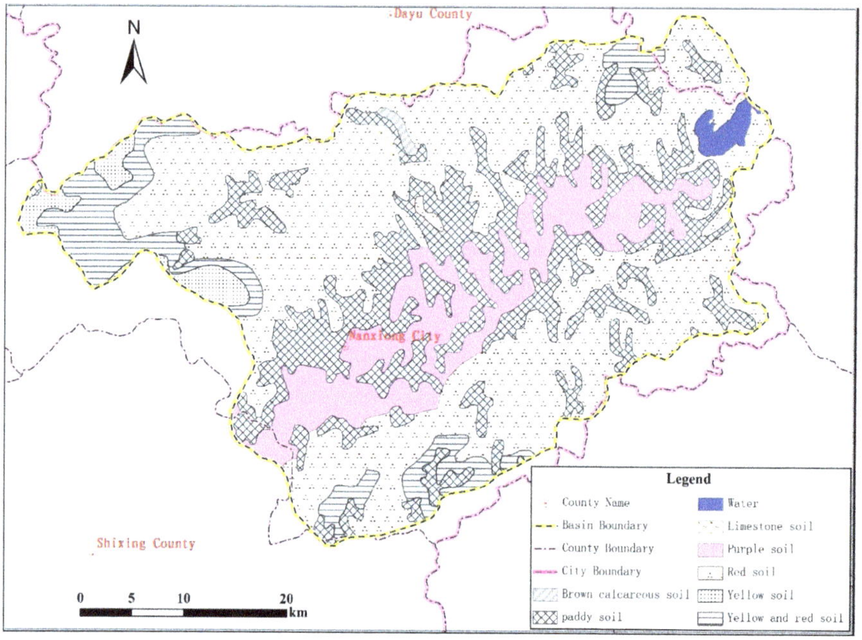

(**a**)

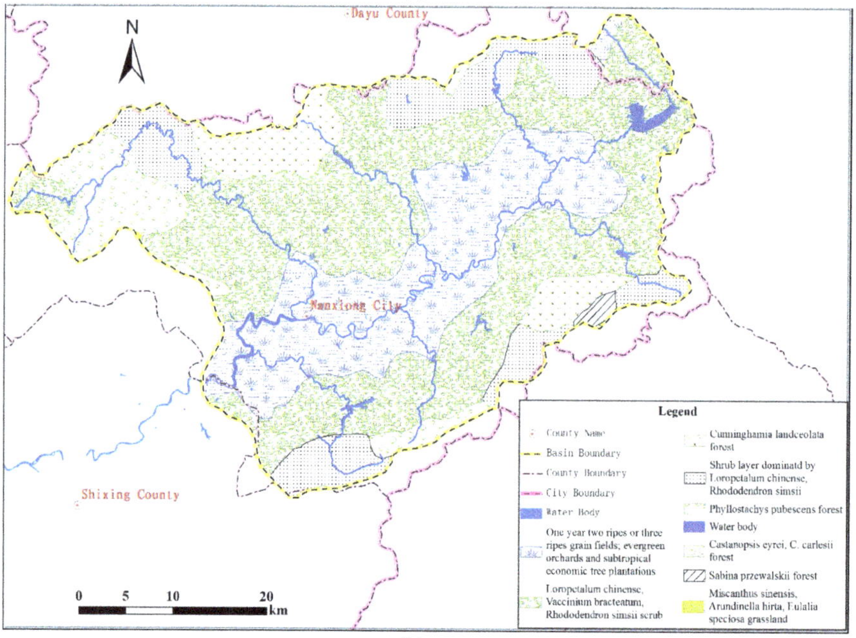

(**b**)

Figure 3. Soil subclasses and vegetation types in the study area: (**a**) soil subclasses; (**b**) vegetation types.

2.3.3. Vegetation

The land use map was obtained from the vegetation dataset of China with a resolution of 1:1,000,000. The vegetation types of the study area are shown in Figure 3b.

The vegetation with the largest area in the basin was Loropetalum chinense, Vaccinium bracteatum, Rhododendron simsii scrub, accounting for 47.33% of the basin area, followed by double-cropping rice and orchards, accounting for 23.98%. We queried the initial value of the attribute information of each vegetation type required by the model in some studies in the literature [27] and made it a CSV-type file named landuse.csv. The file must have columns of data as follows: the first column was ID (for vegetation types); the second column was the surface Manning coefficient of the land use type; columns 3–14 were the vegetation index of January to December, respectively. The index can be inquired in the irrigation and drainage literature of "vegetation water demand" of FAO.

2.3.4. Hydrometeorological Data

(1) Rainfall data and processing

The operation of the model requires a long series of continuous precipitation data (at least 5 years), and we collected the long series of precipitation data of 9 rainfall stations located in the basin. However, there was basically no precipitation data from December to March of the next year of all the rainfall stations.

We selected the forecast period as one hour and processed the precipitation data using the linear interpolation method. We set −9999 when there was no observed data from November to March of the next year. From 8:00 on 1 March 2000, the data of each rainfall station was written into a CSV type file named rainfall.csv, and one column for each station. At the same time, the location of these rainfall stations were written into a CSV file named rainfall.csv.xyz.

(2) Temperature data and processing

TOPKAPI needed temperature data, and there was only one meteorological station named Nanxiong in the basin. The daily average temperature, daily maximum temperature, daily minimum temperature data of the year from 2000 to 2013, and station location of Nanxiong meteorological station was obtained on China Meteorological Data Network (http://data.cma.cn/site/index.html, accessed on 21 April 2020).

It needed the hourly temperature data to obtain the hourly model results; therefore, we converted the daily temperature into the hourly data according to the three characteristic temperature values of the daily average temperature, the daily maximum temperature, and the daily minimum temperature. First, we judged the temperature at a certain time should fall down between which two characteristic values. Then, the linear interpolation method was adopted to obtain the temperature of the time by using these two characteristic values and time intervals. It should be noted that this interpolation may have a deviation from the actual hourly data, and therefore, it is temporarily approximate.

We made a CSV-type file named tmp.csv according to the specified format. The file was more or less the same as rainfall.csv. Similarly, the location of the Nanxiong meteorological station was written into a file named tmp.csv.xyz. The monthly average temperature was also made into a CSV-type file named tmp_m.csv. The model used air temperature data to calculate evaporation, so evaporation data of the basin were not collected.

2.3.5. Flood Data

The flood data mainly refer to the long series flood element data (at least 5 years and the same period with rainfall data) of a hydrological station or water level gauges, the water-level–discharge curve of a forecast section, etc. In the study basin, only Xiaogulu hydrological station had a long series of water level and flow data, and the data of flood elements of the station (water level and flow from March to October) from 2000 to 2013 were collected.

The flood data of Xiaogulu hydrological station were available hour by hour but sometimes failed within the series; on such occasions, the flow would be changed into the hourly data using the linear interpolation method. Then, it was made into a CSV-type file named Qobv.csv according to the format required by the model input. The first two columns of the first line in the file were time and code, and the second line and later lines of the file were the corresponding values.

2.3.6. Other Data

The required water system map was mainly used to refine the drainage network generated by DEM, and that map with the resolution of 1:250,000 of the basin was collected. According to this map, the vector map of the basin boundary was drawn in the GIS software. There were 18 medium and small-sized reservoirs in the basin. According to the model, it needed geographic location, storage capacity curve, discharge curve, long series data of inflow/outflow, reservoir operation data of selected flood events, etc., but we had no data about them except their locations, so the reservoirs in the basin were not simulated, and it was treated as one basin with no reservoir. There was no large-scale external water transfer in and out of the basin, which was not considered in the simulation.

2.4. Model Implementation

The prediction scheme was made based on the river system, hydrological station data compilation, and flood characteristics analysis of the Zhenjiang River basin above Xiaogulu station. First, according to the general situation of the basin, the simulating period was selected, which was from 2000 to 2013, and the time step of the simulation was set to one hour. Then, the data required by the TOPKAPI model were collected and processed. Next, the model parameters were set and the model was run, after which 26 flood events whose peak discharge rates were more than 500 m^3/s were selected according to the actual flood events of the basin, to calibrate, verify, and determine a group of optimal model parameters. Finally, the accuracy and prediction results were evaluated and analyzed.

2.5. Evaluation Indices

In this study, Error of peak discharge (*EPD*), Nash efficiency coefficient (*E*), mean absolute error (*MAE*), root-mean-square error (*RMSE*), coefficient of determination (R^2), explained variance (*EV*), volume control index (*VC*), Chiew and Mcmahon (*CMM*), and index of agreement (*d*) were selected as evaluation indices to analyze the simulation accuracy. Table 2 lists the calculation formula of these evaluation indices.

Table 2. Evaluation indices and the calculation formula.

Index Abbreviation	Index Name	Calculation Formula	Range of Value
EPD	Error of peak discharge	$EPD = \frac{P_i - Q_i}{Q_i}$	$[-\infty, \infty]$
E	Nash efficiency coefficient	$E = 1 - \frac{\sum_{i=i}^{N}(Q_i - P_i)^2}{\sum_{i=i}^{N}(Q_i - \overline{Q})^2}$	$[-\infty, 1]$
MAE	Mean absolute error	$MAE = \frac{\sum_{i=1}^{N}\lvert P_i - Q_i\rvert}{N}$	$[0, \infty]$
RMSE	Root mean square error	$RMSE = \sqrt{\sum_{i=1}^{N}\lvert P_i - Q_i\rvert / N}$	$[0, \infty]$
R^2	Coefficient of determination	$R^2 = \frac{\sum_{i=i}^{N}(Q_i - \overline{Q})(P_i - \overline{P})}{\sqrt{\sum_{i=i}^{N}(Q_i - \overline{Q})^2 \sum_{i=i}^{N}(P_i - \overline{P})^2}}$	$[0, 1]$

Table 2. *Cont.*

Index Abbreviation	Index Name	Calculation Formula	Range of Value
EV	Explained variance	$EV = 1 - \frac{\sum_{i=1}^{N}[(P_i - Q_i) - \frac{1}{N}\sum_{i=i}^{N}(P_i - Q_i)]^2}{\sum_{i=i}^{N}(Q_i - \overline{Q})^2}$	$[-\infty, 1]$
VC	Volume control index	$VC = 1 - \left\lvert \frac{\sum_{i=i}^{N} P_i}{\sum_{i=i}^{N} Q_i} - \frac{\sum_{i=i}^{N} Q_i}{\sum_{i=i}^{N} P_i} \right\rvert$	$[-\infty, \infty]$
CMM	Chiew and Mcmahon	$CMM = 1 - \frac{\sum_{i=i}^{N}\left(\sqrt{P_i} - \sqrt{Q_i}\right)^2}{\sum_{i=i}^{N}\left(\sqrt{Q_i} - \sqrt{\overline{Q}}\right)^2}$	$[-\infty, 1]$
d	Index of agreement	$d = 1 - \frac{\sum_{i=i}^{N}(Q_i - P_i)^2}{\sum_{i=i}^{N}\left(\lvert P_i - \overline{Q}\rvert + \lvert Q_i - \overline{Q}\rvert\right)^2}$	$[0, 1]$

P is the simulated value, and *Q* is the measured value.

3. Simulation Results

When all of the files were prepared and checked, they were used to run and calibrate the TOPKAPI model. Based on the calibrated parameters, the long series hourly simulation results of each year from 2000 to 2013 were obtained, and then, 26 selected floods with peak discharge rates greater than 500 m^3/s were selected to analyze the simulation results of Xiaogulu station.

The TOPKAPI model does not need to divide the simulation period into calibration and verification periods, but according to the needs of model parameter calibration and verification, the first 18 floods were used to calibrate model parameters, and the last 8 floods were used to verify the parameters. Table 3 shows the simulation results of 26 selected flood events. Considering the limit in the length of this article, only some selected flood simulation results are given below (Figure 4).

According to the requirements of the standard for hydrological information and hydrological forecasting (GB/T22482-2008) [28], the evaluation items of flood forecasting accuracy include flood peak discharge (or water level), flood volume (runoff), occurrence time of the flood peak, and process conformance of the flood event. When the flood peak discharge is used to evaluate the accuracy of the prediction scheme (the simulated flood volume in the TOPKAPI model was not calculated yet), the evaluation index is the error of peak discharge: When the pass rate is more than 85.0%, the accuracy class is Level A; when the pass rate is more than 70.0% and less than 85.0%, the accuracy class is Level B; when the pass rate is more than 60.0% and less than 70.0%, the accuracy class is Level C; otherwise, it failed to meet the standard. Table 4 lists the accuracy evaluation results (weather up to the standard) of the TOPKAPI model of 26 selected flood events with peak discharge rates greater than 500 m^3/s, from April 2000 to November 2013.

Based on the test of the selected flood events with peak discharge rates of more than 500 m^3/s, taking 20% of the peak discharge as the allowable error ($EPD \leq 20\%$), the prediction scheme of Xiaogulu station established by the TOPKAPI model had a simulation accuracy as follows: The pass rate was 66.67% in the calibration period and 75% in the validation period, which is Level C and Level B, respectively. In the whole simulation period, taking 20% of the peak discharge as the allowable error, the prediction accuracy was Level C.

When using the Nash efficiency coefficient (E) as the evaluation index, the mean E of the selected 26 flood events was 0.789, and the value of the Nash efficiency coefficient of 21 flood events was larger than 0.7 in the whole period. When using the coefficient of determination (R^2) as the evaluation index, the mean R^2 of the selected 26 flood events was 0.893, and the value of R^2 of 23 flood events was larger than 0.8 in the whole period. When using the index of agreement (d) as the evaluation index, the mean d of the selected

26 flood events was 0.918, and the value of *d* of all flood events was larger than 0.8 in the whole period. Other results of evaluation indices can be seen in Table 3.

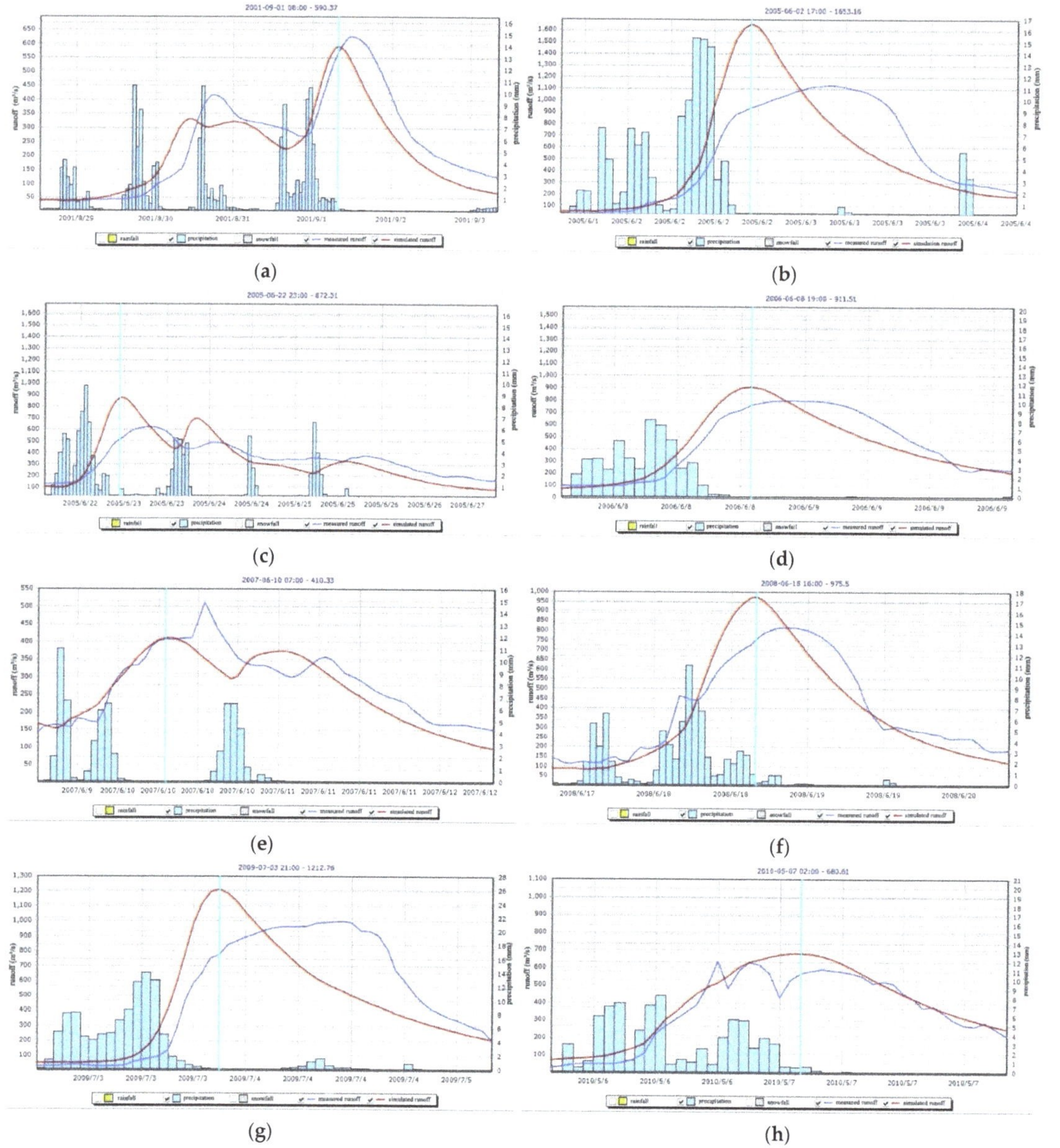

Figure 4. *Cont.*

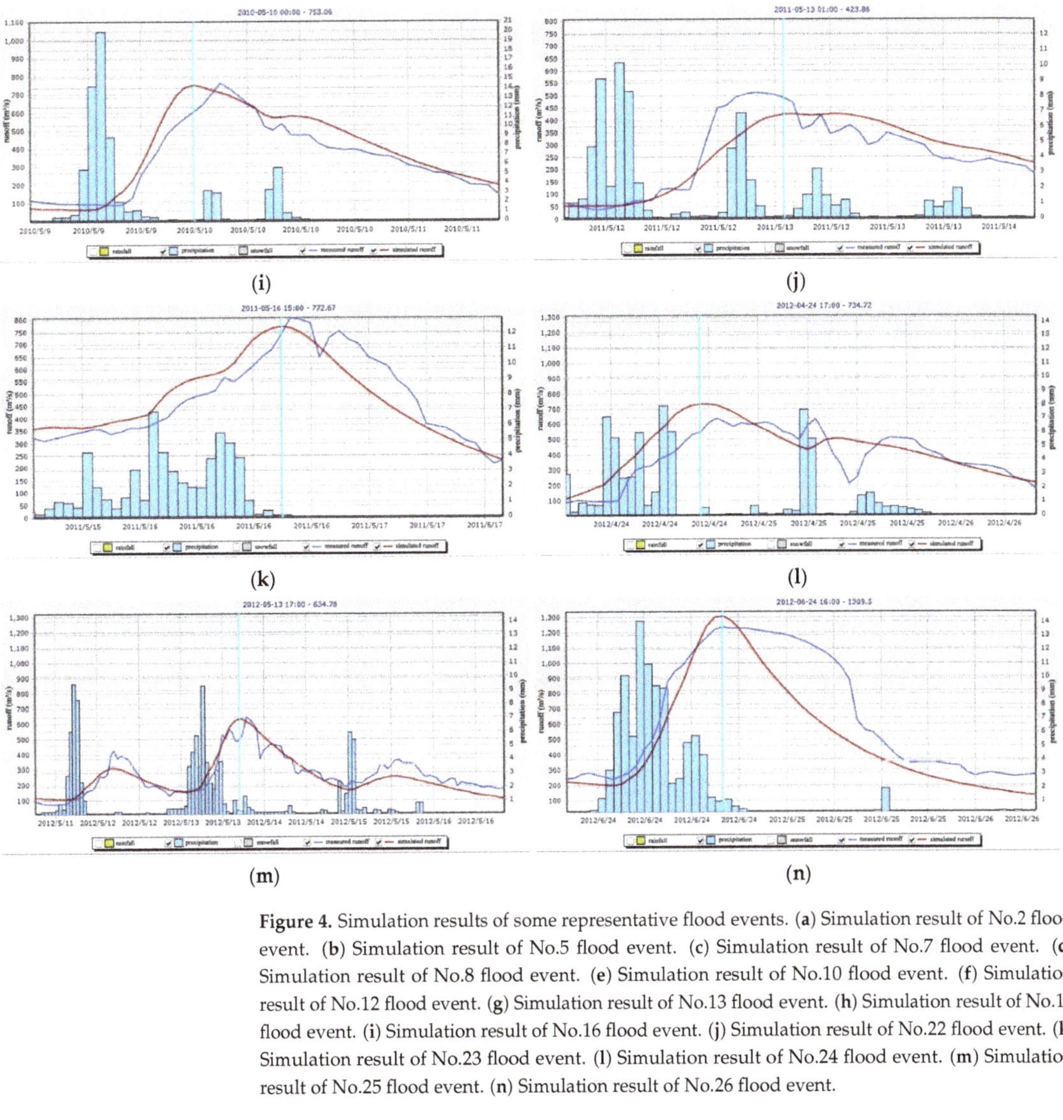

Figure 4. Simulation results of some representative flood events. (**a**) Simulation result of No.2 flood event. (**b**) Simulation result of No.5 flood event. (**c**) Simulation result of No.7 flood event. (**d**) Simulation result of No.8 flood event. (**e**) Simulation result of No.10 flood event. (**f**) Simulation result of No.12 flood event. (**g**) Simulation result of No.13 flood event. (**h**) Simulation result of No.15 flood event. (**i**) Simulation result of No.16 flood event. (**j**) Simulation result of No.22 flood event. (**k**) Simulation result of No.23 flood event. (**l**) Simulation result of No.24 flood event. (**m**) Simulation result of No.25 flood event. (**n**) Simulation result of No.26 flood event.

Table 3. Simulation results of 26 selected flood events with peak discharge rates greater than 500 m^3/s.

NO.	Flood Events	Observed Flood Peak Discharge/(m^3/s)	Simulated Flood Peak Discharge/(m^3/s)	Error of Peak Discharge/%	E	MAE/(m^3/s)	RMSE/(m^3/s)	R^2	EV	VC	CMM	d	Up to the Standard
1	2001-05-09 02:00–2001-05-12 08:00	516	581.82	12.756	0.65	60.57	78.21	0.87	0.66	0.94	0.7	0.92	√
2	2001-08-29 14:00–2001-09-03 08:00	627	590.37	−5.842	0.73	70.07	81.32	0.88	0.78	0.76	0.73	0.92	√
3	2002-10-28 08:00–2002-11-01 20:00	1050	978.73	−6.788	0.77	144.5	180.2	0.92	0.83	0.55	0.83	0.93	√
4	2005-05-24 00:00–2005-05-26 04:00	512	570.78	11.480	0.6	53.89	92.04	0.85	0.65	0.73	0.69	0.91	√
5	2005-06-01 15:00–2005-06-04 06:00	1130	1653.16	46.297	0.46	211.93	301.28	0.78	0.46	0.98	0.75	0.87	×
6	2005-06-12 04:00–2005-06-14 08:00	674	843.65	25.171	0.41	69.16	112.8	0.89	0.56	0.72	0.62	0.89	×
7	2005-06-22 02:00–2005-06-27 08:00	626	872.31	39.347	0.19	94.32	121.7	0.85	0.19	0.96	0.36	0.87	×
8	2006-06-08 02:00–2006-06-09 20:00	801.4	911.51	13.740	0.83	86.62	108.52	0.91	0.83	0.96	0.88	0.96	√
9	2006-07-25 18:00–2006-07-29 08:00	1250	1573.18	25.854	0.73	167.43	233.04	0.88	0.76	0.67	0.83	0.93	×
10	2007-06-09 12:00–2007-06-12 08:00	511	410.33	−19.701	0.74	39.17	49.79	0.89	0.78	0.86	0.74	0.93	√
11	2008-06-12 06:00–2008-06-15 08:00	959	845.65	−11.820	0.84	77.65	113	0.94	0.88	0.75	0.88	0.95	√
12	2008-06-17 08:00–2008-06-20 08:00	815	975.5	19.693	0.82	80.02	102.55	0.94	0.83	0.89	0.86	0.96	√
13	2009-07-03 00:00–2009-07-05 04:00	1000	1212.76	21.276	0.52	199.69	260.37	0.76	0.54	0.82	0.75	0.86	×
14	2009-07-14 16:00–2009-07-16 04:00	555	320.89	−42.182	0.64	86.6	117.87	0.93	0.69	0.56	0.78	0.84	×
15	2010-05-06 02:00–2010-05-07 22:00	637	680.61	6.846	0.9	47.54	64.53	0.97	0.93	0.8	0.92	0.97	√
16	2010-05-09 06:00–2010-05-11 10:00	763.24	753.06	−1.334	0.86	58.59	72.59	0.97	0.91	0.77	0.89	0.97	√
17	2010-05-14 04:00–2010-05-15 20:00	611	668.45	9.403	0.87	57.51	67.07	0.95	0.9	0.8	0.89	0.97	√
18	2010-05-22 15:00–2010-05-24 00:00	601.92	561.85	−6.657	0.42	66.58	88.18	0.72	0.43	0.93	0.42	0.85	√
19	2010-06-14 23:00–2010-06-17 00:00	859	1078.18	25.516	0.8	93.95	116.02	0.94	0.86	0.77	0.87	0.95	×
20	2010-06-24 18:00–2010-06-26 09:00	819	933.42	13.971	0.81	71.99	102.62	0.97	0.88	0.71	0.88	0.96	√
21	2011-05-08 07:00–2011-05-10 04:00	768	452.86	−41.034	0.54	94.63	130.63	0.98	0.78	0.38	0.65	0.83	×
22	2011-05-12 02:00–2011-05-14 04:00	513	423.86	−17.376	0.79	50.75	67.37	0.89	0.79	0.99	0.87	0.94	√
23	2011-05-15 14:00–2011-05-17 14:00	808.9	772.67	−4.479	0.79	64.8	77.29	0.89	0.79	0.99	0.81	0.94	√
24	2012-04-24 01:00–2012-04-26 10:00	638.27	734.72	15.111	0.52	88.16	113.4	0.8	0.59	0.79	0.6	0.87	√
25	2012-05-11 08:00–2012-05-16 20:00	647.86	634.78	−2.019	0.81	44.57	54.65	0.92	0.82	0.91	0.81	0.95	√
26	2012-06-23 20:00–2012-06-26 08:00	1240	1309.3	5.589	0.74	149.38	201.57	0.93	0.86	0.51	0.75	0.93	√

Table 4. Qualification rate statistics of one-hour forecast scheme for Xiaogulu station.

Period	Evaluation Items/Index	Flood Events	Events up to Standard	Pass Rate/%	Assessment Result
Calibration	Peak discharge/*EPD*	18	12(EPD ≤ 20%)	66.67	Level C
Validation	Peak discharge/*EPD*	8	6(EPD ≤ 20%)	75.00	Level B
	Peak discharge/*EPD*	26	18(EPD ≤ 20%)	69.23	Level C
	Process conformance/*E*	26	21(E > 0.7)	80.77	
All periods	Mean of *E*		0.789		
	Mean of the R^2		0.893		
	Mean of the *d*		0.918		

4. Discussion

When using the indices of E, R^2 and d as the evaluation indices, the simulation results are good, but using the *EPD* as the evaluation index, according to the requirements of the standard GB/T22482-2008, some flood events are failed to the standard, and at the same time, the simulated occurrence time of the selected flood events has deviation, too. Therefore, the forecast results using the calibrated parameters can only be used for estimation. In general, the reasons for the deviation between the simulated and observed peak discharge and its occurrence time of 26 flood events can be explained as follows:

(1) The regulation influence of small- and medium-sized reservoirs on the basin's water storage: Since the operation data of these 18 reservoirs had not been collected, the water storage and drainage of these reservoirs had not been simulated and, therefore, these areas were treated as no reservoir in the basin. The total catchment area of these reservoirs was up to 324.04 km^2, accounting for 16.86% of the study area. The five medium-sized reservoirs had regulation capacity, among which Hengjiang reservoir had multi-year regulation capacity. In the flood season, it can store and release water according to the predicted inflow, thus affecting the flooding process of its downstream basin.

Figure 4b,d,g reveal similar simulated and measured processes that floods might have undergone following the reservoirs' storage and regulation. Figure 4g can serve as an example for a simple analysis. Several rainfall stations with the largest rainfall in this flood event were located in the upper and middle reaches, and the total rainfall of Dayuan station with the largest rainfall (within 52 h) was 264.6 mm, and that of Nanpu station with the smallest rainfall is 28.9 mm. It was possible that through the storage of several reservoirs, the measured flood peak flow of Xiaogulu station would be cut off by about 20%, and the occurrence time of the peak would be delayed, according to the simulated results (regarding no reservoirs in the basin).

(2) Deviation in the processing of temperature, precipitation, and flow data: Since the hourly data of meteorological stations in the basin were not available, and only one meteorological station in the basin had daily temperature data, the hourly temperature data obtained from the daily average temperature, daily maximum temperature, and daily minimum temperature might have some deviation. Meanwhile, the hourly precipitation data would also have deviation when the precipitation lasted for a short time, but since it was evenly distributed to an hour, the rainfall intensity would be reduced, affecting the flood peak discharge. Similarly, the measured flow of the Xiaogulu station was processed into a one-hour period, and therefore, the hourly flow obtained by the linear interpolation method would also have some deviation. Figure 4e,h,k, show the processing trace of linear interpolation of measured flow data; nevertheless, the simulated results of these flood events still met the standard.

(3) Deviation of soil and land use attribute information: As the detailed soil and land use attribute information in the basin required by the model was not collected, the same type of soil and land use attribute information in other regions was used instead. The spatial difference in different places was large. Although this aspect had been taken into account in parameter calibration, the deviation would still exist, especially in the parameters of soil thickness and hydraulic conductivity, and they would have some influence on the simulation results. Therefore, simulation results in the calibration period based on

the normal range of calibrated soil attribute parameter values were not ideal, and the corresponding verification results were not satisfactory. The runoff recession process in Figure 4a,c,e shifted to an earlier time, compared with the measured process, and this phenomenon might relate to the soil and land use parameters.

(4) The influence of rainstorm centers and the deviation of processing point rainfall into areal rainfall caused by rainfall non-uniformity: The average rainfall station density was 213.5 km^2/station in this basin, but there was a considerable amount of rainfall in southern China, and the rainstorm center moved frequently, so relatively speaking, the rainfall stations were fewer, and the influence of processing point rainfall into areal rainfall could not be avoided.

(5) The reason attributed to the model itself and the impact of discontinuous rainfall and discharge data: The model is suitable for long series and continuous hydrological simulation. If the rainfall data were missing, the rainfall would be treated as 0, while if the temperature data were missing, the model would continue to calculate according to the monthly average temperature provided. The processing method of missing rainfall data would lead to lower simulated runoff. The model had a long preheating period (3–4 months or even longer) at the beginning of the simulation. In the case of only excerpted precipitation data from March to November were collected, the model had just been successfully preheated, but the precipitation data were interrupted again from November to March of the next year, so the model had to repeatedly preheat, resulting in some deviation between the simulated and the measured runoff. Since there were no rainfall and discharge data collected from November to March of the next year, when the model automatically read the data, the rainfall was treated as 0, resulting in inaccurate flood simulation values of March to May of each year, or even longer.

(6) Influence of initial state: In order to accelerate the preheating of the model, when setting the initial state of the soil, a larger percentage of monthly average saturated water content was set each month, and the same initial state was used in each year, which may be inconsistent with the actual situation.

5. Conclusions

Based on the simulation results and the discussion above, some suggestions are provided when using TOPKAPI in the study area and similar basins with reservoirs: (1) the operation data of medium-sized reservoirs in the basin, especially the water discharge of these reservoirs in case of large floods, should be collected; then, it should be recalibrated, and the model parameters, verified; (2) after collecting the hourly rainfall, temperature, flow, and evaporation data, the processing and integration of measured data need to be studied, to ensure the processing accuracy meets the requirements. Meanwhile, the quality of data used should be improved, with the aim to increase data collection and compilation quality in the dry season; (3) the attribute information of soil and land use and other data needed by the model in the basin need to be collected.

There are 5 medium-sized reservoirs and 13 small-sized reservoirs in this study area. Due to the lack of reservoir operation data, the model parameter calibration and validation can only be based on precipitation data and measured flow data, which increases the difficulty and inaccuracy of the model parameter calibration and flood process analysis. In practice, the real-time flood forecasting should be closely combined with the reservoir dispatching, especially withthe five medium-sized reservoirs in the basin. The soil moisture content of the early stage should be adjusted in real time, to obtain good results.

Author Contributions: Conceptualization, L.L. and Y.H.; methodology, L.L. and K.L. (Kuang Li); data curation, K.L. (Kexin Liu) and H.X.; analysis, L.L., Z.L. and Y.Y.; writing—original draft preparation, L.L.; writing—review and editing, L.L. and X.L.; funding acquisition, L.L. All authors have read and agreed to the published version of the manuscript.

Funding: This research was funded by Project U2040212 supported by the National Natural Science Foundation of China, the R&D Program of China Three Gorges Corporation (NO. WWKY-2021-0081, and NO. WWKY-2021-0411), the National Key R&D Program of China (NO. 2018YFC0406400), and the Special Project of National Science and Technology Basic Resources Investigation (NO. 2021xjkk04005).

Institutional Review Board Statement: Not applicable.

Informed Consent Statement: Not applicable.

Data Availability Statement: Data used for this study can be made available upon request.

Acknowledgments: We gratefully acknowledge financial support from the R&D Program of China Three Gorges Corporation and the National Key R&D Program of China.

Conflicts of Interest: The authors declare no conflict of interest.

Appendix A

Table A1. Basic information of the five medium-sized reservoirs.

NO.	Reservoir Name	Watershed Area/km^2	Total Storage Capacity/10^4 m^3	Normal Storage Capacity/10^4 m^3	Normal Water Level/m
1	Pubu	76.0	3233	2950	346
2	Baojiang	48.1	2350	1914	198.086
3	Zhongping			1432	346
4	Hengjiang	24.25		1282	204
5	Kongjiang	101.44	6522	5828	195.77

Table A2. Basic information of the 11 small-sized reservoirs.

NO.	Reservoir Name	Watershed Area/km^2	Total Storage Capacity/10^4 m^3	Utilizable Capacity/10^4 m^3	Flood Regulation Storage Capacity/10^4 m^3
1	Dayuan	26	681	523	156
2	Luotian	1.02	220.8	185.7	20.8
3	Wuni	3.2	212.2	179	29
4	Sheling	4.6	124.5	80.1	41.9
5	Weibei	2.09	795.3	667.4	80.9
6	Yangmei	11.35	524	406	190
7	Meiling	4.1	153	118	44
8	Zhuhaokeng	1.6	131	124	31.6
9	Songshan	3.98	121	99.6	14
10	Daba	1.2	134.7	113	17.6
11	Zaixia	2.02	103	79.5	18
12	Chishui				
13	Dakengtang				

References

1. Sun, D. Theoretical-technical framework of flash flood disaster prevention. *J. China Inst. Water Resour. Hydropower Res.* **2021**, *19*, 313–317.
2. Guo, L.; Ding, L.Q.; Sun, D.Y.; Liu, C.J.; He, B.S.; Liu, R.H. Key techniques of flash flood disaster prevention in China. *J. Hydraul. Eng.* **2018**, *49*, 1123–1136.
3. Reed, S.; Schaake, J.; Zhang, Z. A distributed hydrologic model and threshold frequency-based method for flash flood forecasting at ungauged locations. *J. Hydrol.* **2007**, *337*, 402–420. [CrossRef]
4. Li, H.; Wang, R.; Huang, Q.; Xiang, J.; Tan, G. Advances on Flood Forecasting of Small-Medium Rivers. *J. China Hydrol.* **2020**, *40*, 16–23.
5. Sivapalan, M.; Takeuchi, K.; Franks, S.W.; Gupta, V.K.; Karambiri, H.; Lakshmi, V.; Liang, X.; McDonnell, J.J.; Mendiondo, E.M.; O'connell, P.E.; et al. IAHS Decade on Predictions in Ungauged Basins (PUB), 2003–2012: Shaping an exciting future for the hydrological sciences. *Hydrol. Sci. J.* **2003**, *48*, 857–880. [CrossRef]
6. Liu, Z.; Li, Z.; Yang, D.; Hou, A.; Ju, X. *Research on Indicator Determination and Forecasting Technology for Flood Warning for Small-Mediun River*; Science Press: Beijing, China, 2016.
7. Borga, M.; Anagnostou, E.; Bloschl, G.; Creutin, J.D. Flash flood forecasting, warning and risk management: The HYDRATE project. *Environ. Sci. Policy* **2011**, *14*, 834–844. [CrossRef]

8. Schmugge, T.J.; Kustas, W.P.; Ritchie, J.C.; Jackson, T.J.; Rango, A. Remote sensing in hydrology. *Adv. Water Resour.* **2002**, *25*, 1367–1385. [CrossRef]
9. Sui, D.Z.; Maggio, R.C. Integrating GIS with hydrological modeling: Practices, problems, and prospects. *Comput. Environ. Urban Syst.* **1999**, *23*, 33–51. [CrossRef]
10. Wen, L.S.; Chao, M.H. A Unified Framework of the Cloud Computing Service Model. *J. Electron. Sci. Technol.* **2013**, *11*, 150–160.
11. Todini, E. Flood forecasting and decision making in the new Millennium. Where are we? *Water Resour. Manag.* **2017**, *31*, 3111–3129. [CrossRef]
12. Zhai, X.; Guo, L.; Liu, R.; Zhang, Y.; Liu, C. Development and application of China flash flood hydrological model: Case study in small and medium-sized catchments of Anhui Province. *J. Basic Sci. Eng.* **2020**, *28*, 1018–1036.
13. Wang, Y.; Liu, R.; Guo, L.; Tian, J.; Zhang, X.; Ding, L.; Wang, C.; Shang, Y. Forecasting and providing warnings of flash floods for ungauged mountainous areas based on a distributed hydrological model. *Water* **2017**, *9*, 776. [CrossRef]
14. Todini, E. The ARNO rainfall-runoff model. *J. Hydrol.* **1996**, *175*, 339–382. [CrossRef]
15. Mazzetti, C. *TOPKAPI Model and Its Application Guide*; Liu, Z.; Weng, M.; Zhao, C., Translators; Hohai University Press: Nanjing, China, 2014.
16. Liu, Z.; Todini, E. Assessing the TOPKAPI non-linear reservoir cascade approximation by means of a characteristic lines solution. *Hydrol. Processes* **2005**, *19*, 1983–2006. [CrossRef]
17. Liu, Z.; Todini, E. Towards a comprehensive physically-based rainfall-runoff model. *Hydrol. Earth Syst. Sci.* **2002**, *6*, 859–881. [CrossRef]
18. Liu, Z. *Toward a Comprehensive Distributed/Lumped Rainfall-Runoff Model: Analysis of Available Phsically-Based Models and Proposal of a New TOPKAPI Model*; The University of Bologna: Bologna, Italy, 2002.
19. Coccia, G.; Mazzetti, C.; Ortiz, E.A.; Todini, E. Application of the TOPKAPI Model within the DMIP 2 Project. 2009. Available online: https://www.researchgate.net/publication/265678071_Application_of_the_TOPKAPI_model_within_the_DMIP_2_project (accessed on 7 November 2021).
20. Wang, T.; Gao, D.; Ma, H.; Zhang, M. Distributed hydrological model for nalinggele river and salt lakes in its lower reaches. *J. Salt Lake Res.* **2006**, *14*, 18–21.
21. Li, Z.; Wang, X.; LV, Y.; Chen, L.; Li, L. Application of TOPKAPI Model and Comparison with Xin'anjiang Model. *Water Power* **2013**, *39*, 6–10.
22. Jian, H.; Liu, H.; Tong, B. Comparative study on construction and application of distributed hydrological model. *Yellow River* **2020**, *42*, 24–29.
23. Mazzetti, C. *TOPKAPI™ Model References*; Progea Srl.: Bologna, Italy, 2015.
24. Liang, L.; Li, K.; Ye, Y.; Xu, H.; Hu, Y. TOPKAPI model and its application of flood early warning and forecasting in China. In *World Environmental and Water Resources Congress 2019: Water, Wastewater, and Stormwater; Urban Water Resources; and Municipal Water Infrastructure*; Scott, G.F., Hamilton, W., Eds.; American Society of Civil Engineers: Pittsburgh, PA, USA, 2019; pp. 113–123.
25. Soil Survey Office of Guangdong Provincial. *Guangdong Soil Records*; Science Press: Beijing, China, 1996.
26. Available online: http://vdb3.soil.csdb.cn/ (accessed on 20 April 2020).
27. Chang, Y.; Yu, Y.; Cui, L.; Hou, J.; Zeng, D. Vegetation absorbed photosynthetically active radiation estimates based on HJ-1A satellite HSI data. *For. Eng.* **2017**, *33*, 22–27.
28. Sun, J.; Zhang, J.; Wang, J.; Liang, J.; Zhang, S.; Chen, S.; Yue, J.; Rui, X.; Li, L.; Liu, Z.; et al. *GB/T 22482-2008*; Standard for Hydrological Information and Hydrological Forecasting. National Standard: Beijing, China, 2009.

water

MDPI

Article

Assessing Numerical Model Skill at Simulating Coastal Flooding Using Field Observations of Deposited Debris and Photographic Evidence

Sean Ferguson [1,*], Mitchel Provan [1], Enda Murphy [1,2], Dominique Bérubé [3], Marc Desrosiers [3], André Robichaud [4] and Joseph Kim [2]

[1] Ocean, Coastal and River Engineering Research Centre, National Research Council Canada, Ottawa, ON K1A 0R6, Canada; mitchel.provan@nrc-cnrc.gc.ca (M.P.); enda.murphy@nrc-cnrc.gc.ca or emurp073@uottawa.ca (E.M.)
[2] Department of Civil Engineering, University of Ottawa, Ottawa, ON K1N 6N5, Canada; jkim245@uottawa.ca
[3] Natural Resources and Energy Development, Geological Surveys Branch, Government of New Brunswick, Bathurst, NB E2A 7B8, Canada; dominique.berube@gnb.ca (D.B.); marc.desrosiers@gnb.ca (M.D.)
[4] Campus de Shippagan, Secteur Administration, Arts et Sciences Humaines, Géographie, Université de Moncton, Shippagan, NB E8S 1P6, Canada; andre.j.robichaud@umoncton.ca
* Correspondence: sean.ferguson@nrc-cnrc.gc.ca

Citation: Ferguson, S.; Provan, M.; Murphy, E.; Bérubé, D.; Desrosiers, M.; Robichaud, A.; Kim, J. Assessing Numerical Model Skill at Simulating Coastal Flooding Using Field Observations of Deposited Debris and Photographic Evidence. *Water* **2022**, *14*, 589. https://doi.org/10.3390/w14040589

Academic Editors: Zhang Cheng, Slobodan P. Simonovic and Subhankar Karmakar

Received: 22 December 2021
Accepted: 11 February 2022
Published: 15 February 2022

Abstract: Despite the growing range and availability of resources to support coastal flood hazard model development, there is often a scarcity of data to support critical assessment of the performance of community-scale coastal inundation models. Even where long-term tide gauge measurements are available in close proximity to the study area, the records provide little insight into the spatial distribution and limits of overland flooding, or the influence of topographic features and structures on flooding pathways. We present methods to support the assessment of model performance using field observations in lieu of, or supplementary to, conventional water-level records. A high-resolution, numerical coastal flood hazard model was developed to simulate storm surge-driven flooding in the Acadian Peninsula region of New Brunswick, Canada. Owing to the remoteness of the study area from tide gauge stations, model performance was assessed based on a comparison with field measurements of deposited wrack and debris, as well as photographic and video evidence of coastal flooding, for two significant storm surge events in recent history. Our research findings illustrate the value of observational and qualitative data for characterizing coastal flood hazards, lending gravity to the importance of non-conventional data sources, particularly in data-scarce regions.

Keywords: flooding; storm surge; numerical modelling; high-water marks; debris

1. Introduction

Approximately 6.5 million Canadians live in the vicinity of Canada's marine coasts, making these regions intrinsically important to the national identity, culture, and economy [1]. Coastal flood hazards and risks in Canada are becoming increasingly exacerbated by climate-driven phenomena such as sea-level rise, and anthropogenic factors that contribute to increased development, population growth, and loss of natural features in the coastal zone [2–5]. The southern Gulf of Saint Lawrence and sections of New Brunswick's coast are particularly sensitive to the impacts of sea-level rise and storm surges [3]. In response to this growing threat, numerous research initiatives have been conducted in New Brunswick and the surrounding area to investigate coastal flooding at both regional [3,6] and local scales [2,7–10].

Hazard assessment is a crucial component of disaster risk management. Flood hazard modelling (and derived mapping and communication products) permits the quantitative assessment of the consequences of flooding and ultimately supports evidence-based decision making and disaster risk management [5,11,12]. A number of analytical methods, with varying complexity, are applied in practice to model and assess coastal flood hazards. Bathtub modelling techniques, where high-water elevations are mapped onto a digital elevation model, are often employed to evaluate overland flood depths and extent [7,10,13–15]. Numerical hydrodynamic modelling techniques, although more computationally demanding, are an attractive alternative owing to their explicit consideration of hydrodynamics, resulting in generally superior accuracy [16], and their provision of additional information regarding spatial and temporal variation of flood hazards, providing insight to water velocities, flood propagation, and duration. Two-dimensional (and in some cases three-dimensional) hydrodynamic modelling techniques are often employed to simulate coastal water levels and flows in response to meteorological and tidal forcing [17–20]. Circulation models are sometimes coupled with other computational models to simulate other physical processes at various scales and resolutions such as wave runup, overtopping, and geomorphological processes [17,21]. To alleviate computational demand, nested modelling techniques and unstructured grids are commonly employed to permit detailed simulation nearshore, while maintaining sufficient domain size to capture synoptic scale atmospheric and meteorological forcing [17,19,22]. Previous studies have employed bathtub modelling approaches to characterize coastal flood hazards in the Acadian Peninsula region of New Brunswick [7,10]. However, to the authors' knowledge, hydrodynamic modelling of storm surges, including overland inundation, at spatial resolutions conducive to damage and risk assessment have not yet been conducted for this region.

The Acadian Peninsula is located at the northeastern extremity of New Brunswick, Canada (Figure 1). The region is bordered to the north by Chaleur Bay and to the south by the Gulf of Saint Lawrence. The region is home to several communities experiencing coastal flooding and erosion, which have contributed to damage to infrastructure, including houses, roads, and wharfs [23]. A number of coastal flood assessment and mapping projects [7,10] and adaptation initiatives [9] have been undertaken for Acadian Peninsula communities. However, one of the biggest challenges hindering accurate flood hazard characterization and development of reliable flood modelling and assessment tools is the scarcity of proximate water level records. Currently, there are no long-term, active tide gauges located on the Acadian Peninsula. The region is approximately equidistant between the two closest long-term tide gauge stations located in Belledune (tide gauge 2145) and Lower Escuminac (tide gauge 2000), approximately 85 km and 75 km away, respectively [24]. Even where tide gauge measurements exist, they provide little insight into the spatial distribution and inland extent of coastal flooding [2]. To overcome these challenges, previous studies have relied on surveyed high-water marks, such as shoreline debris, to support validation of inundation models [10].

Where adequate observational data exist, deposited debris such as driftwood and sea wrack have been used to support analyses related to coastal flooding [8,26–30]. Didier et al. [8] used debris lines to evaluate coastal flood extent, as well as predictive accuracy of wave runup estimations, for the community of Maria, Québec; Maria is located approximately 100 km northwest of the Acadian Peninsula. Similarly, Bernatchez et al. [2] used observational data and photographic evidence in the community of Maria to investigate flood extent for a winter storm event. Often, researchers pursue the acquisition of observational data in a sufficient quantity to support the delineation of the full flood extent [2,8,27]. However, depending on the size of the study area, the effort and resources required to support this level of field data collection can be prohibitively large. In this paper, we describe the assessment of model skill for a numerical hydrodynamic model developed to evaluate storm surge-driven flood hazards for selected locations of the Acadian Peninsula region of New Brunswick, Canada. Model skill was assessed using surveyed high-water marks coinciding with the 6 December 2010 and 21 December 2010 events. High-water marks were comprised of shoreline debris observations as well as pegs placed by the field team marking direct observation of maximum flood extent. In addition, a qualitative assessment of model skill was conducted through a comparison with photographic and video evidence.

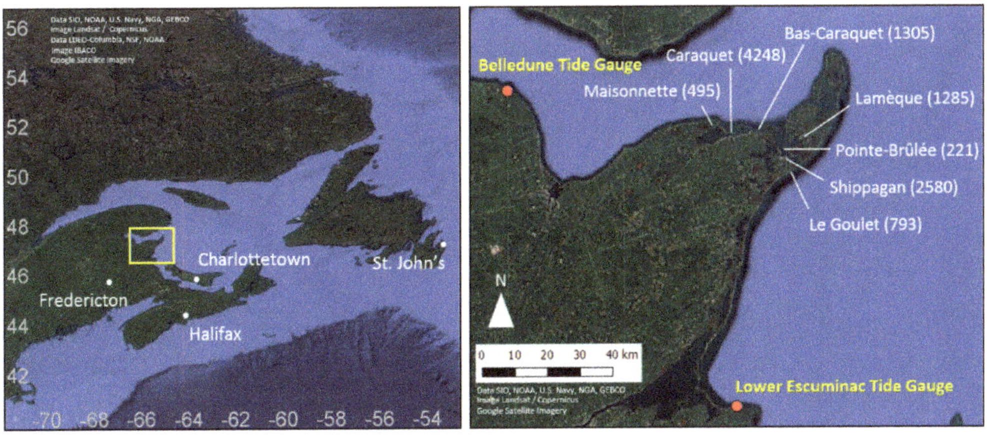

Figure 1. Study location. The image on the left shows the Acadian Peninsula (yellow box) within the broader extent of Atlantic Canada. The image on the right shows selected communities from the Acadian Peninsula where inundation modelling was completed using a fine-resolution, community-scale hydrodynamic model; population estimates are shown in parentheses [25].

2. Storm Surge Modelling

Numerical hydrodynamic modelling was conducted at two scales using TELEMAC-2D, which solves the shallow water (Saint-Venant) equations governing free surface flows in two dimensions (i.e., depth averaged) on unstructured computational meshes [31]. A coarse-resolution, regional-scale model was developed to simulate the generation and evolution of storm surges on the Atlantic Canada continental shelf, including the Gulf of Saint Lawrence, in response to synoptic-scale wind and atmospheric pressure fields [32]. A fine-resolution, community-scale model was developed and driven by output from the regional model to simulate inundation at seven Acadian Peninsula communities. The seven communities of interest included Maisonnette, Bas-Caraquet, Caraquet, Pointe-Brûlée, Shippagan, Le-Goulet, and Lamèque (Figure 1). This paper is focused on assessment of the community-scale model skill; however, some information pertaining to the regional-scale model is provided for context. A detailed summary of the regional-scale model development is presented in Provan et al. [32].

The regional-scale model domain covered a large portion of the continental shelf from Portland, Maine (United States) to Sandwich Bay, Newfoundland and Labrador (Canada). The model uses an unstructured computational mesh with resolution ranging between 30 km near the offshore boundary and 200 m in the vicinity of the Acadian Peninsula. The model bathymetry was constructed using a combination of bathymetric datasets [33,34]. The regional-scale model was forced using wind and atmospheric pressure data from the ERA5 global atmospheric reanalysis dataset [35]. Tidal forcing was applied using the global TPXO database and its regional and local variants [36]. The regional-scale model was calibrated based on measured data available from the Belledune and Lower Escuminac tide gauges. Water level residuals were extracted from the gauge records by computing the difference between measured levels and the predicted astronomical tide [37], and were used as a basis for calibration, assuming storm surges represent the dominant contribution to the residual. Model wind drag formulation/coefficients and bed roughness were iteratively adjusted until satisfactory results were achieved. Ultimately, the regional-scale model was able to predict peak storm surges at the Lower Escuminac gauge to within 15% of measured water level residuals for 24 of the largest 44 storm surge events recorded at the gauge.

The community-scale model encompassed the Acadian Peninsula region of New Brunswick extending from the north coast of New Bandon Parish, eastward to Miscou Island, and southward to Tracadie-Sheila. The model domain extended approximately 20–30 km northward into Chaleur Bay and 30 km southward into the Gulf of Saint Lawrence. The model uses an unstructured (triangular) computational mesh with characteristic element edge lengths on the offshore boundary ranging between approximately 250 m and 2000 m. Characteristic edge lengths of 10 m were specified in the vicinity of the communities of interest to resolve floodplain features, flooding pathways, and hazards, at scales of relevance to support consequence assessments. Model elevations were generated from a combination of bathymetric and topographic datasets [33,38–40], including a hydro-enforced 1 m-resolution digital elevation model (DEM) from 2018 and a 1 m-resolution DEM from 2009 provided by the Government of New Brunswick. The hydro-enforced DEM was used in six of the seven communities. The 2009 DEM was used in the vicinity of Le Goulet to provide a consistent basis for comparing model output to observations, which pre-dated significant coastal engineering works and alterations to the dynamic Le Goulet coastline, including construction of artificial dunes and sand retention structures [9]. Spatially and temporally varying water levels were prescribed at the offshore boundary of the community-scale model based on output from the regional-scale model, which included tidal forcing based on the TPXO tide model and its regional and local variants [36]. Corrections for wave setup were not incorporated into the community-scale model. Surface wind and atmospheric pressure forcing was applied within the model domain, based on linear interpolation of gridded ERA5 reanalysis data to the model mesh [35]. Initial estimates for the wind drag formulation and coefficients were based on calibrated parameters for the regional-scale model. Initial estimates for land roughness were based on land cover data acquired from the Government of New Brunswick [41–43] and typical roughness values by land cover type obtained from literature [44,45].

3. High-Water-Mark Surveys and Photographic and Video Evidence

High-water-mark surveys were conducted by members of the New Brunswick Geological Surveys Branch using a high precision real-time kinematic differential global positioning system (RTK-DGPS) consisting of two rovers and a base station. Leica Viva GS15 GPS receivers were used, which have a reported horizontal and vertical accuracy of 8 mm and 15 mm, respectively [46]. However, duplicate readings on ground-control points revealed that survey accuracy was actually 20 mm to 30 mm in all three axes (i.e., easting, northing, and orthometric height). The surveys were conducted on 9 December 2010 in the vicinity of Le Goulet and Shippagan; 23 June 2011 in the vicinity of Pointe-Brûlée; and 28 June 2011 in the vicinity of Bas-Caraquet. The survey data consisted of point measurements indicating horizontal position (easting and northing) and elevation georeferenced to the Canadian Geodetic Vertical Datum of 1928 (CGVD28). Elevations were converted to the Canadian Geodetic Vertical Datum of 2013 (CGVD2013) using the GPS·H tool [47] to support comparison with the model; model simulations were conducted with elevations referenced to mean water level (MWL), but model results could be easily transformed to CGVD2013. Most of the survey data were based on observations of debris (sea-wrack) deposited onshore by the flood water, primarily consisting of eel grass vegetation. Exceptions included data recorded in Bas-Caraquet (locations BC1 and BC2 on Figure 2), which represent observed gravel deposits, and data recorded in Le Goulet (locations LG1 and LG2), which represent pegs placed by the field team during the flood event marking the maximum flood extent. When debris were used as an indicator of high water, measurements were recorded at ground elevation near the mid-section of the debris lines.

Referring to survey locations shown in Figure 2, most surveyed high-water marks were located either substantially landward of the beach crest (PB1, PB2, PM1, LG1, and LG2), were sheltered within a small bay with fetches less than 3 km (PB3, PB4, PB5, and SG1), or were otherwise sheltered by built structures (SH1 and SG2), such that wave runup contributions to total water levels were expected to be relatively low. This is supported by photographic evidence of the December 2010 flood events, where only small ripples were observed at the flood extent beyond the beach crest in Pointe-Brûlée, Baie-de-Petit-Pokemouche, and Le Goulet (refer to Section 4.2). Exceptions included survey locations in

Bas-Caraquet (BC1 and BC1), where surveyed points are located immediately adjacent to the shoreline, which is exposed to Chaleur Bay.

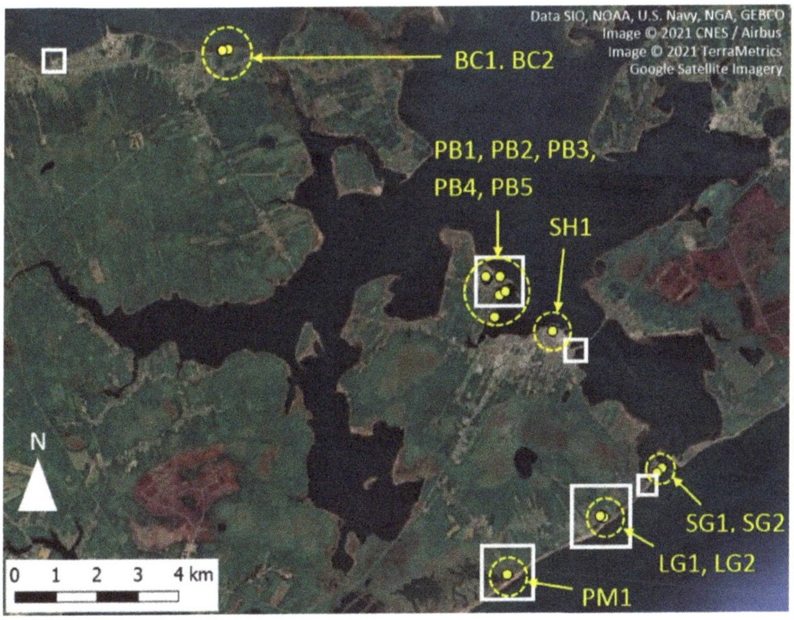

Figure 2. Data available to support model skill assessment. High-water-mark survey data are shown as yellow points. General areas where photographic or video evidence were available are shown with white boxes. Short-form identification labels assigned to high-water-mark survey locations are shown: BC = Bas-Caraquet; PB = Pointe-Brûlée; SH = Shippagan; SG = Shippgan Gully; LG = Le Goulet; PM = Baie-de-Petit-Pokemouche.

The survey data collected on 9 December 2010 (in the vicinity of Baie-de-Petit-Pokemouche, Le Goulet, Shippagan Gully, and Shippagan) reflect peak flood levels and extents produced during the 6 December 2010 event. It was not fully certain whether debris lines surveyed in June 2011 (in the vicinity of Pointe-Brûlée and Bas-Caraquet) reflected peak flood levels associated with the 6 December 2010 event or the 21 December 2010 event; both events produced similar water levels in the Acadian Peninsula region. To address this uncertainty, modelled results for the 6 December 2010 and 21 December 2010 events were carefully compared for each model setup evaluated during the calibration process (see Section 4.1). Modelled results for the 21 December 2010 event consistently showed more extensive flooding in comparison to the 6 December 2010 event in the vicinity of Pointe-Brûlée for all comparable model setups. In the vicinity of Bas-Caraquet, the modelled results for the 21 December 2010 event showed slightly more extensive flooding compared to the 6 December 2010 event. This suggests that the June 2011 survey data in Pointe-Brûlée and Bas-Caraquet represent peak flood levels associated with the 21 December 2010 event. Furthermore, only single debris lines were observed at Pointe-Brûlée and Bas-Caraquet, further suggesting that the debris lines were formed by the 21 December 2010 event (the later of the two events) in these locations. Altogether, survey data corresponding to the 6 December 2010 event were available at six locations and survey data corresponding to the 21 December 2010 event were available at seven locations (Figure 2).

Flooding during the 6 December 2010 and 21 December 2010 events was photographed. Shoreline debris was also photographed following the events. Footage of the 21 December 2010 event was available from a video uploaded to YouTube by a local resident [48]. The video owner was contacted to confirm the locations shown in the video to support

Table 1. Comparison of modelled and measured flood elevations and extents.

Location	Event	Spatial Average Vertical Error (m) (Model-Survey)	Spatial Average Horizontal Error (m)
PM1	6 December 2010	0.01	5.54
LG1	6 December 2010	0.38	11.33
LG2	6 December 2010	0.36	15.43
SG1	6 December 2010	0.67	6.00
SG2	6 December 2010	0.49	4.97
SH1	6 December 2010	0.40	6.94
PB1	21 December 2010	−0.06	2.20
PB2	21 December 2010	−0.06	2.85
PB3	21 December 2010	0.33	4.79
PB4	21 December 2010	0.29	5.45
PB5	21 December 2010	0.10	5.33
BC1	21 December 2010	−0.24	2.73
BC2	21 December 2010	−0.28	7.64

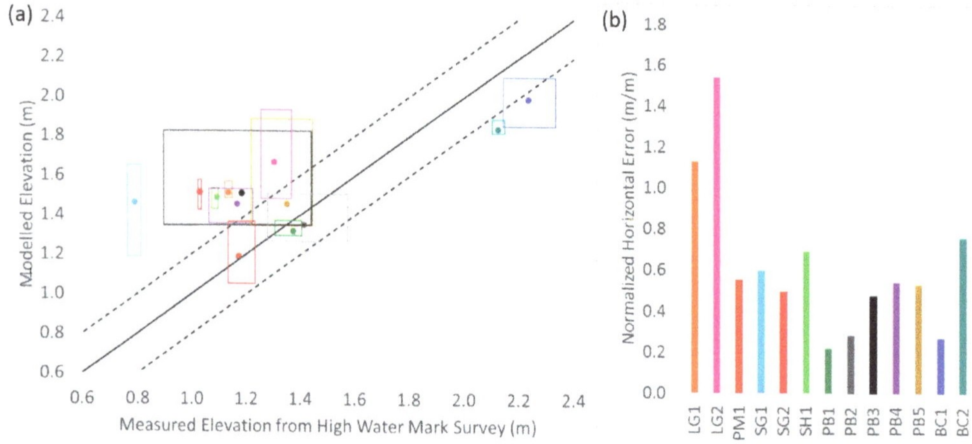

Figure 4. Assessment of (**a**) vertical and (**b**) horizontal model skill. The solid line in (**a**) represents equality of measured and modelled elevations, and the dashed lines represent RMSE of model ground elevations compared to measured point elevations from the high-water-mark survey. Points in (**a**) indicate computed averages, and boxes indicate the range of modelled and measured elevations at each location. Values in (**b**) are normalized by the model element edge length (i.e., 10 m). Colors indicate matching locations.

4.2. Comparison with Photographic and Video Evidence

Qualitative assessment of model skill was conducted by comparing modelled results with photographic and video evidence of flooding captured during and after the 6 December 2010 and 21 December 2010 events. The photographic and video evidence permitted assessment of model skill in areas where survey data were not collected and provided a basis for model validation. In general, modelled results agreed well with photographic and video evidence. However, the model overestimated inundation extents at Le Goulet, as apparent from horizontal error values computed for survey locations LG1 and LG2 (Table 1), which were the highest of all locations evaluated. Figures 5–7 show modelled results in comparison to photographic and video evidence near Pointe-Brûlée, Baie-de-Petit-Pokemouche, and Le Goulet, respectively; where available, surveyed high-water marks are also shown in yellow.

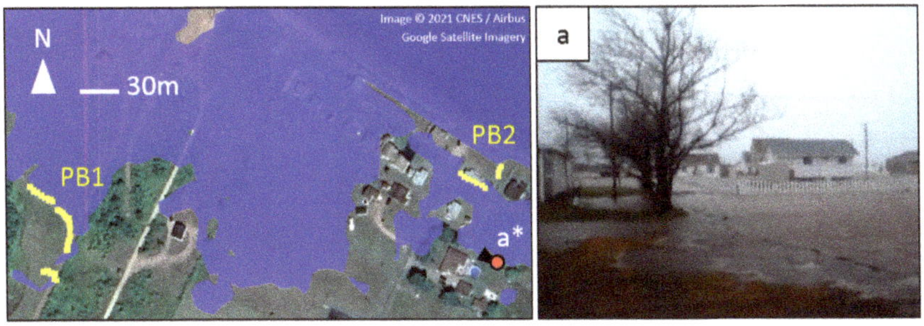

Figure 5. Modelled results for the 21 December 2010 event in Pointe-Brûlée compared to (**a**) video evidence of flooding (video frame extracted from a YouTube video uploaded by a local resident [48]). Approximate observer location and direction of view are shown by the red icon. Video frame (**a**) coincides with observer location (a*). Survey data in the vicinity of the video evidence are shown in yellow.

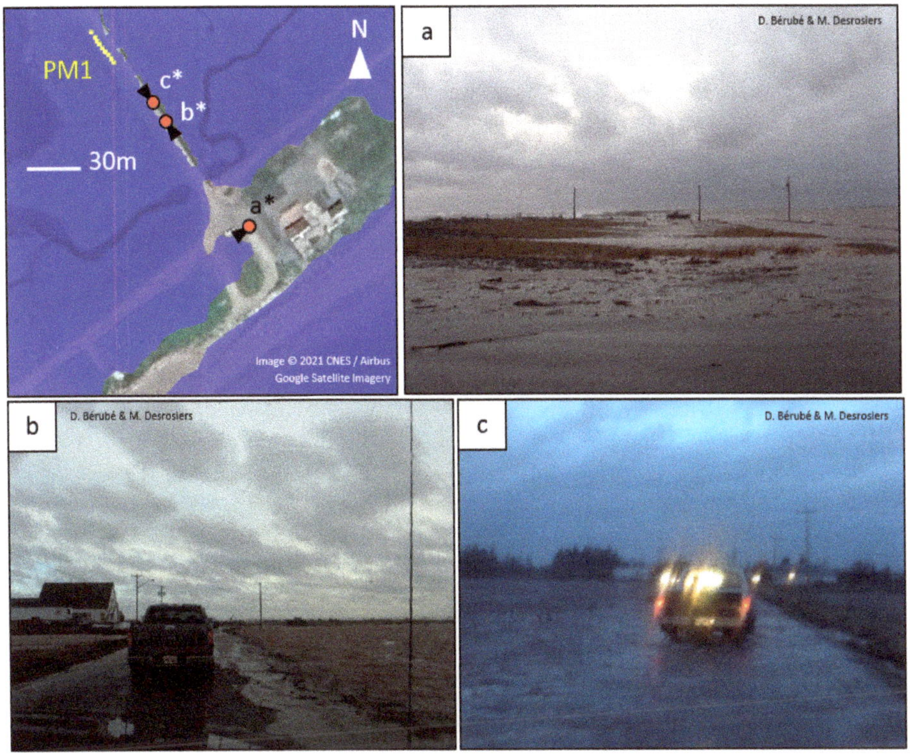

Figure 6. Modelled results for the 6 December 2010 event in Baie-de-Petit-Pokemouche compared to (**a**–**c**) photographic evidence. Approximate observer location and direction of view are shown by the red icons. Photographs (**a**), (**b**) and (**c**) coincide with observer locations (a*), (b*) and (c*), respectively. Survey data in the vicinity of the photographic evidence are shown in yellow.

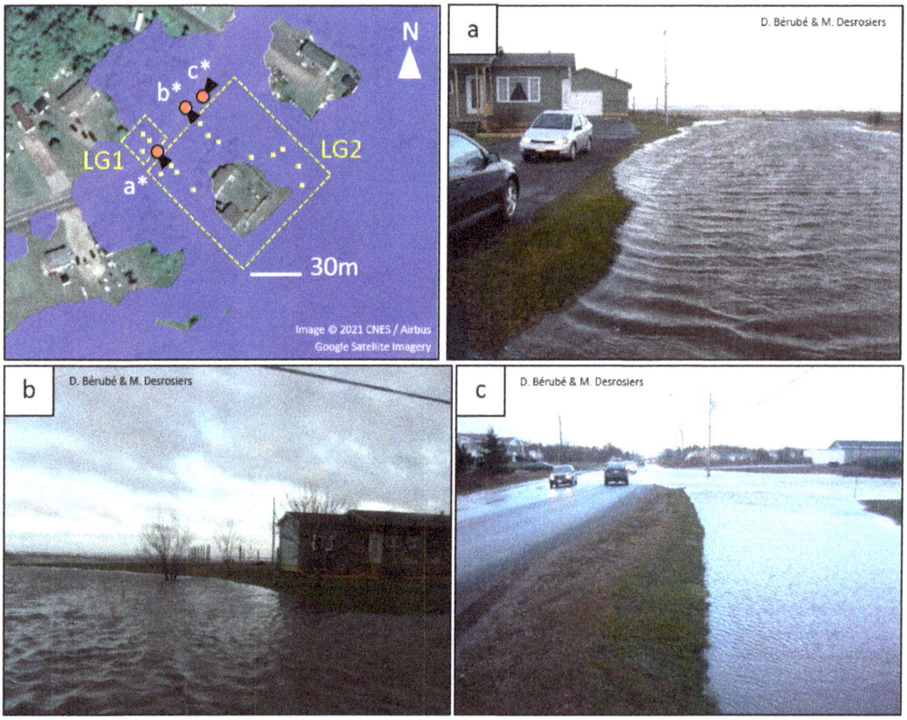

Figure 7. Modelled results for the 6 December 2010 event in Le Goulet compared to (**a**–**c**) photographic evidence. Approximate observer location and direction of view are shown by the red icons. Photographs (**a**), (**b**) and (**c**) coincide with observer locations (a*), (b*) and (c*), respectively. Survey data in the vicinity of the photographic evidence are shown in yellow.

5. Discussion and Conclusions

5.1. Synthesis

A high-resolution, numerical hydrodynamic model was applied to simulate storm surge-driven coastal flooding at communities on the Acadian Peninsula in eastern Canada. Comparisons of model output to surveyed debris and high-water marks demonstrated that the community-scale inundation model generally produces sub-mesh-resolution horizontal accuracy, with vertical accuracy of approximately 0.34 m. Both horizontal and vertical discrepancies may be explained, in part, by wave contributions to observed high-water marks, which were not included in the community-scale modelling. The largest horizontal error was observed in Le Goulet (LG1 and LG2), which is the site most exposed to waves propagating from the longest north-easterly fetches in the Gulf of Saint Lawrence. Furthermore, model ground elevations incorporate interpolation error. In addition, assumptions and uncertainties pertaining to flood depths required to mobilize and deposit debris may have also contributed to discrepancies. Ultimately, considering the model resolution and aforementioned uncertainties, model skill was deemed acceptable to support community-scale flood hazard assessment.

The presented work illustrates methods to assess model skill using surveyed high-water marks, in conjunction with photographic and video evidence, in lieu of, or supplementary to, proximate tide gauge records. Photographic and video evidence provided valuable information to support qualitative assessment of model skill where survey data were not available. The presented methods, which provide insight into model skill and uncertainty in predicting key flood hazard metrics (i.e., the areal extent and depth, or maximum elevation, of flooding), are conducive to community-scale flood disaster risk management where an understanding of such metrics is required to inform assessment of damages and consequences. Although flood extent and elevations may be inferred from point measurements of water level at tide gauges located on the coast (e.g., using bathtub modelling techniques), such approaches leave many open questions regarding skill in predicting overland flooding pathways, and floodwater interactions with coastal floodplain features, which influence hazards.

Unlike area-based methods for assessing model skill at predicting inundation, which require abundant and expansive data collection (e.g., [2,8,27]), modelled horizontal flood extents were assessed using a distance-based method on discrete clusters of survey data. Furthermore, area-based methods require the selection and application of interpolation methods to fully delineate flood extent, introducing further error and uncertainty to model skill assessment [27].

The presented methods lend gravity to the importance of non-conventional data sources in data-scarce regions and provide motivation to advance research focused on coastal debris detection and delineation through field or remote-sensing techniques.

5.2. Limitations and Future Research Needs

The community-scale hydrodynamic model was designed to simulate elevated water levels caused by tides and storm surge, and resulting flood hazards. The model excluded wave effects (i.e., wave runup, overtopping), morphodynamic effects (e.g., dune erosion or breaching), and their contributions to flood hazards, which are sources of model uncertainty in more exposed, open-coast communities. Integration or coupling of wave and morphodynamic models with the hydrodynamic model could enable improved flood hazard assessment and quantify or reduce these sources of uncertainty.

Debris transport and deposition are affected by the physical properties of the debris, shoreline characteristics, and hydrodynamics [50–52]. In this study, sea wrack and shoreline debris were used to support model skill assessment, assuming that they represent a proxy for peak water level elevation and flood extent, following similar assumptions by others [8,26–30]. Future research addressing the processes and mechanisms underlying the transport and deposition of specific debris types in different coastal settings would help to clarify linkages between water levels and deposits of debris, and their utility to support coastal flood hazard assessment.

In this study, photographic and video evidence was used to corroborate the quantitative model skill assessment (relying on debris surveys) in a purely qualitative way. However, recent advancements in video- and imagery-based methods and machine learning techniques to measure bathymetry, depth, and flow velocity offer promise for quantitative assessment [53–56]. Further advancements in these areas, targeted at assessment of hydraulic parameters and interpretation of photographic evidence of historical flooding, would contribute to improved model skill assessment, particularly in data-scarce regions.

The methods and findings presented in this paper offer guidance on assessing model skill in data-scarce regions where coastal debris persist, albeit based on data for a limited number of storm-driven flooding events on Canada's Atlantic coast. Further case studies and validation of the presented methods would enable generalization to a broader range of site conditions.

Author Contributions: Conceptualization, S.F. and E.M.; methodology, S.F. and E.M.; formal analysis, S.F. and M.P.; data curation, D.B., M.D. and A.R.; writing—original draft preparation, S.F.; writing—review and editing, J.K., E.M., D.B., M.D., A.R. and M.P. All authors have read and agreed to the published version of the manuscript.

Funding: This work was funded as part of the Coastal Flood Mitigation Canada project by Defence Research and Development Canada through the Canadian Safety and Security Program, with support from the National Research Council of Canada's Ocean Program.

Data Availability Statement: This study relied on a number of third party datasets, which are referenced in the text; readers are directed to referenced sources for access to data. Some of the data products produced in this study may be available upon request to the authors, subject to intellectual property and data sharing policies of their affiliated institutions. This study relied on data provided by the Canadian Hydrographic Service (CHS), pursuant to CHS MOU No. 2020-0206-1260-NRCC and CHS MOU No. 2021-0805-1260-NRCC. The incorporation of data sourced from CHS in this study shall not be construed as constituting an endorsement by CHS of this study. The products produced in this study do not meet the requirements of the *Charts and Nautical Publications Regulations, 1995* or the *Navigation Safety Regulations, 2020* under the *Canada Shipping Act, 2001*. Official charts and publications, corrected and up-to-date, must be used to meet the requirements of those regulations.

Acknowledgments: The authors thank Nicky Hastings, Jackie Yip, Julie Van de Valk, Robert Capozi, Reid McLean, and Maxime Doucet for their helpful contributions to this research. This paper was improved by comments provided by three critical reviewers.

Conflicts of Interest: The authors declare no conflict of interest.

References

1. Lemmen, D.S.; Warren, F.J.; James, T.S.; Mercer Clarke, C.S.L. (Eds.) *Canada's Marine Coasts in a Changing Climate*; Government of Canada: Ottawa, ON, Canada, 2016; ISBN 978-0-660-05052-2.
2. Bernatchez, P.; Fraser, C.; Lefaivre, D.; Dugas, S. Integrating Anthropogenic Factors, Geomorphological Indicators and Local Knowledge in the Analysis of Coastal Flooding and Erosion Hazards. *Ocean. Coast. Manag.* **2011**, *54*, 621–632. [CrossRef]
3. Environment Canada. *Impacts of Sea-Level Rise and Climate Change on the Coastal Zone of Southeastern New Brunswick*; Environment Canada: Ottawa, ON, Canada, 2006; ISBN 0-662-43947-3.
4. Murphy, E.; Lyle, T.; Wiebe, J.; Hund, S.V.; Davies, M.; Williamson, D. *Coastal Flood Risk Assessment Guidelines for Building and Infrastructure Design*; National Research Council Canada: Ottawa, ON, Canada, 2020; ISBN 978-0-660-36745-3.
5. Apollonio, C.; Bruno, M.F.; Iemmolo, G.; Molfetta, M.G.; Pellicani, R. Flood Risk Evaluation in Ungauged Coastal Areas: The Case Study of Ippocampo (Southern Italy). *Water* **2020**, *12*, 1466. [CrossRef]
6. Daigle, R. *Sea-Level Rise and Flooding Estimates for New Brunswick Coastal Sections Based on IPCC 5th Assessment Report*; R.J. Daigle Enviro: Moncton, NB, Canada, 2017.
7. Aubé, M.; Hébert, C.; Simard, I. *Analyse Des Risques d'inondation et d'érosion Dans Les Régions de Grande-Anse, Maisonnette, Bertrand et Caraquet*; Institut de Recherche sur les Zones Côtières Inc. & Geomediatix Innovations Inc.: Shippagan, NB, Canada, 2016.
8. Didier, D.; Bernatchez, P.; Boucher-Brossard, G.; Lambert, A.; Fraser, C.; Barnett, R.; Van-Wierts, S. Coastal Flood Assessment Based on Field Debris Measurements and Wave Runup Empirical Model. *J. Mar. Sci. Eng.* **2015**, *3*, 560–590. [CrossRef]
9. Hébert, C.; Aubé, M. *Évaluation d'options d'adaptation Aux Changements Climatiques: Restauration Des Dunes à Le Goulet*; Institut de Recherche sur les Zones Côtières Inc.: Shippagan, NB, Canada, 2015.
10. Robichaud, A.; Simard, I.; Doiron, A.; Chelbi, M. Infrastructures à Risque Dans Trois Municipalités de La Péninsule Acadienne. 2011. Atlantic Climate Adaptation Solutions Association. Available online: https://atlanticadaptation.ca/en/islandora/object/acasa%253A653 (accessed on 27 May 2021).
11. Natural Resources Canada. *Federal Flood Damage Estimation Guidelines for Buildings and Infrastructure-Version 1.0*; Government of Canada: Ottawa, ON, Canada, 2021.
12. Rangel-Buitrago, N.; Anfuso, G. *Risk Assessment of Storms in Coastal Zones: Case Studies from Cartagena (Colombia) and Cadiz (Spain)*; Springer: Heidelberg, Germany, 2015; ISBN 978-3-319-15843-3.
13. Webster, T.L.; Forbes, D.L.; MacKinnon, E.; Roberts, D. Flood-Risk Mapping for Storm-Surge Events and Sea-Level Rise Using Lidar for Southeast New Brunswick. *Can. J. Remote Sens.* **2006**, *32*, 194–211. [CrossRef]
14. Creach, A.; Chevillot-miot, E.; Mercier, D.; Pourinet, L. Vulnerability to Coastal Flood Hazard of Residential Buildings on Noirmoutier Isalnd (France). *J. Maps* **2016**, *12*, 371–381. [CrossRef]
15. Aucelli, P.P.C.; Di Paola, G.; Incontri, P.; Rizzo, A.; Vilardo, G.; Benassai, G.; Buonocore, B.; Pappone, G. Coastal Inundation Risk Assessment Due to Subsidence and Sea Level Rise in a Mediterranean Alluvial Plain (Volturno Coastal Plain–Southern Italy). *Estuar. Coast. Shelf Sci.* **2017**, *198*, 597–609. [CrossRef]
16. Didier, D.; Baudry, J.; Bernatchez, P.; Dumont, D.; Sadegh, M.; Bismuth, E.; Bandet, M.; Dugas, S.; Sévigny, C. Multihazard Simulation for Coastal Flood Mapping: Bathtub versus Numerical Modelling in an Open Estuary, Eastern Canada. *J. Flood Risk Manag.* **2019**, *12*, e12505. [CrossRef]
17. Barnard, P.L.; van Ormondt, M.; Erikson, L.H.; Eshleman, J.; Hapke, C.; Ruggiero, P.; Adams, P.N.; Foxgrover, A.C. Development of the Coastal Storm Modeling System (CoSMoS) for Predicting the Impact of Storms on High-Energy, Active-Margin Coasts. *Nat. Hazards* **2014**, *74*, 1095–1125. [CrossRef]
18. Resio, D.T.; Westerink, J.J. Modeling the Physics of Storm Surges. *Phys. Today* **2008**, *61*, 33–38. [CrossRef]
19. Muis, S.; Verlaan, M.; Winsemius, H.C.; Aerts, J.C.J.H.; Ward, P.J. A Global Reanalysis of Storm Surges and Extreme Sea Levels. *Nat. Commun.* **2016**, *7*, 11969. [CrossRef]

20. Ferrarin, C.; Roland, A.; Bajo, M.; Umgiesser, G.; Cucco, A.; Davolio, S.; Buzzi, A.; Malguzzi, P.; Drofa, O. Tide-Surge-Wave Modelling and Forecasting in the Mediterranean Sea with Focus on the Italian Coast. *Ocean. Model.* **2013**, *61*, 38–48. [CrossRef]
21. Postacchini, M.; Lalli, F.; Memmola, F.; Bruschi, A.; Bellafiore, D.; Lisi, I.; Zitti, G.; Brocchini, M. A Model Chain Approach for Coastal Inundation: Application to the Bay of Alghero. *Estuar. Coast. Shelf Sci.* **2019**, *219*, 56–70. [CrossRef]
22. Federal Emergency Management Agency. *Guidance for Flood Risk Analysis and Mapping: Coastal Water Levels*; Federal Emergency Management Agency: Washington, DC, USA, 2016.
23. Valorēs the Reality of the Acadian Peninsula. Available online: https://adaptationpa.ca/en/adaptation-pa/the-reality-of-the-acadian-peninsula (accessed on 6 May 2021).
24. Fisheries and Oceans Canada Canadian Tides and Water Levels Data Archive. Available online: https://www.meds-sdmm.dfo-mpo.gc.ca/isdm-gdsi/twl-mne/index-eng.htm# (accessed on 16 July 2019).
25. Statistics Canada Census Profile. 2016 Census. Statistics Canada Catalogue No. 98-316-X2016001. Available online: https://www12.statcan.gc.ca/census-recensement/2016/dp-pd/prof/index.cfm?Lang=E (accessed on 7 May 2021).
26. Kim, J.; Murphy, E.; Nistor, I.; Ferguson, S.; Provan, M. Numerical Analysis of Storm Surges on Canada's Western Arctic Coastline. *J. Mar. Sci. Eng.* **2021**, *9*, 326. [CrossRef]
27. Melton, G.; Gall, M.; Mitchell, J.T.; Cutter, S.L. Hurricane Katrina Storm Surge Delineation: Implications for Future Storm Surge Forecasts and Warnings. *Nat. Hazards* **2010**, *54*, 519–536. [CrossRef]
28. Jardine, D.E.; Wang, X.; Fenech, A.L. Highwater Mark Collection after Post Tropical Storm Dorian and Implications for Prince Edward Island, Canada. *Water* **2021**, *13*, 3201. [CrossRef]
29. Harper, J.R.; Henry, R.F.; Stewart, G.G. Maximum Storm Surge Elevations in the Tuktoyaktuk Region of the Canadian Beaufort Sea. *Arctic* **1988**, *41*, 48–52. [CrossRef]
30. MacLeod, R.F.; Dallimore, S.R. Assessment of Storm Surge History as Recorded by Driftwood in the Mackenzie Delta and Tuktoyaktuk Coastlands, Arctic Canada. *Front. Earth Sci.* **2021**, *9*, 698660. [CrossRef]
31. Hervouet, J.M. *Hydrodynamics of Free Surface Flows: Modelling with the Finite Element Method*; John Wiley & Sons, Ltd.: Chichester, West Sussex, UK, 2007; ISBN 978-0-470-03558-0.
32. Provan, M.; Ferguson, S.; Murphy, E. Storm Surge Contributions to Flood Hazards on Canada's Atlantic Coast. *J. Flood Risk Manag.* **2022**. Submitted. Available online: https://www.mdpi.com/journal/water/instructions#references (accessed on 27 May 2021).
33. Canadian Hydrographic Service Non-Navigational (NONNA-100) Bathymetric Data. Available online: https://open.canada.ca/data/en/dataset/d3881c4c-650d-4070-bf9b-1e00aabf0a1d (accessed on 15 July 2019).
34. GEBCO Compilation Group GEBCO 2019 Grid. 2019. Available online: https://doi.org/10.5285/836f016a-33be-6ddc-e053-6c86abc0788e (accessed on 23 July 2019).
35. Hersbach, H.; Bell, B.; Berrisford, P.; Biavati, G.; Horányi, A.; Muñoz Sabater, J.; Nicolas, J.; Peubey, C.; Radu, R.; Rozum, I.; et al. ERA5 Hourly Data on Single Levels from 1979 to Present. Copernicus Climate Change Service (C3S) Climate Data Store (CDS). Available online: https://doi.org/10.24381/cds.adbb2d47 (accessed on 9 December 2019).
36. Egbert, G.D.; Erofeeva, S.Y. Efficient Inverse Modeling of Barotropic Ocean Tides. *J. Atmos. Ocean. Technol.* **2002**, *19*, 183–204. [CrossRef]
37. Foreman, M.G.G. *Manual for Tidal Heights Analysis and Prediction*; Department of Fisheries and Oceans, Institue of Ocean Sciences: Sidney, BC, Canada, 1977.
38. Canadian Hydrographic Service Nautical Charts. Available online: https://www.charts.gc.ca/charts-cartes/nautical-marines-eng.html (accessed on 17 August 2021).
39. Government of New Brunswick 2018 Digital Elevation Model. Available online: https://geonb.snb.ca/nbdem/ (accessed on 5 May 2020).
40. Government of New Brunswick SNB 2018 LiDAR AOI 2. Available online: https://geonb.snb.ca/li/ (accessed on 1 June 2020).
41. Government of New Brunswick Wetlands. Available online: http://www.snb.ca/geonb1/e/DC/RW.asp (accessed on 19 August 2020).
42. Government of New Brunswick Forest. Available online: http://www.snb.ca/geonb1/e/DC/forest.asp (accessed on 5 December 2019).
43. Government of New Brunswick Non-Forest. Available online: http://www.snb.ca/geonb1/e/DC/non-forest.asp (accessed on 5 December 2019).
44. Chin, D.A. *Water Resources Engineering*, 3rd ed.; Pearson Education Inc.: Upper Saddle River, NJ, USA, 2013; ISBN 0-13-283321-2.
45. Houghtalen, R.J.; Akan, A.O.; Hwang, N.H.C. *Fundamentals of Hydraulic Engineering Systems*, 4th ed.; Prentice Hall: Upper Saddle River, NJ, USA, 2010.
46. Leica Geosystems Leica Viva GS15 Data Sheet. Available online: https://leica-geosystems.com/-/media/files/leicageosystems/products/datasheets/leica_viva_gs15_ds.ashx?la=de-de (accessed on 21 January 2022).
47. Government of Canada GPS·H. Available online: https://webapp.geod.nrcan.gc.ca/geod/tools-outils/gpsh.php (accessed on 5 November 2021).
48. Doucet, M. Inondation Du 21 Décembre 2010 (Le Goulet) (Haut-Shippagan). YouTube. Available online: https://www.youtube.com/watch?v=GaL_tjc6sbM (accessed on 14 May 2021).
49. Braudrick, C.A.; Grant, G.E. When Do Logs Move in Rivers? *Water Resour. Res.* **2000**, *36*, 571–583. [CrossRef]

50. Orr, M.; Zimmer, M.; Jelinski, D.E.; Mews, M. Wrack Deposition on Different Beach Types: Spatial and Temporal Variation in the Pattern of Subsidy. *Ecology* **2005**, *86*, 1496–1507. [CrossRef]
51. Oldham, C.; McMahon, K.; Brown, E.; Bosserelle, C.; Lavery, P. A Preliminary Exploration of the Physical Properties of Seagrass Wrack That Affect Its Offshore Transport, Deposition, and Retention on a Beach. *Limnol. Oceanogr. Fluids Environ.* **2014**, *4*, 120–135. [CrossRef]
52. Murphy, E.; Nistor, I.; Cornett, A.; Wilson, J.; Pilechi, A. Fate and Transport of Coastal Driftwood: A Critical Review. *Mar. Pollut. Bull.* **2021**, *170*, 112649. [CrossRef]
53. Brodie, K.L.; Palmsten, M.L.; Hesser, T.J.; Dickhudt, P.J.; Raubenheimer, B.; Ladner, H.; Elgar, S. Evaluation of Video-Based Linear Depth Inversion Performance and Applications Using Altimeters and Hydrographic Surveys in a Wide Range of Environmental Conditions. *Coast. Eng.* **2018**, *136*, 147–160. [CrossRef]
54. Perugini, E.; Soldini, L.; Palmsten, M.L.; Calantoni, J.; Brocchini, M. Linear Depth Inversion Sensitivity to Wave Viewing Angle Using Synthetic Optical Video. *Coast. Eng.* **2019**, *152*, 103535. [CrossRef]
55. Ansari, S.; Rennie, C.D.; Jamieson, E.C.; Seidou, O.; Clark, S.P. Machine Learning Based Surface Velocimetry. In Proceedings of the EGU General Assembly 2021, Online, 19–30 April 2021.
56. Ansari, S.; Rennie, C.D.; Venditti, J.G.; Kwoll, E.; Fairweather, K. Shore-Based Monitoring of Flow Dynamics in a Steep Bedrock Canyon River. *E3S Web Conf.* **2018**, *40*, 06025. [CrossRef]

water

MDPI

Review

Wastewater System Inflow/Infiltration and Residential Pluvial Flood Damage Mitigation in Canada

Dan Sandink [1,*] and Barbara Robinson [2]

[1] Institute for Catastrophic Loss Reduction, 20 Richmond Street East, Suite 210, Toronto, ON M5C 2R9, Canada
[2] Norton Engineering Inc., 243 Glasgow St., Kitchener, ON N2M 2M3, Canada; nortonengineeringinc@gmail.com
* Correspondence: dsandink@iclr.org

Abstract: Pluvial flooding in urban areas is one of the most significant drivers of disaster loss in Canada. Damages during pluvial flood events are associated with overwhelmed urban drainage (stormwater and wastewater) systems. During the period from 2013 to 2021, Canadian property and casualty insurers reported approximately CAD 2 billion in personal property (residential) pluvial sewer backup claims during flood catastrophes. There has been growing interest in managing pluvial urban flood risk, notably through newly funded national programs focused on climate change adaptation. These programs have included the development of new guidelines and standards focused on managing the underlying factors contributing to urban and basement flooding. Inflow and infiltration (I/I) has received limited attention in the pluvial flood literature, however. Informed by significant engagement with practitioners in Canada, this paper provides a review of the issue of I/I into wastewater systems and its relation to pluvial flooding. The paper will address concerns related to private property engagement in I/I and urban pluvial flood reduction programs. Both improved technical standards and administrative support are needed to ensure that wastewater infrastructure is less susceptible to I/I over its lifecycle.

Keywords: extreme rain; pluvial flood; basement flood; wastewater; inflow and infiltration; Canada

Citation: Sandink, D.; Robinson, B. Wastewater System Inflow/Infiltration and Residential Pluvial Flood Damage Mitigation in Canada. *Water* **2022**, *14*, 1716. https://doi.org/10.3390/w14111716

Academic Editor: Marco Franchini

Received: 22 April 2022
Accepted: 23 May 2022
Published: 27 May 2022

1. Introduction

Pluvial flooding in urban areas associated with short-duration, high intensity (SDHI) rainfall events is a chronic cause of property damage and disruption in urban areas around the globe [1–8]. Pluvial flooding is a routine experience in many urban municipalities in Canada and results in hundreds of millions of dollars in insurance and uninsured disaster losses each year. During the period from 2013 to 2021, property and casualty insurers reported approximately CAD 2 billion in personal property sewer backup claims during flood catastrophes (i.e., catastrophe events where insured losses exceeded CAD 25 million) [9]. The flooding of residential properties, associated with SDHI events exceeding stormwater and wastewater systems' capacities, results in floodwater entering buildings via multiple flood mechanisms. These mechanisms include overland flow, infiltration/seepage, sewer surcharge/backup, and internal/building-specific plumbing and drainage system failures [10]. Sewer backup typically contributes more than half of the total insured losses during major urban pluvial flood events in Canada [9,11–13].

Aside from property damage, SDHI events may also drive sanitary and combined sewer overflow events with negative implications for surface water, and SDHI rain events result in significant damage and operation costs associated with municipal infrastructure [5,14]. Increasing urban development, aging public and private-side infrastructure, and sewer construction quality issues, among other factors, are expected to intensify the impacts of pluvial flood events in Canada [15–19]. While SDHI events drive major, damaging flood events, basement flooding associated with overtaxed or poorly maintained sewer infrastructure may occur during less intense rainfall events or even in dry conditions as a result of blockages in pipe systems.

Climate change impacts in many regions of Canada, which are anticipated to include the increased frequency and intensity of SDHI rainfall events, are expected to intensify pluvial flood risk [19,20]. Climate change impacts associated with urban pluvial flood are increasingly recognized in national climate change assessment reports and climate change adaptation-related guidance documents [15–21]. Maintaining and recapturing capacity in wastewater systems is also increasingly important, as urban municipalities across Canada emphasize the increased density of development, infill development, and affordable housing [22,23].

Vulnerable residents, including those occupying basement apartments in flood-prone urban areas [24], may suffer significant impacts during flood events. These residents, typically renters, are unlikely to have insurance coverage for any type of flood or property damage [25]. In the context of affordable housing, it is important to provide adequate basement flood and sewer backup protection for basement apartments [24].

Considerable international literature exists on the topic of urban pluvial flooding—for example, [1–8,26–36]. While it is recognized that extreme rainfall in urban areas causes excess flow of water in wastewater systems via rainfall-derived inflow and infiltration (RDII) [14,26,37], and that this excess flow results in the flooding of buildings via buried sanitary sewer conveyance systems [10,19,36], the specific role of wastewater systems and inflow and infiltration (I/I) in urban pluvial flood and options to mitigate I/I risk in the context of pluvial flooding have received limited attention in the literature.

This paper contributes to the literature on urban pluvial flooding by providing detail with respect to the role of wastewater (sanitary) systems in pluvial flood hazards, with a focus on Canada. The paper will specifically review practical approaches to improving I/I management in wastewater systems in Canadian communities. The discussion is informed by significant consultation with wastewater system practitioners associated with the development of new practical guidance documents [13,38–43], including the development of National Standards of Canada (NSCs). NSCs provide a basis for infrastructure system design, construction, maintenance, and operation and are developed by, and oriented toward, the application by the infrastructure management community in Canada.

The discussion is organized as follows: Section 2 provides an overview of pluvial flood in Canada, including a discussion of common approaches to urban drainage, examples of recent pluvial flood events, and flood mechanisms in urban areas. Section 3 provides an overview of the issue of I/I, its impacts, and examples of I/I occurrence in pluvial flood events. Section 3 further reviews common sanitary sewer characteristics and discusses challenges associated with managing I/I that are attributed to sewer type and local governance characteristics. Section 4 provides detail on the factors that affect the occurrence of I/I in sewer infrastructure, with an emphasis on the administrative factors that lead to its occurrence. Section 5 provides a review of new NSCs that have been developed to respond to climate change impacts in Canada, including the potential increase in pluvial flood risk, focusing specifically on how I/I in wastewater systems can be addressed. Section 6 provides a discussion of the opportunities and next steps for I/I management, including the improved understanding of property owners' behavior and the methods to engage property owners in I/I reduction, recommendations for improved technical standards and administrative approaches to managing I/I, and improved consideration of the overall costs and benefits of I/I for the purpose of informing its management. Section 6 also addresses the study limitations and future research. A summary of the conclusions of the review is provided in Section 7.

2. Pluvial Urban Flood in Canada

Urban pluvial flood is defined as events where "[. . .] rain-driven ponding or overland flow [. . .] results from the exceedance of natural or engineered drainage capacity [. . .]" in urban areas [8]. In Canada, urban pluvial flooding includes scenarios where SDHI rain events exceed the "[. . .] combined hydraulic capacity of [an] area's storm sewers, ditches, and catch basins and water flows from the streets onto properties" [44]. This type of flooding can occur anywhere, including in areas that are not vulnerable to flooding associated with an overflowing water body [27,45].

Urban stormwater management systems may include minor drainage systems (storm sewers, catch basins, inlets, inlet control devices, gutters, ditches, and swales) and major drainage systems (streets, channels, ponds, natural streams, and valleys) which convey stormwater away from urban areas during and after storm events [44]. In Canada, the capacity of these systems may differ depending on the "era" of construction, corresponding to the approximate intervals of 1880–1970, 1970–1990, and 1990 to the current day (Table 1).

Table 1. Eras of stormwater management in Canada [1].

Era	Summary
~1880–1970	• Sewer networks provided to quickly convey stormwater from upstream urban areas to downstream receiving waters • Design capacity typically 1:2 to 1:10 year, occasionally 1:25 year
~1970–1990	• Stormwater detention facilities and overland systems incorporated to convey stormwater when minor system capacities are exceeded • Minor system to manage 1:2 to 1:10 peak flows; major system to manage 1:50 to 1:100 flows
~1990 to present	• Stormwater quality considered along with quantity; application of Best Management Practices to manage stormwater pollution in receiving waters.

[1] [46–48].

Until the 1970s, urban stormwater management emphasized underground (minor) systems with design capacities ranging from 1:2 to 1:10 year return period flows, with limited consideration of overland (major) systems. From the 1970s to the 1990s, major systems were incorporated into urban drainage design. These systems conveyed stormwater when minor system capacity was exceeded and were typically designed to manage 1:50 to 1:100 year return period flows. Conventional approaches to urban stormwater management included "[collecting] and [conveying] water as quickly as possible while maximizing dry land area for urban development" [8]. By the 1990s, however, stormwater management practices also increasingly incorporated considerations concerning water quality for smaller storms [46–50]. Pluvial flood events may occur when rainfall intensity/runoff exceeds the design capacities of these systems, even when they are functioning correctly [8].

Recent pluvial flood events in Canada demonstrate the intensity of extreme rainfall events that exceed system capacity. Each of the events listed in Table 2 were classified as an insurance loss "catastrophe" with significant residential pluvial flood damage. Further, each of the total accumulations presented in Table 2 exceeded local 1:100 year return period short-duration rain events.

Table 2. Select SDHI/pluvial flood catastrophes in Canada with significant sewer backup loss components.

Event	Rainfall Accumulation
Peterborough, ON, 15 July 2004	~80 mm in 1 h, ~260 mm in 24 h [51]
Toronto/GTA, ON, 19 August 2005	132 mm in 2 h, 149 mm in 12 h [52]
Hamilton, Ottawa, ON, July 2012	116 mm to 140 mm in 3 h in the Hamilton area [9]
Toronto, ON, 8 July 2013	102 mm in 2 h, 126 mm in 6 h [52]
Greater Toronto Area, ON, August 2014	150 to 200 mm total accumulation in Burlington [9]
Windsor/Tecumseh, ON, 28 September 2016	220 mm over 24 h, 110 mm in 2 h in Tecumseh, 115–230 mm in Windsor (24 h) [53,54]
Windsor/Tecumseh/Essex, ON, 28–29 August 2017	290 mm in LaSalle, +220 mm in Windsor, 190 mm in Essex [9]

Buildings may be affected by a variety of mechanisms during pluvial flood events in urban areas, such as those outlined in Table 2. These mechanisms, reviewed in detail elsewhere [10,55], include:

- Seepage of ground and surface water (water seeps into the ground adjacent to the foundation walls, causing water to enter buildings through cracks, loose joints, etc. in basements and foundations, and/or groundwater enters homes through cracks the in foundations);
- Sewer backup (surcharging of sewers, resulting in the backup of storm, sanitary, combined, and partially separated systems into buildings, including backflow into foundation drainage systems);
- Overland flow of stormwater (stormwater surface flow enters buildings through aboveground openings).

Further to the above, a variety of property- and building-specific plumbing and drainage factors may exacerbate flood occurrence at the property and building scales, including poor lot grading and drainage, the poor installation and lack of maintenance of building sewer connections, and limited knowledge and maintenance of key interior and exterior sewer and drainage systems by property owners [55]. For example, sewer laterals that have become blocked due to a lack of maintenance or poor installation may drive isolated sewer backup events. With respect to public and private sewage conveyance systems, excessive water entering wastewater systems may drive regional-scale sewer backup events. Sewer backup events also result from site- and/or regional-scale factors affecting sanitary sewer systems. Public-side sewers are generally in a state of less-than-ideal repair, exacerbating flood risk during intense rainfall events [56].

3. Inflow/Infiltration (I/I)

An important mechanism driving the occurrence of sewer backup in wastewater systems during pluvial flood events is excessive water entering sanitary sewer conveyance systems, resulting in reduced system capacity and an increased vulnerability to the surcharge reversal of flow into buildings. This excess flow is referred to as "Inflow and Infiltration" or I/I [57,58]. In general, I/I is defined as any rain or groundwater in the sanitary sewer that should not be there. Infiltration includes "water other than sanitary wastewater that enters a sewer system from the ground through defective pipes, pipe joints, connections, or manholes." Inflow includes "water other than sanitary wastewater that enters a sewer system from sources such as roof leaders, cellar/foundation drains, yard drains, area drains, drains from springs and swampy areas, manhole covers, cross connections between storm sewers and sanitary sewers, and catch basins" [59].

I/I is a chronic issue in regions across North America and internationally [2–7,60], with the negative impacts of I/I exacerbated by increasing urban populations, increasing urban density/infill development, and aging infrastructure. It has been reported that roughly half of all wastewater volume may be attributed to I/I [7,59,61,62].

While this review focuses on the role of I/I in sewer backup flooding as a component of pluvial flood events in urban areas, it is important to note that I/I results in multiple negative impacts. From a public policy perspective, the role of I/I in driving pluvial flood damage via sewer backup should be considered in the context of its multitude of negative impacts [4–7,15,55].

I/I results in a lack of capacity at pumping stations and trunk sewer systems, limiting the potential for urban intensification and additional development in urban communities [42]. I/I also increases the lifecycle costs for wastewater systems and may reduce the years of service for new sewer infrastructure. For example, I/I causes the erosion of bedding and haunching, compromising pipe performance and resulting in early failure for sewer systems [63]. Wastewater treatment plants (WWTPs) may experience negative impacts associated with wet weather peaking, and overflow bypasses at pumping stations and secondary bypasses at WWTPs present significant risks to surface water systems associated with sanitary sewer overflows [3,42]. With respect to the direct financial implications of I/I, the US EPA reported [59]:

Wastewater collection and treatment cost can range from $2 to $5 per thousand gallons [$0.50 to $1.30/m^3]. An annual [I/I] volume of 150 million gallons [567, 800 m^3] would cost between $300,000 and $750,000 per year to transport and treat. For many older collection systems, infiltration can be quite substantial and has been calculated as high as fifty percent of the flow.

I/I is expected to be affected by climate change. In general, rainfall-derived I/I (RDII) is expected to increase with higher rainfall intensity/accumulation [8,27,64–69], and SDHI rainfall events are expected to increase in terms of frequency and severity under changing climate conditions in many regions of Canada [15,16]. Further, reduced periods of frozen ground in northern climates due to higher temperatures may result in increased infiltration during the winter [70]. Coastal regions also face an increasing risk of I/I, as sea level rises increase groundwater levels and saltwater intrusion, leading to compromises in system integrity and contributing directly to extraneous flows in wastewater systems [3,71]. Changing climate conditions may also affect antecedent conditions (i.e., rainfall and moisture conditions before/between SDHI events), with further implications for RDII [37].

3.1. *I/I and Pluvial Flood Events*

Engineering studies commissioned by local governments following pluvial flood events provide insight into the role of I/I in the flooding of buildings. This section provides examples of recent major pluvial flood events in urban municipalities. An emphasis is placed here on the role of wastewater systems, how I/I contributed to these flood events, and proposed solutions that focus on mitigating I/I.

The Binbrook community of Hamilton, Ontario experienced an SDHI rainfall event on 22 July 2012. The total rainfall accumulation over a four-hour period was estimated at 90–140 mm (depending on the rain station location), exceeding local 1:100 year events for the region. Roughly 100 residents reported flooding to the City of Hamilton after the event. Flow monitoring indicated that the sanitary sewer system responded rapidly to the SDHI event, and the flows exceeded the capacity of the local pumping station. Flow monitoring further indicated the reversal of flow within the sanitary sewer pipe. Surveys of the affected residential subdivisions indicated the backing up of water via basement floor drains and basement shower drains, indicative of sanitary sewer surcharging [72].

The City of Peterborough, Ontario experienced a severe SDHI event on 14–15 July 2004. The total accumulations exceeded 225 mm in 24 h and 75 mm in 1 h, exceeding local 1:100 year return period events. The following factors were identified as contributors to significant regional residential flooding during the storm in the City's flood reduction masterplan [51]:

- Extreme rainfall;
- Impervious surfaces in areas where intense rainfall was centered;
- Insufficient storm system capacity;
- Poorly design overland flow routes;
- Excess I/I in the sanitary sewer system.

With respect to the separated sanitary sewer systems, accidental interconnections between storm and sanitary systems, inflow through maintenance hole covers, foundation drain connections to sanitary sewers (permitted until 1991), roof downspout connections to sanitary systems, and groundwater infiltration into the sanitary system (attributed to damaged or misaligned sanitary sewer pipes) contributed to flood damage. Previous monitoring at the City's wastewater treatment plant indicated chronic I/I in the City's sanitary sewer system before the flood event. In dry weather, the plant received twice as much water relative to the domestic water supplied by the City's domestic water treatment plant. In wet weather, the flows were six times higher than expected. The post-flood assessment further indicated higher rates of I/I in high groundwater areas and lower rates of I/I in high elevation areas [51].

On 19 August 2005, in Toronto, Ontario, rainfall accumulation of 132 mm in two hours and 149 mm of accumulation over 12 h [52] resulted in significant regional basement flooding (4200 basement flood complaints were recorded by the City) [73]. Thirty-four basement flood protection areas were created to assist in identifying the causes and solutions for recurring basement flood occurrences [73]. While separated storm sewers were commonly constructed in the areas affected by the flood (designed to accommodate 1:2–5 year return period events), affected areas were constructed before major drainage systems were common practice [74]. Multiple factors drove basement flooding in the affected areas, including both the capacity limitations in storm systems and I/I in sanitary systems.

System modelling in study area 30 indicated that sanitary systems were vulnerable to surcharging under historical storm events and design storm events. Rainfall-derived I/I (RDII) rates were estimated to be 1–3 times the accepted design value for the area (0.26 L/s/ha). It was further observed that overloaded sanitary sewers, along with overloaded minor and poorly defined major storm sewer systems, contributed to flooding [75]. Post-flood flow monitoring (conducted in 2006) in study area 29 indicated sewer flow rates indicative of leaky sanitary systems or sanitary systems with significant inflow from residential foundation drain connections. The recommendations for flood remediation included continued operations and maintenance programs to reduce I/I in the sanitary sewer system. These programs included the sealing of maintenance hole covers, maintenance hole rehabilitation, sanitary sewer relining, and cross connection elimination [76]. Managing I/I to control flood risk was recommended throughout study area 28. The factors contributing to flood occurrence in study area 28 included:

- I/I in the sanitary systems;
- High groundwater tables;
- Surface runoff accumulation in low-lying areas;
- The existence of reverse-slope driveways in residential buildings (directing surface flows into buildings and into sanitary sewers via basement plumbing fixtures);
- Overflow depths above street right-of-way elevations;
- Undersized storm systems;
- Blocked or broken sanitary sewers, manholes, and catch basins.

Many of the study areas affected by the 2005 pluvial flood event in Toronto were affected again by an SDHI rainfall event on 8 July 2013. The July 2013 event included 102 mm of rainfall accumulation over two hours and 126 mm over six hours, again exceeding the local 1:100 year return period events [52]. Following the event, 4759 flood complaints were received by the City. Similar to the 2005 event, the factors that drove flooding during the SDHI event included an overloaded sanitary sewer system attributed to excessive I/I, as well as overloaded storm sewer systems and surface flooding [73].

In May 2009, the Sherwood Forest area of the city of London, Ontario experienced an SDHI rain event, with maximum intensities exceeding 100 mm/hr. High rates of I/I in the separated sanitary sewer system were observed (flow rates reaching 50 L/s—twice the capacity of the sewer system) and contributed to regional basement flooding. In this region of the city, it was common to connect foundation drainage to separated sanitary sewers until the year 1985, and significant inflow into the sanitary sewer was directly attributed to residential foundation drain connections. Specifically, the analysis by Jiang et al. [77] indicated that 85% of RDII was attributed to foundation drain connections in the sanitary sewer and that disconnection could result in a 78% reduction in RDII. The municipality established a source-control approach, focusing on the subsidization of foundation drain disconnection by private residential building owners. Observations following the implementation of a pilot foundation drain disconnection program with partial uptake indicated that flows in the sanitary sewer were halved during wet weather events [77–80].

3.2. Sewer Types and Implications for I/I

In Canada, municipalities are serviced by variety of sewer system types (Table 1). New construction in Canada is serviced by separated sewer systems, while older areas may also be served by legacy "combined" systems. In many regions of Canada, subdivisions constructed during the mid-20th century are served by systems that are separate within the municipal right-of-way but include building downspout and foundation drain connections to sanitary laterals. These older systems may experience significant I/I problems, as exemplified in the cases discussed in Section 3.1. Due to the differing eras of development in any given urban region, communities may be served by a combination (or hybrid) of the systems listed in Table 3.

Table 3. Common types of wastewater conveyance systems in Canada: Sanitary, combined, and semi-combined/partially separated systems [1].

Type	Description
Separated	Sewer systems comprised of fully separated storm systems (which convey stormwater to surface water features and stormwater-management infrastructures such as stormwater ponds) and sanitary sewer systems (sewers that collect domestic wastewater and direct it to a central location for treatment and discharge to the environment).
Combined	Legacy systems that convey both stormwater and sanitary sewage.
Semi-combined/ partially separated	Legacy systems: Separated systems are present on the municipal side of the property line, but household/private-side foundation drainage systems, eavestrough downspouts, area drains, etc. are connected to sanitary sewers on the private-side of the property line.
3rd pipe	Separated storm conveyance systems designed for the dedicated management of building and private property discharges, including foundation drainage and/or downspout drainage.

[1] [10,13,16,55].

Where urban areas are serviced by separated storm and sanitary sewer systems, stormwater and groundwater should be discharged to underground storm sewer conveyance systems and overland flow routes. Very limited amounts of "clean" storm or groundwater should enter the separated sewer systems [5], and, indeed, there exist clear allowable I/I values (e.g., leakage) at acceptance in all construction specifications on the municipal side and in building construction codes in Canada [81]. It is expected, however, that sanitary sewer systems will deteriorate with age, and I/I rates will increase over time [7,13]. The factors influencing deterioration include physical defects, design flaws, illicit connections, root penetration, poorly adjusted manholes, corrosion, soil conditions, and the location of systems below groundwater levels.

Sanitary sewers are typically assumed to have a design life of 75 years, at which time the system is presumably taken out of service and replaced. At the end of the 75 year period, the sewer should still be capable of conveying peak domestic flow and the peak long-term I/I allowance. For this reason, when designing sanitary sewers, an allowance for peak, long-term I/I is included in the calculations for pipe sizing. This value is essentially a safety factor. While local standards vary [81], a value of 0.28 L/s/ha (also expressed in other units) is commonly used across Canada (note that sewer managers in the Province of British Columbia, Canada use half this value) [13]. Sewers demonstrating leakage at or below amounts permitted at construction are deemed "leak acceptable" [42]. Field experience and extensive consultation with sewer managers in Canada, however, has indicated that the occurrence of excessive I/I in "brand new" separated sewer systems is prevalent, indicating issues with the design, construction, and acceptance that have implications for the long-term operation and I/I rates in sewer systems. Specifically, flow monitoring data collected between 2015 and 2017 by Norton Engineering Inc. revealed that 34 of 35 new subdivisions in Ontario, Canada were experiencing I/I rates far exceeding the expected values [38]. Extensive consultation with municipalities across Canada has indicated that excessive I/I in new sewer construction is an issue experienced nationally [39,40,42].

While new combined sewers are no longer constructed in Canada under normal conditions [16], and semi-combined/partially separated systems are no longer permitted in most new construction, sewer systems in service within municipalities in Canada frequently include combinations of the types outlined in Table 3 [13,82]. Third pipe systems, which may include buried gravity pipe systems dedicated to managing building foundation drainage and/or roof drainage, are less common due to their additional expense, but they serve as an effective means of managing these flows by gravity without discharging directly to sanitary systems.

3.3. *Distinguishing between Public- and Private-Side I/I*

Further complicating the management of I/I in Canada are jurisdictional boundaries with respect to the design, construction, operation, and maintenance of sewer infrastructure. In most regions of Canada, an important distinction can be made between the "public" and "private" sides of the property line (Table 4) [10,42].

Table 4. Jurisdictional challenges associated with managing Inflow/Infiltration in Canada [1].

Jurisdiction	Management Implications
Private property ("private side" of the property line)	• The factors driving flood risk and I/I in *existing* buildings are under the control of households/property owners • Addressing private-side flood and I/I risk requires complex, iterative private-side remediation activities in buildings and on private properties • The factors driving building and site-level flood risk and I/I for *new* construction are largely under the jurisdiction of provincial, territorial, and/or municipal authorities responsible for building and plumbing construction codes and building bylaws • With some exceptions (including sewer use by-laws), municipalities/local authorities have limited jurisdiction and control over private-side actions to reduce I/I
Public property ("public side" of the property line)	• Local governments have significant control over the design standards, guides, and local practices governing the design of sanitary sewer conveyance systems, as well as the operation, maintenance, and repair of existing infrastructure • Managing I/I risk in new construction requires engagement and coordination with sectors typically not involved in the design, maintenance, or operation of wastewater systems, including construction code authorities and building and plumbing code inspections staff

[1] [10,13,42,83,84].

In addition to the issues outlined in Table 4, the two-tier local government systems in Canada can be problematic when working to resolve I/I. Where two-tier local governments exist, in general, the upper tier municipality is responsible to the regulator (i.e., provincial government authority) for meeting the I/I and overflow targets, while the lower tier municipality owns and operates the pipes where the I/I occurs, including on the private-side. To some extent, this situation introduces a degree of conflict when the upper and lower tier local governments work to mitigate I/I [13].

4. Factors Affecting the Occurrence of Public and Private-Side I/I

As part of recent efforts to develop practical guidance documents concerning the management of urban pluvial flood and I/I, extensive consultation with local government staff, sewer pipe and appurtenance manufacturers, and building and infrastructure regulators across Canada occurred in 2015–2021, documented in [13,38–43]. The consultation focused on multiple issues driving urban pluvial flood risk, including I/I. A particular emphasis of the consultation was the occurrence of I/I in new sewer construction [42]. Based on this consultation, the factors identified that drive the occurrence of excessive I/I in new construction include:

- General lack of understanding by the industry of the various factors during construction that represent I/I risk;
- Failure to construct private- and public-side sewers according to construction codes, standards, and guidelines in force at the local level;
- Failure to apply testing, quality assurance, and acceptance practices, as outlined in codes, standards, guides, and specifications;
- Conflicts of interests related to which party performs the site inspection (e.g., developer vs. municipal representatives);
- Limited inspection of the private side of the pipe system (these include prescribed notices only);
- Lack of clarity and guidance in construction codes with respect to constructing leak-acceptable sewer infrastructure;
- Jurisdictional issues and silos [42].

Surveys of municipal staff managing wastewater systems in communities with new subdivisions reporting high rates of I/I indicated that many manufacturer-recommended practices and construction code requirements concerning sewer system construction and inspection were not being conducted [38]. Ongoing surveying of the municipal respondents indicated no testing or very low rates of feeler gauge testing for installed pipe gaskets (0% of ~100 municipalities). The respondents further indicated that, though mandatory, sewer testing methods, including leak testing, mandrel, and CCTV inspections, were conducted infrequently (Figure 1).

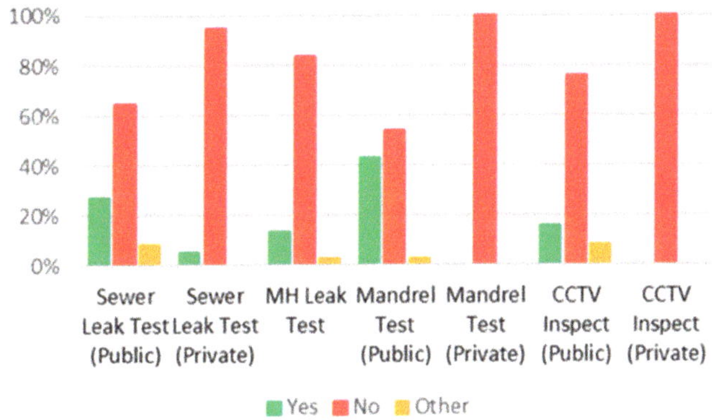

Figure 1. Performance of required sewer inspection/testing. Based on surveys conducted by Norton Engineering Inc. between 2015 and 2019 [40].

On the public side, maintenance hole infiltration or exfiltration tests may be rarely conducted (17% of 35 municipalities reported the performance of these tests), and CCTV inspections of public-side lateral sewers and connections were rarely conducted (14% of 35 respondents reported this type of test) [40]. Additional factors affecting the occurrence of I/I in new construction are outlined in Table 5.

Municipal staff across Canada have reported multiple factors driving the limited application of recommended or required inspection practices. These include: actual or perceived pressure from the development industry and/or pressure from the senior management in local governments to approve new sewage works quickly; compartmentalization (existence of "silos") between the staff/departments/organizations responsible for building and sewer construction and inspections [42,84]; lack of experience in the construction and inspections sectors with respect to appropriate practices for the construction, inspection, and acceptance of public- and private-side sewer systems; and resource limitations, including limited staffing for inspection and oversight [42]. While more experienced municipal engineering staff may oversee the inspection and testing of public-side sewer infrastructure, sewer and drainage systems on the private side of the property line are only "spot-checked" by local building code inspection staff, who may lack plumbing system expertise and who may have limited consideration for the lifecycle issues associated with private-side sewer system performance and the implications of I/I [42].

Table 5. Factors affecting I/I in new sewer construction, municipal-side [1].

Planning, design, and construction	• Location of sewer systems, including buried pipe systems in areas prone to surface flooding and/or high groundwater tables.
Construction	• Quality control and construction issues leading to pipe defects and cross connections.
Inspections and acceptance	• Limited knowledge by inspections staff of the factors driving I/I and their implications for the lifecycle costs of operating and maintaining sewer infrastructure and related systems. • Lack of experience of municipal inspectors. • Lack of appropriate inspections to verify the leak-acceptable status of buried pipe infrastructure, including visual inspections to verify construction practices at different stages of construction (e.g., embedding and haunching), the infiltration and/or exfiltration of pipes after backfill, CCTV inspections, and mandrel testing. • Failure to refer to the performance-based standards for sewers stated in sewer regulations governing the design, construction, and inspection of sewer systems. • Conflicts of interest between the proponent's desire to complete construction quickly and the owners' (e.g., municipality) desire for long-term pipe performance.

[1] [42].

A further jurisdictional issue relates to the responsibility for appropriate inspections of the connection between the municipal-side sanitary sewer "stub" (public-side lateral that extends toward the private property) and the private-side sewer lateral. This public/private-side connection has been identified as being at a high risk for I/I related defects due to the differential settlement, as the municipal-side infrastructure is constructed and buried before the private-side infrastructure is built [42]. Furthermore, although the same underlying construction standards specify the installation of the public- and private-side sewer pipe (e.g., PVC-pipe-manufacturer-recommended practices) [41,42], different regulatory requirements exist on the private and public sides of the property line (e.g., municipal guidelines for public-side construction and provincial construction codes for private-side construction), resulting in inconsistent construction practices for what is essentially a continuous pipe of consistent material [41].

With respect to new construction, flow monitoring allows municipalities to ensure conformance to performance-based specifications, providing real data that can be used in the process of verifying the acceptability of new infrastructure [13,42]. Providing notification to developers and contractors that flow monitoring will be in place and will be considered as part of the approval of new sewer infrastructure provides a strong incentive to ensure that recommended and required practices are in place and that I/I is managed to the extent possible. Flow monitoring, however, has not been routinely conducted for new sewer construction in Canada [42].

Private-Side Issues

The effective management of I/I should include considerations of managing sources of I/I on the private side of the property line. For example, [85] indicated that, based on a response of 26 reporting agencies, private-side contributions of I/I range from 7–80%, with an average contribution of 24%. Additional estimates of private-side contributions of I/I include 40% [86], 55% [87], and 35% [88]. Pawlowski et al. 2014 [88] further estimated that 98% of private-side I/I in a US case-study municipality were associated with foundation drain and downspout connections to sanitary sewers.

Multiple factors drive flood risk and I/I on the private side of the property line. These factors typically relate directly to the failure of property owners to maintain private sewers (e.g., building drains and laterals), as well as protective plumbing equipment (including pumped foundation drainage discharge systems and sewer backwater protection devices), lot grading, and the internal and external drainage features of buildings (see Table 6). As discussed above, partially separated/semi-combined systems are particularly problematic, and municipalities across Canada continue to report issues associated with the connection of foundation drainage and downspouts to sanitary sewers as significant drivers of I/I and the resulting urban pluvial flood/sewer surcharge risk [89].

Table 6. Factors affecting I/I in new construction, private-side [1].

Element	Risk Factor
Planning and design	• Location of buildings and sewer connections in areas prone to surface flooding and/or high groundwater (inc. location of foundation drainage systems, floor slabs, footings, etc. below high groundwater tables) and/or failure to protect buildings from flood risks • Building and drainage design that increase I/I and flood risk (e.g., reverse slope driveways, external area drains that may be cross connected to sanitary sewers, window wells, exterior basement stairwells, appropriate discharge of downspouts) • Site factors that increase the risk of I/I (e.g., plantings that increase the risk of root penetration, the blockage of sewer connections and foundation systems, stormwater infiltration features that increase the risk of infiltration flooding, infiltration into sewer connections) • Failure to incorporate backflow protection in pipe trench design
Construction	• Lack of application of industry-accepted bedding, haunching, and backfilling practices, inadequate pipe materials, improper pipe jointing practices • Failure to seal water entry points in building foundations, basement floor slabs, utility penetrations
Inspection, testing, acceptance	• Limited application of industry accepted pipe testing approaches for leak testing, including the air pressure, infiltration, and exfiltration testing of private-side sewer connections in compliance with consensus-based standards (e.g., NSCs) • Limited application of flow monitoring, consideration of flow monitoring results in the acceptance of new construction
Operation and maintenance	• Lack of maintenance of plumbing system and drainage systems, leading to failures (inc. sump pump system, foundation drain, sewer lateral connection pipe failures) • Lack of inspection, maintenance, and repair of critical drainage features over the life of the building, including inspection and repair and the replacement of sewer connections and sump pump systems • Lack of reporting of pertinent information to local authorities responsible for sewer systems—e.g., backups due to failed pipes, observations concerning flood occurrence and the mechanisms of flood (surface, seepage, sewer backup, other) • Interference with critical drainage features on the property (e.g., alteration of lot grading and drainage)

[1] [10,42,55].

In existing/older sewer systems, building owners have significant control over the lot and building drainage characteristics that may affect the flood risk for the building and property, including foundation drain and downspout connections to sanitary sewers. Reflecting the influence of private property owners on system-level urban pluvial flood risk, municipalities across Canada have implemented multiple strategies to increase the engagement of residents in managing urban pluvial flood risk and I/I on the private side of the property line, including voluntary education and incentive programs. Successful engagement in these actions, however, remains elusive, with few programs resulting in the significant uptake of private-side action to control I/I [10].

New private-side construction is dictated by building construction code requirements (including building and plumbing codes), which are issued, in the majority of circumstances in Canada, by federal, provincial, and territorial governments. Local authorities may also regulate new private-side drainage works via local drainage and sewer use bylaws. In particular, sewer use bylaws regulate the common causes of significant inflow into sanitary systems, including restricting the connection of foundation drainage and downspouts to sanitary sewer connections [83]. Further, in many jurisdictions in Canada, provincial construction codes take precedence over local building bylaws. These provincial construction requirements may not clearly articulate requirements for the discharge of lot-side storm and groundwater and may, for example, permit the connection of foundation drains to sanitary sewers depending on local code enforcement and interpretation [39,90].

While foundation drain and downspout connections to sanitary sewers are typically restricted in Canada, it is widely reported that building owners make changes to their basement plumbing to avoid the use of sump pumps or to reduce existing flooding (including connecting or draining foundation drains to sanitary sewers) [90]. Further, municipalities face significant difficulty in monitoring and enforcing private-side I/I sources, as property owner permission may be required to access buildings and properties in order to evaluate cross connections. Post-inspection recommendations for the remediation of cross connections are typically disruptive and costly for households, further limiting households' involvement in I/I reduction [83].

Additional private-side factors that drive I/I and sewer surcharge risk include sewer connections that are in poor repair (e.g., poorly jointed, cracked, or with clean-out caps removed). The limited adoption of practices that reduce risk on the private side of the property line can drive I/I, as flood waters that enter buildings may then enter sanitary sewers via basement floor drains [55]. Further, foundation drainage system failures, the failure of sump pump systems due to power interruptions and/or the lack of maintenance, and the backing up of municipal stormwater systems into private-side drainage systems (e.g., where foundation drains discharge by gravity to municipal stormwater systems) may also drive private-side I/I risk [10,55]. The use of "reverse slope" driveways also represents a higher flood risk, as they direct surface water directly into buildings and then into sanitary systems via floor drains [91–94].

With respect to new construction, pipe manufacturers and construction guidelines and codes recommend specific installation and testing requirements to ensure the proper performance of installed sewer pipe. Specifically, the National Plumbing Code of Canada (NPCC) includes leak testing provisions (air and water testing, ball tests, and final tests for private-side drainage systems) and also includes specific pipe bedding practices to reduce the risk of penetration and the poor grading of pipes [95]. Construction codes also reference standards that are applied in public-side sewer construction [39]. Pipe manufacturers commonly reference consensus-based standards to outline the appropriate construction and inspection practices for buried pipe systems, including CAN/CSA B182.11 and ASTM D2321. These standards provide detailed information concerning the construction, installation, and testing of buried pipes for sewer conveyance systems. On the private side, however, these requirements are largely ignored [42].

5. Development of New Standards and Resources in Canada

As a result of significant losses derived from urban pluvial flood impacts, a number of new NSCs have been developed to help guide risk reduction associated with pluvial flooding and potential climate change impacts. Each of these standards includes specific guidance with respect to wastewater systems to manage the sewer surcharge and I/I components of urban pluvial flood risk. A high-level summary of the I/I considerations and requirements offered in these standards is provided in Table 7.

Two of the above standards focus on managing I/I in the context of urban pluvial flood risk on the public and private sides of the property line. Specifically, detailed measures required to manage urban pluvial flooding and I/I on the private side of the property line are outlined in CSA Z800-18: Basement Flood Protection and Risk Reduction Guideline. The guideline was developed by a technical committee comprised of local and national wastewater, stormwater, and buildings experts and includes details concerning the design and maintenance of private-side residential plumbing and drainage features that serve to mitigate urban pluvial flood and I/I risk. At the time of writing, a new NSC (BNQ 3682-320) concerning the management of inflow/infiltration in new sewer construction is under development [42,97]. This upcoming standard focuses on the most important drivers of I/I in new sewer construction, which have been identified by a wide cross section of municipal, provincial, and federal government professionals involved in sewer and building design and construction.

Table 7. Consideration of wastewater systems and I/I in the National Standards of Canada targeted at urban pluvial flood risk reduction and climate change adaptation [1].

Standard	Wastewater I/I System Considerations
CSA Z800-18	• Comprehensive private-side I/I mitigation options for new and existing construction: Design, construction, maintenance, operation, and risk management considerations for private properties. • Focus is on private properties and low-rise residential/small buildings.
CSA W204-19	• New construction guidance to reduce the risk of urban flood events, with a focus on pluvial flood/stormwater management and other flood causes (e.g., riverine). • Compliance requires: fully separated sewer systems, that sanitary sewers convey extraneous I/I without surcharging, and that surcharging should not occur during the design event specified by the authority with jurisdiction. Under extreme (1 in 100 year) I/I conditions, hydraulic grade lines shall be 0.3 m below the underside building footings. • Additional compliance requirements include: the hydraulic design of sanitary sewer systems complying with storm sewer system requirements and that public-side maintenance holes not to be located in sags or be sealed when located in sags where stormwater ponding may occur.
CSA S900.1-18	• Guidance for adapting wastewater treatment plants to the potential impacts of climate change, including vulnerability and risk assessments for plants under changing climate conditions and the development of adaptation options based on these assessments. • Includes considerations for extreme precipitation events and implications for I/I and sensitivity analyses that consider the potential impacts of extreme precipitation events on I/I and the resulting plant design flows and peaking factors.
CSA W210-21	• Urban flood risk management standard for older, built-up (mature) communities for application by local authorities with differing levels of resources/data available for assessments. • Assessment of I/I factors depending on "[. . .] technical maturity, staff capacity, and available resources [. . .]," ranging from "foundational" or basic assessments using the era of construction to "advanced" assessments that include considerations such as sanitary sewer systems modelling and inspections of private laterals and cross connections. • Comprehensive consideration of the factors that drive I/I and multiple types of sewer backup risk that affect buildings (e.g., storm, sanitary, combined, partially separated sewer backup risk) on the municipal and private sides of the property line for flood risk assessment.
CSA W211-21	• Sanitary sewage collection, conveyance, and treatment systems "out of scope" for this standard. • Includes some consideration of sanitary sewer systems in relation to flood (e.g., impacts on sanitary sewers where storm sewers are not maintained).
BNQ 3682-320	• Under development at the time of writing, the standard is to provide comprehensive private- and municipal-side mitigation options for new and existing construction, including factors related to the design, construction, maintenance, and operation of public and private sanitary sewers. • Considers the integration of flood types and that the protection of buildings from surface and infiltration flood reduces the risk of I/I (e.g., restricting use of reverse slope driveways to mitigate I/I during pluvial flood events). • To include the consideration of the interaction between stormwater management features, such as features designed for stormwater infiltration, and sanitary sewer systems.

[1] [44,45,55,69,96,97].

6. Discussion

Protecting sewer systems from I/I aligns directly with protecting buildings from urban pluvial flood impacts. As presented in Table 7, new NSCs concerning urban pluvial flooding include topics and considerations related to the management of I/I and sewer surcharge risk. Similarly, programs aimed at private-side flood protection simultaneously focus on the reduction of private-side contributions of I/I [10]. However, discussion concerning the management of I/I in the pluvial flood management literature is limited.

With respect to policy formulation, this discussion focuses on addressing key barriers in I/I management in both existing/older and new construction, as identified in consultation with practitioners across Canada [13,38–43,90] and in Technical Committee discussions leading to the development of new NSCs. Notably, the actions taken by building owners (on the private side) remain an ongoing impediment to the effective management of pluvial flooding and I/I specifically. There is also an identified need to strengthen the capacity of local governments to manage new construction in a manner that better addresses I/I. The improved education and coordination of the municipal staff involved in the inspection and approval of buildings and sewer infrastructure, backed by coordinated technical standards and regulations that clearly specify consistent standards for sewer systems and drainage on both the private and public sides of the property line, should be implemented. Improved technical regulation should rest upon improved administrative support, including the capacity for inspections and improved understanding on the part of building officials of the importance of appropriate sewer inspections and testing before the acceptance of new infrastructure. Improved technical and administrative requirements must also be backed by a better assessment of the benefits and costs of I/I management.

6.1. Public Risk Perception and Risk Reduction Behavior

The private side of the property line is an important contributor to I/I in both new and existing construction. Local authorities across Canada have developed and implemented voluntary homeowner engagement programs to assist in the reduction of I/I and pluvial flood risk, including programs that provide direct financial incentives for sewer backflow protection and downspout and foundation drain disconnection.

The experience of wastewater system and urban pluvial flood managers in Canada to date suggests that it is difficult and expensive to address the factors resulting in I/I and urban pluvial flood risk post-construction. For example, the enforcement of I/I measures on the private side of the property line is politically unpopular [83], and voluntary programs targeted at property owners do not typically experience high uptake [10].

There exists considerable literature on household and public risk perceptions and risk reduction behavior related to natural hazards [98–111], with a substantial amount of the literature focusing on flooding specifically [112–123]. Further, studies have focused on the household response to pluvial flood events, including mitigation actions [27,124]. However, there exist few studies that comprehensively characterize the propensity of households and private property owners to engage in I/I and basement flood risk reduction, as these issues relate to pluvial flood [10]. While authors [27] have provided useful discussion concerning residents' propensity to engage in "medium" and "high" cost flood protection activities, these activities were consistent with protecting buildings from the direct impact of flooding (e.g., sewer backflow protection, relocating electrical systems to higher floor levels, flood barriers) rather than addressing the private-side drivers of sanitary sewer I/I that may contribute to flooding, including downspout and foundation drain connections.

It should not be expected that private property owners will choose to engage in risk reduction behavior in existing homes, especially in the case of significant drainage improvements (disconnecting foundation drains, replacing leaking buried sewer pipes, etc.), even where significant financial incentives are made available. Engaging private property owners, including households, in sewer maintenance and repair will require innovative practices that include inspections and maintenance requirements that are triggered during key windows of opportunity, including when permits are issued for private-side work concerning sanitary building sewers and drains and when private land parcels are redeveloped [83,125,126]. New sewer and drainage systems should be installed according to the best practices and manufacturer recommendations. The design, construction, and inspection of new systems should be conducted in a manner that will limit or eliminate the need to make significant changes to building drainage systems post-construction. Investments in the inspection and enforcement of new sewer construction can help offset the long-term issues with private-side systems that prove exceedingly difficult to address.

6.2. Improve Technical Standards & Regulation

In general, concerted effort to manage urban pluvial flood risk, including elements of risk related to wastewater systems, will require the regulation of I/I management practices on both the municipal and private sides of the property line. Through regulation, risk reduction methods concerning planning, design, construction, inspection, testing, and acceptance must become standard practice. Specific items that may require regulation on both the public and private sides of the property line are outlined in Table 8. Regulatory approaches to improving private-side construction are provided in Table 9.

With respect to the interface between private and municipal systems, there exist inconsistencies between private- and public-side technical requirements for sewer installation. In many instances, installation and inspection requirements for municipal-side sewers are more comprehensive, while key documents governing the private-side installation of sewers and drainage systems lack technical clarity with respect to restricting cross connections and ensuring proper installation procedures (e.g., gasketed connections for buried pipe, proper bedding, and backfilling procedures adhering to the manufacturer's requirements). Notably, with respect to sewer laterals extending onto private property, different local regulatory requirements may be in place for a continuous length of pipe. The requirements for the public- and private-side systems should be better aligned.

Table 8. Best practices concerning the design, construction, inspection, and acceptance of new sewer systems to reduce I/I risk, municipal-side [1].

Element	Approach
Planning and Pre-Design	• Direct new sewer construction away from surface flood hazard areas, groundwater, or areas with site conditions that exacerbate I/I risk, including stormwater infiltration features. • Where construction in flood-prone areas is unavoidable, incorporate I/I and flood risk reduction mitigation options (e.g., apply more restrictive, leak-proof standards for sewer pipe design to reduce the risk of infiltration over its lifespan).
Design	• Ensure that systems can achieve minimum design flushing velocity. • Design sewage pumping stations to operate under all flow conditions. • Locate maintenance holes away from surface ponding areas, use riser rings in manholes. • Design according to site conditions: e.g., leak-proof joints are required in areas exposed to groundwater. • Appropriate differentiation of storm and sanitary sewer pipe materials to reduce the risk of cross connections (e.g., with respect to the size, color, and placement of pipes). • Accommodate flow monitoring in system design (e.g., place manholes at the downstream end of new sewer systems serving subdivisions).
Construction	• Install flow monitors at the downstream end of new subdivisions when trunk systems are established so that conformance with existing performance-based standards can be confirmed. • Increase education concerning I/I for municipal and consultant inspectors and enforce the generally required full-time inspection of pipe construction by the designated inspector.
Inspection and acceptance	• Inspect all new sewers using CCTV, compare written reports to CCTV recording, sign off on comparison. • Provide written reports and report interpretation for sewer and maintenance hole leak testing and pipe deflection tests. • Apply third party leak testing for manholes and sewers. • Visually inspect manholes prior to acceptance. • Provide acceptance packages that include all the items required by the standards, specifications, and regulations. • Apply flow monitoring results to inform the acceptance of new sewers for all subdivisions.

[1] [39].

Table 9. Best practices concerning the design, construction, inspection, and acceptance of new sewer systems to reduce I/I risk, private-side [1].

Element	Approach
Planning and Pre-Design	• New buildings should not be located in areas exposed to surface flood risks, groundwater, or areas with site conditions that exacerbate I/I risk, including stormwater infiltration features. Where construction in these areas is unavoidable, apply flood and I/I mitigation options (e.g., construct buildings without basements). • Private-side connections into public-side municipal stubs should be located above seasonally high groundwater tables. • Building foundation drains, floor slabs, and foundation footings should be located above seasonally high groundwater tables.
Design	• Apply surface, groundwater, and infiltration flood hazard protection to reduce the risk of building flooding that can contribute to I/I, including site grading and drainage and all aspects of lot grading (reverse slope driveways are prohibited, basement windows, exterior stairwells are used only where necessary). • Do not permit the cross-connections of private-side drainage features to sanitary sewers (including downspouts, area drains, foundation drainage systems). • Use appropriate sewer pipe materials for buried applications, gasketed pipe joints, and appropriate pipe strength. • Sanitary sewer slope, protection of backfill through pipe trenches.
Construction	• Ensure appropriate application of construction practices for pipes (including bedding, haunching, and backfilling according to accepted construction standards), address common/recurring issues identified in new construction (including ensuring building air barrier systems remain intact and addressing infiltration and surface flood hazards).
Inspection and acceptance	• Visual inspections to ensure site grading and drainage complies with design requirements. • Thorough inspections of the sewer connection (pre-backfill visual inspection of jointing, bedding, and haunching) and post-backfill (CCTV, pressure, infiltration and/or exfiltration testing, deflection testing). • Acceptance depending on the inspection results and flow monitoring data.

[1] [127].

NSCs concerning pluvial flood management and I/I, specifically CSA Z800-18 and the upcoming BNQ 3682-320 NSC (Table 7), emphasize restricting construction in areas known to be at risk of flooding, including regions prone to high groundwater and overland/surface flooding. In several regions of Canada, land use planning is in place to restrict development in known flood-prone areas [128]. Specifically, land use planning regulations in Ontario, Canada's most populous province, state that "development shall generally be directed [. . .] to areas outside of [. . .] hazardous lands adjacent to river, stream and small inland lake systems which are impacted by flooding hazards [. . .]" [129] (p. 32). However, land-use planning restrictions across Canada are inconsistent, with several jurisdictions permitting development in flood prone areas. Further, non-river flooding, including surface stormwater flooding attributed to sags/topographical lows outside of riverine flood hazard areas, is typically not represented on official maps that are used to guide land use planning in Canada [130]. Thus, despite the recommendations in new NSCs that development be directed away from flood-prone areas, it is likely that many new sewer systems will be constructed in areas that are at risk of surface flooding. These eventualities are recognized in new NSCs, and the standards include accommodations with respect to building and sewer design to reduce the risk of damage should development be located in these areas (including constructing buildings without basements and applying additional sewer design methods to ensure that systems remain leak-acceptable).

6.3. Improve Administrative Capacity for Local Authorities

With respect to new construction, guidance for authorities with jurisdiction over the construction of sewer systems should extend beyond technical requirements, special provisions and drawings in construction documents, and sewer use bylaws, and it should consider administrative and enforcement provisions to ensure that existing technical requirements are adhered to. Administrative guidance should therefore be incorporated into technical standards concerning the design, construction, inspection, and operation of sewer infrastructure. Other municipal guidance documents not directly related to sewer design and construction should be written with a view to reducing I/I risk, including agreements with property developers and local government official plans. Best practices in the local regulation of sewer systems include ensuring that required inspection practices take place. Additional administrative approaches to managing I/I include education and awareness for those responsible for the installation, inspection, and management of sewer systems, including both professionals and private property owners (Table 10).

Table 10. Non-structural/administrative strategies, municipal- and private-side.

Approach
• Education and information for private-side building and plumbing codes inspection staff concerning I/I, its causes, and how proper construction and inspection techniques serve to limit I/I risk over the lifespan of sewer infrastructure.
• Education for public-side engineering staff working in I/I to help improve the understanding of private-side construction codes (i.e., building and plumbing codes) and the sections therein related to I/I risk.
• Application of local information concerning sewer system services as part of the maintenance, alteration, etc. of private-side drainage features.
• Provide information to local authorities concerning flood occurrence and its causes.

6.4. Improve the Asessment of the Costs of I/I and the Benefits of Interventions to Reduce I/I

Allocating the budget and resources to correct I/I remains challenging for sewer managers, as buried wastewater conveyance systems are out of sight, and I/I is typically invisible to the public and to decisionmakers. It is therefore necessary to ensure that available financial resources and existing data are used wisely to develop I/I management programs. As discussed above, the real, direct costs of I/I include sewer overflows at plants and pumping stations and the loss of sewer capacity that could be used to allow for additional development. Additional costs associated with I/I may include the need to upsize existing pipes/infrastructure to accommodate I/I flows, loss of the lifespan of the sewer, flooding of buildings and homes, treatment costs, the need to expand sewage treatment to help manage extraneous I/I flows, and legal risk to the municipality associated with flood and environmental damage, among additional negative impacts.

Traditional engineering studies (including Environmental Assessments or EAs) that determine whether to expand wastewater treatment plants, however, focus only on chemical and power costs when assessing the benefits of reducing I/I [13]. This approach is inadequate and does not account for the real costs of I/I, such as indirect societal, environmental, and economic costs. This approach has been applied because many of the known costs of I/I are difficult to calculate, as they are complex and depend on multiple factors, and limited resources are availed to practitioners (finances, time, and expertise) for detailed benefit–cost assessments. Nevertheless, a more comprehensive and systematic approach to estimating the costs of I/I is required, especially in light of increasing urban development and the risks posed by changing climate conditions.

A practical method to better assess the overall costs is to rely on the "user fees" charged by local authorities as a proxy for the overall costs of I/I. User fees are used to cover the costs associated with buried municipal systems, upsizing pipes due to lost capacity, replacing pipes that have not reached their design capacity, municipal liability for flood damage, and treatment plant expansion costs. These municipal costs are reflected in the rates charged to property owners, and these values may be more readily available to municipalities to assess the overall costs of managing I/I [13].

6.5. Limitations, Future Research

The limitations of this review include the lack of quantitative information on the relative contribution of different factors to I/I and how these factors directly or indirectly contribute to flood damage in buildings. Authors have identified private-side factors that drive damage [131] and have studied household perceptions and behavior related to pluvial flooding [27], but there has been a limited focus on physical flood mechanisms (i.e., the relative contribution to damage of different flood mechanisms). The relative contribution of private-side contributors to I/I has been studied [6,77,88], but a comprehensive understanding of public- and private-side factors that drive I/I during intense rainfall events is elusive. For example, quantitative information on the I/I benefits of properly installed/leak-acceptable sewer laterals is not typically sufficient to provide detailed benefit–cost assessment studies.

This review focused on providing an overview and discussion on the role of wastewater systems and I/I in the management of pluvial flood risk in Canada and on new approaches with respect to policy and regulation to reduce I/I, both as it relates to pluvial flooding in urban areas and to the multitude of additional negative impacts of I/I. This discussion did not focus on the mitigation of pluvial flood risk directly, including early public warning systems and blue-green or nature-based infrastructure, but rather on managing of the risk of I/I associated with pluvial flood events. Further, focusing on the wastewater component of pluvial flooding specifically, this review did not compressively address the issues associated with urban flooding, including flash flooding in urban basins [132,133]. The emphasis on wastewater and I/I reflects the input from practitioners across Canada that have struggled to manage this problem both in existing/older construction and in new construction and the dearth of discussion on wastewater systems and I/I in the pluvial flooding literature.

The authors here focused on the outcomes of consultations and opportunities for the better integration of new guidance documents and NSCs concerning sewer system and pluvial flood management in Canada. The consultation conducted since 2015 involved practitioners from local, provincial, and federal agencies concerning the management of sewer systems, and thus, the recommendations and discussion provided here focused on the needs and interests of sewer practitioners.

There is a growing literature and increased emphasis on the topic of climate change adaptation in urban areas, including a particular emphasis on urban stormwater management and pluvial flooding [19,134–137]. The administrative and behavioral aspects of infrastructure design, construction, inspection, and acceptance, however, remain a gap in the literature. As discussed above, recurring barriers to effective pluvial flood protection, including wastewater and I/I components, include property owners' behavior, the activities and behavior of those responsible for the installation and inspection of new sewer infrastructure, and limited administrative and regulatory capacity to ensure that new sewer infrastructure is installed according to industry requirements and best practices. "Siloization," or the compartmentalization of responsibilities for urban water services, identified in previous studies, results in "[. . .] misaligned strategic goals" [136] (p. 13). This review further identified the compartmentalization of responsibilities for municipal- and private-side sewer infrastructure, contributing to the inconsistent application of technical standards and best practices, as well as to the inconsistent inspection and acceptance requirements for sewer infrastructure. Future research should explore practical methods for ensuring appropriate infrastructure governance that supports the consistent application of technical standards and best practices.

7. Conclusions

Urban pluvial flooding, resulting in flooded residential buildings, is one of the most significant drivers of disaster loss in Canada. As a result of historical impacts and an expected increase in the intensity of urban pluvial flood events, practical guidance documents and resources have been developed for infrastructure and property managers in Canada to help mitigate risk. Increasingly, the role of wastewater systems and their contribution to both flood damage and additional negative flood impacts (social, environmental, and financial impacts to communities) are recognized and are being incorporated into the standards oriented toward urban pluvial flood management.

The conclusions of this review include:

1. Though there is extensive literature on pluvial flooding and I/I, the literature on managing I/I as part of pluvial flood risk management is limited. The experience in Canada indicates that I/I is a significant contributor to pluvial flood risk, and greater emphasis should be placed on managing I/I as part of pluvial flood risk management. I/I's role in urban pluvial flooding should be factored into the multitude of additional negative impacts of excessive, chronic I/I to better motivate the management of I/I.
2. I/I is an ongoing issue in existing/older construction. In Canada, regions served by "partially separated" or "semi-combined" systems are particularly vulnerable to high inflow rates.
3. Extensive consultation across Canada has revealed that I/I is an issue in new sewer construction. The lack of administrative capacity to inspect and enforce sewer design requirements and the limited application of best practices in sewer design and construction—notably on the private-side—are key factors that affect the occurrence of I/I in new construction.
4. The engagement of property owners in I/I management and pluvial flood risk reduction, including the application of resource-intensive risk reduction options, remains an ongoing issue that has not been addressed through the engagement methods identified in the perception and behavior literature. New construction must be made as resilient as possible to avoid scenarios where the local authorities must revert to the education/voluntary compliance of property owners to mitigate I/I and pluvial flood risk.
5. New NSCs have been developed that focus both on the management of flood risk in urban areas (including pluvial flood) and on the management of I/I. The implementation of the practices outlined in these NSCs will require both adjustments to the technical aspects of the design, construction, inspection, and acceptance guidelines (e.g., as outlined in new NSCs) as well as administrative support to comply with and enforce improved standards. Improved understanding of the benefits and costs of reducing I/I, accounting for its myriad negative impacts, should support the implementation of the technical and administrative best practices.

The consultation of the relevant stakeholders across Canada has revealed the importance of I/I as a significant contributing factor to urban pluvial flood risk and that I/I can occur in both new and existing sewer construction. Increased emphasis on the administrative aspects of managing I/I—including the collaboration between the practitioners involved in monitoring and regulating sewer construction on the public and private sides of the property line, aligning the regulatory requirements for sewer construction, the improved accounting of the costs and benefits of managing I/I, and applying innovative means of engaging private property owners in I/I management—will contribute to the management of urban pluvial flood risk in Canada.

Author Contributions: Conceptualization, D.S. and B.R.; investigation, B.R. and D.S.; writing—original draft preparation, D.S. and B.R.; writing—review and editing, D.S. and B.R. All authors have read and agreed to the published version of the manuscript.

Funding: This research received no external funding.

Acknowledgments: The authors acknowledge the assistance of Mark Elliot in the preparation of the disaster loss figures included in the manuscript.

Conflicts of Interest: The authors declare no conflict of interest.

References

1. Zhu, Z.; Chen, X. Evaluating the effects of low impact development practices on urban flooding under different rainfall intensities. *Water* **2017**, *9*, 548. [CrossRef]
2. Mobini, S.; Becker, P.; Larsson, R.; Berndtsson, R. Systemic inequity in urban flood exposure and damage compensation. *Water* **2020**, *12*, 3152. [CrossRef]
3. Cahoon, L.B.; Hanke, M.H. Inflow and infiltration in coastal wastewater collection systems: Effects of rainfall, temperature, and sea level. *Wat. Environ. Res.* **2019**, *91*, 322–331. [CrossRef] [PubMed]
4. Pagliacci, F.; Defrancesco, E.; Bettella, F.; D'Agostino, V. Mitigation of Urban Pluvial Flooding: What Drives Residents' Willingness to Implement Green or Grey Stormwater Infrastructures on Their Property? *Water* **2020**, *12*, 3069. [CrossRef]
5. Sola, K.J.; Bjerkholt, J.T.; Lindholm, O.G.; Ratnaweera, H. Analysing consequences of infiltration and inflow water (I/I-water) using cost-benefit analyses. *Water Sci. Technol.* **2020**, *82*, 1312–1326. [CrossRef]
6. Yap, H.T.; Ngien, S.K. Assessment on inflow and infiltration in sewerage systems of Kuantan, Pahang. *Water Sci. Technol.* **2017**, *76*, 2918–2927. [CrossRef]
7. Beheshti, M.; Sægrov, S.; Ugarelli, R. Infiltration/Inflow Assessment and Detection in Urban Sewer System. *Vann* **2015**, *1*, 24–34.
8. Rosenzweig, B.R.; McPhillips, L.; Chang, H.; Cheng, C.; Welty, C.; Matsler, M.; Iwaniec, D.; Davidson, C.I. Pluvial flood risk and opportunities for resilience. *Wiley Interdiscip. Rev. Water* **2018**, *5*, e1302. [CrossRef]
9. Catastrophe Indices and Quantification. Catastrophe Bulletins. Available online: www.CatIQ.com (accessed on 22 April 2022).
10. Sandink, D.; Binns, A. Reducing urban flood risk through building-and lot-scale flood mitigation approaches: Challenges and opportunities. *Front. Water* **2021**, *3*, 86. [CrossRef]
11. Friedland, J.; Cheng, H.; Peleshok, A. *Water Damage Risk and Property Insurance Pricing*; Canadian Institute of Actuaries: Ottawa, ON, Canada, 2014.
12. Sandink, D.; Kovacs, P.; Oulahen, G.; Shrubsole, D. Public relief and insurance for residential flood losses in Canada: Current status and commentary. *Can. Water Resour. J.* **2016**, *41*, 220–237. [CrossRef]
13. Robinson, B.; Sandink, D. *Developing an Efficient and Cost-Effective Inflow and Infiltration (I/I) Reduction Program*; Institute for Catastrophic Loss Reduction: Toronto, ON, Canada; Standards Council of Canada: Ottawa, ON, Canada, 2021.
14. Tan, P.; Zhou, Y.; Zhang, Y.; Zhu, D.Z.; Zhang, T. Assessment and pathway determination for rainfall-derived inflow and infiltration in sanitary systems: A case study. *Urban Water J.* **2019**, *16*, 600–607. [CrossRef]
15. Amec Foster Wheeler and Credit Valley Conservation. National Infrastructure and Buildings Climate Change Adaptation State of Play Report. Prepared for the Infrastructure and Buildings Working Group, Part of Canada's Climate Change Adaptation Platform. 2017. Available online: http://www.ibwgsop.org/ (accessed on 22 April 2022).
16. Canadian Standards Association. *CSA PLUS 4013-2019: Development, Interpretation, and Use of Rainfall Intensity-Duration-Frequency (IDF) Information: Guideline for Canadian Water Resources Practitioners*; Technical Guide; Canadian Standards Association: Toronto, ON, Canada, 2019.
17. Nirupama, N.; Simonovic, S.P. Increase of flood risk due to urbanisation: A Canadian example. *Nat. Hazards* **2007**, *40*, 25–41. [CrossRef]
18. Denault, C.; Millar, R.; Lence, B. Assessment of possible impacts of climate change in an urban catchment. *J. Am. Water Resour. Assoc.* **2006**, *42*, 685–697. [CrossRef]
19. Brown, C.; Jackson, E.; Harford, D.; Bristow, D. Cities and Towns. In *Canada in a Changing Climate: National Issues Report*; Warren, F.J., Lulham, N., Eds.; Government of Canada: Ottawa, ON, Canada, 2021; Chapter 2.
20. Bush, E.; Lemmen, D.S. (Eds.) *Canada's Changing Climate Report*; Government of Canada: Ottawa, ON, Canada, 2019; 444p.
21. Credit Valley Conservation and Zizzo Strategy. *Developing a Stormwater Quality Management Standard (QMS) in Light of a Changing Climate*; Standards Council of Canada: Ottawa, ON, Canada, 2018.
22. Province of Ontario. *Bill 140, Strong Communities through Affordable Housing Act*; Legislative Assembly of Ontario: Toronto, ON, Canada, 2011.
23. Province of Ontario. *Promoting Affordable Housing Act*; Legislative Assembly of Ontario: Toronto, ON, Canada, 2016.
24. City of Toronto. City of Toronto's First Resilience Strategy. City of Toronto. Available online: https://www.toronto.ca/services-payments/water-environment/environmentally-friendly-city-initiatives/resilientto/ (accessed on 22 April 2022).
25. Insurance Canada. Nearly Half of Tenants Have No Renter's Insurance. 2018. Available online: https://www.insurance-canada.ca/2018/04/11/kanetixtenants-renters/ (accessed on 12 January 2021).
26. Zhang, Z. Estimating rain derived inflow and infiltration for rainfalls of varying characteristics. *J. Hydraul. Eng.* **2007**, *133*, 98–105. [CrossRef]
27. Rözer, V.; Müller, M.; Bubeck, P.; Kienzler, S.; Thieken, A.; Pech, I.; Schröter, K.; Buchholz, O.; Kreibich, H. Coping with pluvial floods by private households. *Water* **2016**, *8*, 304. [CrossRef]
28. Douglas, I.; Garvin, S.; Lawson, N.; Richards, J.; Tippett, J.; White, I. Urban pluvial flooding: A qualitative case study of cause, effect and nonstructural mitigation. *J. Flood Risk Manage.* **2010**, *3*, 112–125. [CrossRef]
29. Maksimović, Č.; Prodanović, D.; Boonya-Aroonnet, S.; Leitao, J.P.; Djordjević, S.; Allitt, R. Overland flow and pathway analysis for modelling of urban pluvial flooding. *J. Hydraul. Res.* **2009**, *47*, 512–523. [CrossRef]

30. Skougaard Kaspersen, P.; Høegh Ravn, N.; Arnbjerg-Nielsen, K.; Madsen, H.; Drews, M. Comparison of the impacts of urban development and climate change on exposing European cities to pluvial flooding. *Hydrol. Earth Syst. Sci.* **2017**, *21*, 4131–4147. [CrossRef]
31. Palla, A.; Colli, M.; Candela, A.; Aronica, G.T.; Lanza, L.G. Pluvial flooding in urban areas: The role of surface drainage efficiency. *J. Flood Risk Manage.* **2018**, *11*, S663–S676. [CrossRef]
32. Van Dijk, E.; van der Meulen, J.; Kluck, J.; Straatman, J.H. Comparing modelling techniques for analysing urban pluvial flooding. *Water Sci. Technol.* **2014**, *69*, 305–311. [CrossRef]
33. Schmitt, T.G.; Scheid, C. Evaluation and communication of pluvial flood risks in urban areas. *Wiley Interdiscip. Rev. Water.* **2020**, *7*, e1401. [CrossRef]
34. Bruwier, M.; Maravat, C.; Mustafa, A.; Teller, J.; Pirotton, M.; Erpicum, S.; Archambeau, P.; Dewals, B. Influence of urban forms on surface flow in urban pluvial flooding. *J. Hydrol.* **2020**, *582*, 124493. [CrossRef]
35. Olsen, A.S.; Zhou, Q.; Linde, J.J.; Arnbjerg-Nielsen, K. Comparing methods of calculating expected annual damage in urban pluvial flood risk assessments. *Water* **2015**, *7*, 255–270. [CrossRef]
36. Hamidi, A.; Grossberg, M.; Khanbilvardi, R. Evaluation of Urban Drainage Infrastructure: New York City Case Study. In *AGU Fall Meeting Abstracts*; American Geophysical Union: Washington, DC, USA, 2017; Volume 2017, p. H33O-02.
37. Mikalson, D.; Guo, Y.; Adams, B. Rainfall derived inflow and infiltration modeling approaches. *J. Water Manag. Model.* **2012**. [CrossRef]
38. Robinson, B. *Project to Address Unacceptable Inflow and Infiltration (I/I) in New Subdivisions: Phase 1 Final Report, 2015 to 2017*; Norton Engineering Inc.: Kitchener, ON, Canada, 2017.
39. Robinson, B. *Manual of Best Practices to Reduce Risk of I/I on the Public Side*; Norton Engineering Inc.: Kitchener, ON, Canada, 2019.
40. Robinson, B. *Working Together to Build Better Sewers and Build Sewers Better: Preparation for Climate Change*; Water Environment Association of Ontario: Mississauga, ON, Canada, 2019.
41. Robinson, B. *Building Code Regulations and Engineering Standards as They Relate to I/I*; Norton Engineering Inc.: Kitchener, ON, Canada, 2018.
42. Robinson, B.; Sandink, D.; Lapp, D. Reducing the Risk of Inflow and Infiltration (I/I) in New Sewer Construction. In *A National Foundational Document for the Development of a National Standard of Canada*; Institute for Catastrophic Loss Reduction: Toronto, ON, Canada; Standards Council of Canada: Ottawa, ON, Canada, 2019.
43. Sandink, D.; Robinson, B.; Dale, N.; Okrutny, P. *Practical Guidance for Private-Side Drainage Systems to Reduce Basement Flood Risk: Addressing Critical Information Gaps*; National Research Council of Canada: Ottawa, ON, Canada, 2021. [CrossRef]
44. Canadian Standards Association. *W204-19: Flood Resilient Design of New Residential Communities*; Canadian Standards Association: Toronto, ON, Canada, 2019.
45. Canadian Standards Association. *CSA W210:21 Prioritization of Flood Risk in Existing Communities*; Canadian Standards Association: Toronto, ON, Canada, 2021.
46. Bolivar-Phillips. Adaptive Approaches in Stormwater Management. 2013. Prepared for the City of Ottawa, Ottawa. Available online: https://documents.ottawa.ca/sites/documents/files/documents/stormwater_management_en.pdf (accessed on 22 May 2022).
47. Hulley, M.; Watt, E.; Zkovs, G. Potential impacts of climate change on stormwater management. In Proceedings of the WaterTech2008 Conference, Lake Louise, AB, Canada, 16–18 April 2008.
48. Watt, E.; Marsalek, J. Critical review of the evolution of the design storm event concept. *Can. J. Civ. Eng.* **2013**, *40*, 105–113. [CrossRef]
49. Pour, S.H.; Abd Wahab, A.K.; Shahid, S.; Asaduzzaman, M.; Dewan, A. Low impact development techniques to mitigate the impacts of climate-change-induced urban floods: Current trends, issues and challenges. *Sustain. Cities Soc.* **2020**, *62*, 102373. [CrossRef]
50. Ministry of the Environment, Conservation and Parks. *Low Impact Development Stormwater Management Guidance Manual (Draft for Consultation)*; Ministry of the Environment, Conservation and Parks: Ottawa, ON, Canada, 2022.
51. Worsely, B. *City of Peterborough Flood Reduction Master Plan*; UMA/AECOM; City of Peterborough: Peterborough, ON, Canada, 2005.
52. Environment and Climate Change Canada. Engineering Climate Datasets, via Simonovic; et al. 2019. IDF_CC Tool. 2017. Available online: https://www.idf-cc-uwo.ca/home.aspx (accessed on 17 April 2022).
53. City of Windsor. *Special Meeting of City Council*; City of Windsor: Windsor, ON, Canada, 2017.
54. Town of Tecumseh. Public Works and Environmental Services Report No. 40/16: Rainfall Event of 29 September 2016. Available online: http://173.209.52.99/sites/default/files/PW%20Report%20No%2040-16_Rainfall%20Event%20of%20September%2029%2C%202016_16Nov10.pdf (accessed on 22 May 2022).
55. Canadian Standards Association. *CSA Z800-18: Basement Flood Protection and Risk Reduction Guideline*; Canadian Standards Association: Toronto, ON, Canada, 2019.
56. Federation of Canadian Municipalities. *The Canadian Infrastructure Report Card*; Canadian Federation of Municipalities: Ottawa, ON, Canada, 2019.
57. Region of Niagara. Inflow and Infiltration. Available online: niagararegion.ca/living/sewage/inflow-infiltration.aspx (accessed on 18 March 2022).

58. Metro Vancouver. *Inflow and Infiltration Allowance Assessment–Integrated Liquid Waste Management Plan*; Metro Vancouver: Vancouver, BC, Canada, 2014.
59. US Environmental Protection Agency. *Guide for Estimating Infiltration and Inflow*; U.S. EPA Water Infrastructure Outreach: Washington, DC, USA, 2014.
60. Gustafsson, L.G. Alternative drainage schemes for reduction of inflow/infiltration-prediction and follow-up of effects with the aid of an integrated sewer/aquifer model. In Proceedings of the 1st International Conference on Urban Drainage via Internet, online, 21 September 2000; pp. 21–37.
61. Langeveld, J.G.; De Haan, C.; Klootwijk, M.; Schilperoort, R.P.S. Monitoring the performance of a storm water separating manifold with distributed temperature sensing. *Water Sci. Technol.* **2012**, *66*, 145–150. [CrossRef]
62. Thapa, J.B.; Jung, J.K.; Yovichin, R.D. A Qualitative Approach to Determine the Areas of Highest Inflow and Infiltration in Underground Infrastructure for Urban Area. *Adv. Civ. Eng.* **2019**, *2019*, 2620459. [CrossRef]
63. Water Environment Federation (WEF); American Society of Civil Engineers (ASCE); Water, Environmental & Resources Institute (EWRI). *Existing Sewer Evaluation and Rehabilitation: WEF Manual of Practice No. FD-6 ASCE/EWRI Manuals and Reports on Engineering Practice No. 62*, 3rd ed.; McGraw-Hill: New York, NY, USA, 2019.
64. Semadeni-Davies, A.; Hernebring, C.; Svensson, G.; Gustasfsson, L. The impact of climate change and urbanization on drainage in Helsingborg, Sweden: Combined sewer system. *J. Hydrol.* **2008**, *350*, 100–113. [CrossRef]
65. GENIVAR Inc. *Climate Change Vulnerability Assessment of the Town of Prescott's Sanitary Sewage System*; PIEVC Program: Toronto, ON, Canada, 2011.
66. Nasrin, T.; Sharma, A.; Muttil, N. Impact of short duration rainfall events on sanitary sewer network performance. *Water* **2017**, *9*, 225. [CrossRef]
67. Torres, M. Infiltration and Inflow in Oslo Municipality: Assessment of I/I Volumes, Sources, and Relationship with Measurable Wastewater System's Characteristics. Master's Thesis, Norwegian University of Life Sciences, Ås, Norway, 2013.
68. Zhang, M.; Liu, Y.; Cheng, X.; Zhu, H.; Shi, H.; Yuan, Z. Quantifying rainfall-derived inflow and infiltration in sanitary sewer systems based on conductivity monitoring. *J. Hydrol.* **2018**, *558*, 174–183. [CrossRef]
69. Canadian Standards Association. *S900.1-18. Climate Change Adaptation for Wastewater Treatment Plants*; CSA: Toronto, ON, Canada, 2018.
70. Semadeni-Davies, A. Urban water management versus climate change: Impacts on cold region waste water inflows. *Clim. Change* **2004**, *6*, 103–126. [CrossRef]
71. Flood, J.; Calhoon, L. Risk to coastal wastewater collection systems from sea-level rise and climate change. *J. Coast. Res.* **2011**, *27*, 652–660. [CrossRef]
72. City of Hamilton. *Binbrook Sanitary and Stormwater System Performance*; City of Hamilton: Hamilton, ON, Canada, 2012.
73. City of Toronto. *Staff Report: Impact of 8 July 2013 Storm on the City's Sewer and Stormwater Systems*; City of Toronto: Toronto, ON, Canada, 2013.
74. Genivar and Clarifica. *Investigation of Chronic Basement Flooding, Sewershed Area 28–Final Project File*; City of Toronto: Toronto, ON, Canada, 2008.
75. XCG Environmental Consulting Services. *Flood Remediation Plan: Environmental Assessment Project File Report, Sewershed Study Area 30*; City of Toronto: Toronto, ON, Canada, 2008.
76. Stantec. *Sewershed Area 29 Chronic Basement Flooding Class EA: Alternative Solutions*; City of Toronto: Toronto, ON, Canada, 2008.
77. Jiang, A.Z.; McBean, E.A.; Binns, A.; Gharabaghi, B. Quantifying rainfall-derived inflow from private foundation drains in sanitary sewers: Case study in London, Ontario, Canada. *J. Hydrol. Eng.* **2019**, *24*, 05019023. [CrossRef]
78. Chambers, K. *Weeping Tile Disconnection to Reduce the Impact of Basement Flooding—London, Ontario*; Institute for Catastrophic Loss Reduction Basement Flood Symposium: Toronto, ON, Canada, 2013.
79. Sandink, D. *Involving Homeowners in Urban Flood Risk Reduction a Case Study of the Sherwood Forest Neighborhood, London, Ontario*; Institute for Catastrophic Loss Reduction: Toronto, ON, Canada, 2011.
80. City of London. *Update on Results of Sherwood Forest Weeping Tile Disconnect Pilot Project*; City of London: London, ON, Canada, 2015.
81. Federation of Canadian Municipalities and National Research Council. *Infiltration/Inflow Control/Reduction for Wastewater Collection Systems: A Best Practice by the National Guide to Sustainable Municipal Infrastructure*; Federation of Canadian Municipalities and National Research Council: Ottawa, ON, Canada, 2003.
82. Metro Vancouver Liquid Waste Services Department. *Private Lateral Foundation Drains and Semi-Combined Sewers as an Inflow and Infiltration Source*; Metro Vancouver: Burnaby, BC, Canada, 2016.
83. Kyriazis, J.; Zizzo, L.; Sandink, D. *Assessing Local Mandatory Measures to Reduce Flood Risk and Inflow & Infiltration in Existing Homes*; Institute for Catastrophic Loss Reduction: Toronto, ON, Canada, 2017.
84. Kesik, T. *Management of Inflow and Infiltration in New Urban Developments*; Institute for Catastrophic Loss Reduction: Toronto, ON, Canada, 2015.
85. Water Environment Research Foundation. *Methods for Cost-Effective Rehabilitation of Private Lateral Sewers*; Water Environment Research Foundation: Alexandria, VA, USA, 2006.
86. Pearlman, S. *Minimizing Municipal Costs for Inflow and Infiltration Remediation: A Handbook for Municipal Officials*; Neponset River Watershed Association: Canton, OH, USA, 2017.

87. Nelson, D.; Cantrell, C.; Gross, J. Columbus Private Source Infiltration and Inflow Identification Pilot Program. In Proceedings of the Water Environment Federation, WEFTEC 2005, Session 81 through Session 90, Orleans, LA, USA, 29 September–3 October 2012; pp. 6845–6859.
88. Pawlowski, C.; Rhea, L.; Shuster, D.; Braden, G. Some factors affecting inflow and infiltration from residential sources in a core urban area: Case study in Columbus, Ohio, neighborhood. *J. Hydrol. Eng.* **2014**, *140*, 105–114. [CrossRef]
89. Sandink, D. *Urban flooding in Canada*; Institute for Catastrophic Loss Reduction: Toronto, ON, Canada, 2013.
90. Robinson, B. *Inflow/Infiltration (I/I) in New Construction: A Huge Issue in Ontario and Across Canada*; Ontario Building Officials Association: Woodbridge, ON, Canada, 2018.
91. Kovacs, P.; Guilbault, S.; Sandink, D. *Cities Adapt to Extreme Rainfall*; Institute for Catastrophic Loss Reduction: Toronto, ON, Canada, 2014.
92. City of Hamilton. *Lot Grading Policy, Criteria and Standards for Single and Semi-Detached Dwelling Units Created through Development Applications*; City of Hamilton: Hamilton, ON, Canada, 2011.
93. City of Toronto. *City of Toronto Zoning By-Law 569-2013*; City of Toronto: Toronto, ON, Canada, 2013.
94. City of Vaughan. *City of Vaughan's Zoning By-Law 1–88 (Comprehensive Zoning By-Law)*; City of Vaughan: Vaughan, ON, Canada, 2012.
95. National Research Council of Canada. *National Plumbing Code of Canada*; National Research Council: Ottawa, ON, Canada, 2020.
96. Canadian Standards Association. *CSA W211: 21. Management Standard for Stormwater Systems*; Canadian Standards Association: Toronto, ON, Canada, 2021.
97. Bureau de Normalisation du Québec–BNQ. *BNQ 3682-320 Mitigation of the Risks of Inflow and Infiltration in New Sewer Networks*; Under Development, Bureau de Normalisation du Québec: Quebec, QC, Canada, 2022.
98. Joffe, H.; Perez-Fuentes, G.; Potts, H.W.; Rossetto, T. How to increase earthquake and home fire preparedness: The fix-it intervention. *Nat. Hazards* **2016**, *84*, 1943–1965. [CrossRef]
99. Lindell, M.K.; Perry, R.W. Household adjustment to earthquake hazard: A review of research. *Environ. Behav.* **2000**, *32*, 461–501. [CrossRef]
100. Mulilis, J.P.; Lippa, R. Behavioral change in earthquake preparedness due to negative threat appeals: A test of protection motivation theory. *J. Appl. Soc. Psychol.* **1990**, *20*, 619–638. [CrossRef]
101. Tanes, Z.; Cho, H. Goal setting outcomes: Examining the role of goal interaction in influencing the experience and learning outcomes of video game play for earthquake preparedness. *Comput. Hum. Behav.* **2013**, *29*, 858–869. [CrossRef]
102. Martin, I.M.; Bender, H.; Raish, C. What motivates individuals to protect themselves from risks: The case of wildland fires. *Risk Anal. Int. J.* **2007**, *27*, 887–900. [CrossRef]
103. Mozumder, P.; Helton, R.; Berrens, R.P. Provision of a wildfire risk map: Informing residents in the wildland urban interface. *Risk Anal. Int. J.* **2009**, *29*, 1588–1600. [CrossRef] [PubMed]
104. Penman, T.D.; Eriksen, C.; Horsey, B.; Green, A.; Lemcke, D.; Cooper, P.; Bradstock, R.A. Retrofitting for wildfire resilience: What is the cost? *Int. J. Disaster Risk Reduct.* **2017**, *21*, 1–10. [CrossRef]
105. Shafran, A.P. Risk externalities and the problem of wildfire risk. *J. Urban Econ.* **2008**, *64*, 488–495. [CrossRef]
106. Adame, B.J.; Miller, C.H. Vested Interest, Disaster Preparedness, and Strategic Campaign Message Design. *Health Commun.* **2015**, *30*, 271–281. [CrossRef] [PubMed]
107. Glik, D.C.; Eisenman, D.P.; Zhou, Q.; Tseng, C.; Asch, S.M. Using the Precaution Adoption Process Model to Describe a Disaster Preparedness Intervention among Low-Income Latinos. *Health Educ. Res.* **2014**, *29*, 272–283. [CrossRef]
108. Jassempour, K.; Shirazi, K.K.; Fararooei, M.; Shams, M.; Shirazi, A.R. The impact of educational intervention for providing disaster survival kit: Applying precaution adoption process model. *Int. J. Disaster Risk Reduct.* **2014**, *10*, 374–380. [CrossRef]
109. Thomas, T.N.; Sobelson, R.K.; Wigington, C.J.; Davis, A.L.; Harp, V.H.; Leander-Griffith, M.; Cioffi, J.P. Applying instructional design strategies and behavior theory to household disaster preparedness training. *J. Public Health Manag. Pract.* **2018**, *24*, e16–e25. [CrossRef]
110. Semenza, J.C.; Ploubidis, G.B.; George, L.A. Climate change and climate variability: Personal motivation for adaptation and mitigation. *Environ. Health* **2011**, *10*, 1–12. [CrossRef]
111. Song, J.; Peng, B. Should We Leave? Attitudes towards Relocation in Response to Sea Level Rise. *Water* **2017**, *9*, 941. [CrossRef]
112. Babcicky, P.; Seebauer, S. The Two Faces of Social Capital in Private Flood Mitigation: Opposing Effects on Risk Perception, Self-Efficacy and Coping Capacity. *J. Risk Res.* **2017**, *20*, 1017–1037. [CrossRef]
113. Botzen, W.J.W.; Kunreuther, H.; Czajkowski, J.; de Moel, H. Adoption of Individual Flood Damage Mitigation Measures in New York City: An Extension of Protection Motivation Theory. *Risk Anal.* **2019**, *39*, 2143–2159. [CrossRef] [PubMed]
114. Bubeck, P.; Botzen, W.J.W.; Aerts, J.C.J.H. A Review of Risk Perceptions and Other Factors That Influence Flood Mitigation Behavior. *Risk Anal.* **2012**, *32*, 1481–1495. [CrossRef] [PubMed]
115. Dittrich, R.; Wreford, A.; Butler, A. The Impact of Flood Action Groups on the Uptake of Flood Management Measures. *Clim. Change* **2016**, *138*, 471–489. [CrossRef]
116. Erdlenbruch, K.; Bonte, B. Simulating the Dynamics of Individual Adaptation to Floods. *Environ. Sci. Policy* **2018**, *84*, 134–148. [CrossRef]
117. Fox-Rogers, L.; Devitt, C.; O'Neill, E.; Brereton, F.; Clinch, J.P. Is There Really 'nothing You Can Do'? Pathways to Enhanced Flood-Risk Preparedness. *J. Hydrol.* **2016**, *543*, 330–343. [CrossRef]

118. Grothmann, T.; Reusswig, F. People at risk of flooding: Why some residents take precautionary action while others do not. *Nat. Hazards* **2006**, *38*, 101–120. [CrossRef]
119. Haer, T.; Botzen, W.J.W.; Aerts, J.C.J.H. The Effectiveness of Flood Risk Communication Strategies and the Influence of Social Networks-Insights from an Agent-Based Model. *Environ. Sci. Policy* **2016**, *60*, 44–52. [CrossRef]
120. Koerth, J.; Jones, N.; Vafeidis, A.T.; Dimitrakopoulos, P.G.; Melliou, A.; Chatzidimitriou, E.; Koukoulas, S. Household adaptation and intention to adapt to coastal flooding in the Axios–Loudias–Aliakmonas National Park, Greece. *Ocean Coast. Manag.* **2013**, *82*, 43–50. [CrossRef]
121. Richert, C.; Erdlenbruch, K.; Figuières, C. The determinants of households' flood mitigation decisions in France-on the possibility of feedback effects from past investments. *Ecol. Econ.* **2017**, *131*, 342–352. [CrossRef]
122. Terpstra, T. Emotions, trust, and perceived risk: Affective and cognitive routes to flood preparedness behavior. *Risk Anal. Int. J.* **2011**, *31*, 1658–1675. [CrossRef]
123. Terpstra, T.; Lindell, M.K.; Gutteling, J.M. Does communicating (flood) risk affect (flood) risk perceptions? Results of a quasi-experimental study. *Risk Anal. Int. J.* **2009**, *29*, 1141–1155. [CrossRef]
124. Dillenardt, L.; Hudson, P.; Thieken, A.H. Urban pluvial flood adaptation: Results of a household survey across four German municipalities. *J. Flood Risk Manag.* **2021**. [CrossRef]
125. City of Toronto. *Update on the Engineering Review Addressing Basement Flooding*; Staff Report to Council; City of Toronto: Toronto, ON, Canada, 2008.
126. City of Surrey. *Surrey Stormwater Drainage Regulation and Charges, By-Law, 2008, No. 16610*; City of Surrey: Surrey, UK, 2008.
127. Robinson, B. *Manual of Best Practices to Reduce Risk of I/I on the Private Side*; Norton Engineering Inc.: Kitchener, ON, Canada, 2020.
128. Thistlethwaite, J.; Henstra, D. Municipal flood risk sharing in Canada: A policy instrument analysis. *Can. Water Resour. J.* **2017**, *42*, 349–363. [CrossRef]
129. Province of Ontario. *Provincial Policy Statement*; Queen's Printer for Ontario: Toronto, ON, Canada, 2020.
130. MMM Group Limited. *National Floodplain Mapping Assessment–Final Report*; Public Safety Canada: Ottawa, ON, Canada, 2014.
131. Spekkers, M.H.; Kok, M.; Clemens, F.H.; Ten Veldhuis, J.A. Decision tree analysis of factors influencing rainfall-related building damage. *Nat. Hazards Earth Syst. Sci. Discuss.* **2014**, *2*, 2263–2305.
132. Islam, A.R.; Talukdar, S.; Mahato, S.; Kundu, S.; Eibek, K.U.; Pham, Q.B.; Kuriqi, A.; Linh, N.T. Flood susceptibility modelling using advanced ensemble machine learning models. *Geosci. Front.* **2021**, *12*, 101075. [CrossRef]
133. Ruidas, D.; Chakrabortty, R.; Islam, A.R.; Saha, A.; Pal, S.C. A novel hybrid of meta-optimization approach for flash flood-susceptibility assessment in a monsoon-dominated watershed, Eastern India. *Environ. Earth Sci.* **2022**, *81*, 1–22. [CrossRef]
134. Henstra, D.; Thistlethwaite, J.; Vanhooren, S. The governance of climate change adaptation: Stormwater management policy and practice. *J. Environ. Plan. Manag.* **2020**, *63*, 1077–1096. [CrossRef]
135. Andrey, J.; Kertland, P.; Warren, F.J. Water and Transportation Infrastructure. In *Canada in a Changing Climate: Sector Perspectives on Impacts and Adaptation*; Warren, F.J., Lemmen, D.S., Eds.; Government of Canada: Ottawa, ON, Canada, 2014; pp. 233–252.
136. Bohman, A.; Glaas, E.; Karlson, M. Integrating sustainable stormwater management in urban planning: Ways forward towards institutional change and collaborative action. *Water* **2020**, *12*, 203. [CrossRef]
137. Hager, J.K.; Mian, H.R.; Hu, G.; Hewage, K.; Sadiq, R. Integrated planning framework for urban stormwater management: One water approach. *Sustain. Resilient Infrastruct.* **2021**, *23*, 1–22. [CrossRef]

MDPI
St. Alban-Anlage 66
4052 Basel
Switzerland
Tel. +41 61 683 77 34
Fax +41 61 302 89 18
www.mdpi.com

Water Editorial Office
E-mail: water@mdpi.com
www.mdpi.com/journal/water

www.ingramcontent.com/pod-product-compliance
Lightning Source LLC
Chambersburg PA
CBHW061454210926
51850CB00008B/1712
* 9 7 8 3 0 3 6 5 5 7 3 5 9 *